Environmental Pollution and the Brain

Environmental Pollution and the Brain

Sultan Ayoub Meo

MBBS, PhD, M Med Ed, FRCP

Professor and Consultant in Clinical Physiology
Department of Physiology, College of Medicine,
King Saud University, Riyadh, Saudi Arabia

CRC Press
Taylor & Francis Group
Boca Raton London New York

CRC Press is an imprint of the
Taylor & Francis Group, an **informa** business

First edition published 2022
by CRC Press
6000 Broken Sound Parkway NW, Suite 300, Boca Raton, FL 33487–2742

and by CRC Press
2 Park Square, Milton Park, Abingdon, Oxon, OX14 4RN

© 2022 selection and editorial matter, Sultan Ayoub Meo.

CRC Press is an imprint of Taylor & Francis Group, LLC

ISBN: 978-1-032-06509-0 (hbk)
ISBN: 978-1-032-08003-1 (pbk)
ISBN: 978-1-003-21246-1 (ebk)

DOI: 10.1201/9781003212461

Typeset in Minion Pro
by Apex CoVantage, LLC

This book is dedicated to my **Parents, Wife, and Children**
For their continuous encouragement in my life

Contents

Preface

Since the existence of life on this planet, the environment has played a vital role in the survival of all living organisms. The four classical elements, earth, water, fire, and air, which form the basics of matter, all combine to make up our surroundings, each one of them contributing in a unique way to keep the circle of life on this planet going. Marking the dawn of mankind from the stone age to the increasing competition of the 21st century that reaffirms Darwin's theory of "Survival of the Fittest," mankind has evolved to make today's world so technologically advanced and comfortable for the common person, yet simultaneously dangerous to thrive in.

The History of Medicine has witnessed various milestones, such as discovering the first antibiotic, penicillin, in 1928, and artificial insulin. These breakthroughs saved the lives of millions and changed the face of modern medicine, all thanks to the advancement in technology and research that made them possible. However, despite advancements in the scientific field, it is not surprising to note how diseases and the microbes causing them are, with time, becoming harsher, stronger, and more resistant, equally combatting the efforts of human beings in defeating them.

The industrial and technological revolution has undoubtedly enhanced the global economy and improved the health care system and quality of life. Over the past three decades, occupational and environmental pollution has become an emerging global public health problem in developing and developed countries. Environmental pollution has caused changes in the composition of air, water, and soil, and similar changes have been observed in the pattern of health and diseases. Being induced by coarse, fine, and ultrafine particulate matters, environmental pollution-related diseases are influenced by the nature, type, duration of exposure, and the concentration and size of airborne particulate matters in the breathing zone. Worldwide, people are facing multiple acute and chronic health problems due to various types of environmental pollution.

However, beyond the smog, soot, toxic gases, and waste that are generated by industries globally, one highly neglected form of pollution, invisible to the naked eye, that is found in the environment and surrounds us in all aspects is the proliferation of electromagnetic fields (EMFs), also known as "electromagnetic smog." With the widespread use of the internet, Wi-Fi, mobile phones, and smart devices, non-ionizing radiation generated from these and other electronic devices raise significant concerns about human health's adverse effects.

The human mind, highly complex to be able to come up with the most significant innovations, unfortunately also happens to be the same organ to be directly affected by man's creations, especially by environmental pollution and radiofrequency electromagnetic field radiation (RF-EMFR). At a time when the environmental health science community is seriously discussing the global issues associated with the environment such as climate change there is an acute need to address the RF-EMFR pollution "electromagnetic smog."

While writing this book, it aimed to cover the fundamental aspects of environmental pollution, RF-EMFR, brain biology, essential areas for biological, environmental sciences, faculty, researchers, students, and policymakers. Another valuable aspect of this book is that the entire text is replete with references up to date with the current research literature, which, being parallel to the text, provide tips that ultimately lead to a grip on the concept of the subject.

I am very thankful to my wife and my children for all their support while writing this book. I also express my gratitude to my mentors, friends, and students for their help in bringing out the first edition of this book that I wrote as a sole author with great effort and enthusiasm.

I also thank my publisher, Taylor and Francis, CRC Press Boca Raton, FL, and Abingdon, Oxon, for their continuous support. I truly hope the information found in this book will help enlighten its readers on how environmental pollution and human-made EMR affect various body organs and systems.

Let's make sure that before our comfort comes, our world and its people's health, safety, and security are more important!

Sultan Ayoub Meo

Acknowledgments

I begin by extending gratitude to the Almighty for giving me the time, peace of mind, ability, health, and strength to write this book and further spread the knowledge I possess.

First, I am grateful to my parents for their prayers, wishes, and their support during my entire academic career.

I am very thankful to my wife, Nadra Sultan Meo, and children: Dr. Anusha Sultan Meo, Army Medical College, National University of Medical Sciences (NUMS), Rawalpindi, Pakistan; Tehreem Sultan Meo, Lincoln's Inn, UK; Muhammad Zain Sultan Meo; Muhammad Omair Sultan Meo, College of Medicine, Alfaisal University, Riyadh; and Maheen Sultan Meo, PISES, Riyadh, Saudi Arabia for their immense assistance to concentrate on my work with enthusiasm and dedication while writing this book. They were also involved in literature review and designing some figures.

I am also very grateful to the President, King Saud University, Dean, College of Medicine, King Saud University, Chairman, and all my colleagues in the Department of Physiology, College of Medicine, King Saud University, Riyadh, Kingdom of Saudi Arabia.

I further extend gratitude to my mentors Prof. Muhammad Abdul Azeem, United Medical School, Karachi, Pakistan; Prof. David C. Klonoff, Clinical Professor of Medicine, School of Medicine, University of California, San Francisco, USA; and Prof. Fahad Abdullah Al-Zamil, College of Medicine, King Saud University, Riyadh, Saudi Arabia for their guidance and support.

For assistance in the literature review, I would like to mention Dr. Naseer Ahmed, Aga Khan University, Karachi Pakistan, and Mr. Adnan Mahmood Usmani, at College of Medicine, King Saud University, Riyadh, Kingdom of Saudi Arabia. The graphical and illustrations work in this book is credited to Mr. Niazudin Ahmed, to whom I am also very thankful.

To all my other mentors, friends, and students who might have been missed here, I believe that this might not have been possible without you all for your support in bringing out the first edition of my book.

Thank you.

Sultan Ayoub Meo

About the Author

Prof. Sultan Ayoub Meo is a medical graduate (MBBS) with higher postgraduate degrees (MPhil) and Doctorate (PhD) in Physiology. Prof. Meo received Fellowships (FRCP) from the Royal College of Physicians of Dublin, Ireland, Royal College of Physicians of London, Royal College of Physicians and Surgeons of Glasgow, and Royal College of Physicians of Edinburgh, UK. In addition to garnering MBBS, MPhil, PhD, and four fellowships of the highly respectable Royal Colleges of the United Kingdom and Ireland, he also had obtained a postgraduate master's degree in Medical Education (MMedEd), from the University of Dundee, Scotland, UK.

Prof. Meo has 25 years of teaching experience and has been actively involved in undergraduate and postgraduate teaching in physiology and supervision of Masters, PhD, and Fellowship students. Prof. Meo has been appointed as a PhD supervisor and examiner in many Universities in Saudi Arabia and Malaysia. He has also been involved in the clinical division to provide clinical physiology services to the community.

Prof. Meo has been credited with ten books and over 200 research articles in leading science journals. He has been named an Associate Editor and board member of many international peer-reviewed journals.

Prof. Meo has been invited to deliver lectures at more than 125 international conferences in various countries, including Saudi Arabia, United Arab Emirates, Kingdom of Bahrain, Pakistan, Indonesia, Turkey, China, UK, and the USA.

In 2017, the honorable governor Riyadh Prince Faisal bin Bandar bin Abdulaziz Al Saud honored Prof. Meo with the KSU-Saudi Arabia Excellency award in Medicine.

Biology of the Brain

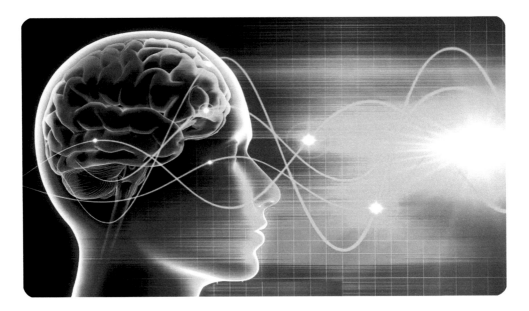

1.1 Introduction

Being the main organ on which all other body systems depend, the human brain is a highly complex organ. It comprises 85–100 billion neurons, of which approximately 20 billion are found in the cerebral cortex [1, 2]. Each cortical neuron has, on average, 7,000 synaptic connections, linking different neurons and resulting in an arrangement of 0.15 quadrillion synapses and more than 150,000 km of myelinated nerve fibers [3]. The complexity of the brain is due to its biological, neuronal, hormonal, mechanical, and electrical connectivity with each cell (75–100 trillion cells) in the human body. Neuronal connections provide and control the entire physiological and biochemical actions, reactions, thoughts, feelings, memory, and experiences by coding and processing information [4].

The nervous system is the chief controlling and coordinating system that regulates all body functions and systems. It has two main divisions: "(i) the central nervous system (CNS), which consists of two parts, the brain and spinal cord, and (ii) the peripheral nervous system (PNS), which consists of cranial and spinal nerves." The brain's sensory part collects the signal information from the body, that is, the receptors, whereas the motor part controls the body's responses, that is, the effectors. The higher cognitive functions of the body, including voluntary and involuntary actions, are robust and controlled by various brain centers [5–7].

Subsequently, the narrative of the neurons, receptors, and synapses is the first step toward a solicitous understanding of the basic biological and physiological principles of the CNS and PNS (Figure 1.1).

1.2 Neurons

Neurons, also known as nerve cells or nerve fibers, are the principal excitable cells of the nervous system, playing a major role in receiving, processing, and transmitting the information.

DOI: 10.1201/9781003212461-1

1

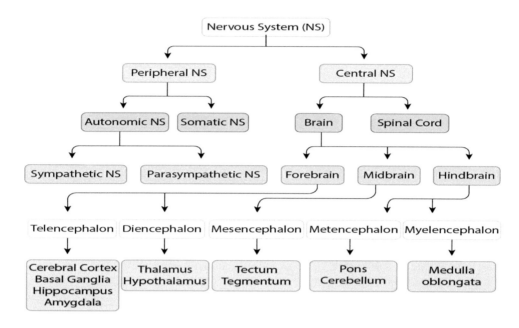

Figure 1.1   Nervous system: central and peripheral nervous system.

Neurons are the basic biological units of the brain, spinal cord, and PNS. Being the primary functional and structural unit of the nervous system, they are supported by neuroglia and astrocytes. They are the principal elements of perception, memory, thought, and actions and the units of consciousness [8]. The two main classes of cells that make up nervous tissue are neurons and neuroglia.

The neurons consist of three parts: the axon, dendrite, and cell body called soma (Figure 1.2)—and are classified into unipolar, bipolar, and multipolar neurons. On the basis of the anatomical features, neurons are also classified into two types: Golgi type 1 neuron with a long axon and Golgi type 2 neuron with a short axon. However, physiologically, "neurons are classified into three types: sensory neurons, interneurons, and motor neurons." Neurons vary in number, size, and shape. The number of neurons in the body ranges from 85 billion to 100 billion, which comprise approximately 10% of the entire brain structure [1, 2].

1.2.1 Cell Body (Soma)

The cell body (soma) of a neuron houses the nucleus, nucleolus, and other minute organelles that perform essential chemical, biological, and metabolic functions and sustain cell survival. The size of a neuronal cell body is approximately 0.005–0.1 mm. The essential minute organelles found in the neuronal cytoplasm include mitochondria, Golgi apparatus, endoplasmic reticulum, ribosomes, lysosomes, endosomes, peroxisomes, centrioles, microtubules, and microfilaments. These organelles are also found in other body cells (Table 1.1; Figure 1.3). The organelles found in a nerve cell are fundamental for the expression of genetic information that controls the synthesis of cellular proteins, energy synthesis, growth, and replacement [9, 10].

Recent studies have revealed that proteins are synthesized in the dendrites, but axons and axon terminals could not synthesize these proteins, as they do not contain

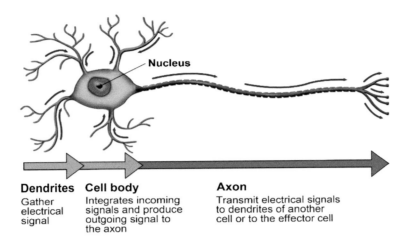

Figure 1.2 Neuron.

ribosomes. Through the mechanism of "anterograde transport," tiny substances are transported via microtubules along the entire length of the axon to the terminals, where they insert into the plasma membrane and other organelles and material are degraded by lysosomes [10].

1.2.2 Dendrites

The receiving end of the neuron or input part consists of tree-like branches of the neuronal cell body (soma), known as dendrites. The word "dendrite" was coined by William His, a Swiss anatomist, in 1889, derived from the Greek words "dendr" or "dendro," meaning tree. The dendrite spines are small outgrowths, which further increase the receptive surface area of a neuron. They control the number, dispersal, and incorporation of inputs into a neuron and regulate the generation of the various dendrite branching patterns with distinct features of assorted types of neurons [11]. To maintain the typical characteristics of the neurons and perform function appropriately, the neurons must connect through dendrites to receive information [11]. The normal morphology of dendrites provides a physiological promise with vital functional implications in determining the types, nature, severity of signals, and their integration received by neurons. The receptor relations, signaling pathways, and cytoskeletal elements have been recognized as significant elements that contribute significantly to the organization of dendrites of different neurons [12].

The morphological and physiological characteristics of dendrites vary from those of axons [13]. Dendrites have specialized structures, such as spines, which act as the primary excitatory synaptic sites, and these structures differentiate dendrites from axons. Axons and dendrites contain additional minute organelles, including the Golgi body [14–15]. The arrangements of several cytoskeletal structures influence the process in which minute organelles and molecules are transported along the axons and dendrites [12].

1.2.3 Axons

Almost all neurons consist of a single axon, the diameter of which varies from a micrometer to a meter, in certain nerves of the human brain. Axons are usually less branched, straighter, and smoother in appearance than dendrites and have microtubules, neurofilaments, and scattered

Table 1.1 Structural and Functional Characteristics of Organelles of the Neurons		
Organelles of Neuron	**Structural Characteristics**	**Functional Characteristics**
Plasma cell membrane	Similar to other cells, neuronal cells are also surrounded by a plasma cell membrane made of lipids bilayer and proteins. It is composed of an inner and an outer layer. The lipid layer is made up of two rows of phospholipid molecules	The neuronal plasma membrane has a semipermeable membrane. It serves as a barrier to encompass the cytoplasm and organelles, has a selective role for the diffusion of certain ions and substances, and restricts others. It stores nutritional and eliminates injurious substances, and catalyzes biochemical reactions. It generates an electrical potential and conduct nerve impulse
Nucleus	Large, rounded with one clear nucleolus, widely scattered in the neuronal cell body (soma)	It contains genetic material; DNA synthesizes RNA from DNA and transports it through pores toward the cytoplasm for protein synthesis. It controls genetic and all neuronal activities
Mitochondria	Spherical or rod-shaped in appearance, covered with a double membrane. Largely scattered throughout the cell body, axons and dendrites contain enzymes and form energy, hence known as the powerhouse of the cell	The brain is a highly metabolically active organ; neurons require more energy, mitochondria synthesize energy
Golgi apparatus	Membrane-bound folded structure, closely located near to nucleus, with clusters of flattened cisternae and small vesicles	Involve in packages process, augment carbohydrate to protein molecules for transport to nerve terminals
Rough endoplasmic reticulum	Ribosomes are located in the cytoplasm and proximal part of dendrites, absent in the axon	Synthesize protein
Lysosomes	Small membrane-bound vesicles with a diameter of about 8 nm; three types: primary, secondary, and residual	Engulfs and digests the debris and invading substances
Microtubules	Are present between neurofibrils, 20–25 nm in diameter, scattered in dendrites, cell body, and axon	Neuronal transport mechanism. Allow the organelles to move precisely and track other structures

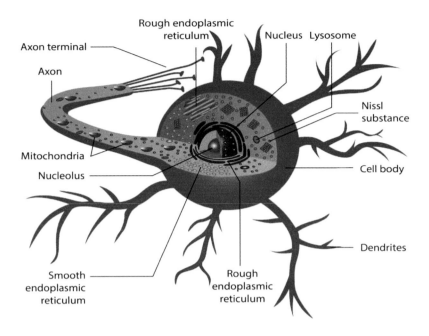

Figure 1.3   Organelles of neuron.

mitochondria. They play a major role in the conduction of an electrochemical impulse, called an action potential, propagating outward or away from the cell body toward the axon terminus [16].

Action potentials move swiftly at a speed of approximately 100 meters per second. The transmission speed of impulses also depends on the presence of the myelin sheath, fatty insulation around the axon made up of "Schwann cells" that prevents the loss of electrical impulse to the neuron's surrounding, aiding the faster transmission of the impulse along with the extent of the axon. Axons are more than a meter long, and an action potential takes a few milliseconds to transfer along the entire length of the axons. An action potential is generated in the axon hillock and conducted down toward axon terminals. Axon terminals are small branches that form synapses with other neurons, cells, muscles, or glands (Table 1.2) [9–10]. However, some exceptions exist. For instance, in few neurons, for example, peripheral sensory neurons, the input occurs in the axons, and some dendrites also serve as output systems [16].

Neurons are mainly categorized by four functional properties: electrical excitability, secretion (mainly vesicular and peptide), molecular synthesis (mostly proteins), and growth and plasticity [16]. Neurons have three main types: sensory neurons, interneurons, and motor neurons.

1.3 Types of Neurons
1.3.1 Sensory Neurons
Sensory neurons are mainly located in the receptors in organs and tissue, such as the skin, tongue, eyes, ears, and nose. They are activated by sensory input or a stimulus, that is, a

Table 1.2	Biological Characteristics of Axons and Dendrites	
Characteristics	**Axons**	**Dendrites**
Anatomical features	Large-sized, may be between 0.11 mm and 1–2 meters. The axons of many neurons are covered in a myelin sheath	Short sized, 1.5 mm, irregular surface, covered in dendritic spines
Branch	Branched at the terminal or distal ends	Tree-like appearance, highly branched, branches arise from the cell body
Number per neuron	Usually, one axon per neuron	Many dendrites per neuron
Myelin sheath	It may be myelinated or non-myelinated	Unmyelinated and are not insulated
Knobs	Synaptic knobs	No knobs at the tip of the branch
Neurofibrils and Nissl's granules	Contain neurofibrils but no Nissl's granules	Contain both neurofibrils and Nissl's granules
Transport of impulse	The axons mainly conduct impulses away from the cell body	Function as an "antennae" of the neuron, covered with hundreds to thousands of synapses. Dendrites receive signals from other nerve cells and conduct impulses toward the cell body

change in the external environment of an organism, which, once detected, activates the receptor that transmits signals from sensory organs to the spinal cord (enter into the spinal cord) and brain in the form of neuronal messages. Once the impulses arrive in the brain, they are translated into a form known as "sensations," such as touch, temperature, taste, smell, vision, and hearing [17].

Sensory neurons are classified based on their cell body size, axon diameter, degree of myelination, and axonal conduction velocities as follows: Aβ, Aδ, or C types. C-type sensory neurons are the smallest and most abundant, with unmyelinated axons, and have the slowest conduction velocity (0.2–2 m/s). Aδ- and Aβ-type sensory neurons have medium- and large-cell body sizes with light and heavy myelinated processes, thus exhibiting intermediate and rapid conduction velocities, respectively. The conduction velocities in Aδ-type sensory neurons vary from 5 m/s to 30 m/s, whereas those in Aβ-type sensory neurons range from 16 m/s to 100 m/s. Most Aβ fibers have low mechanical thresholds, indicating that Aβ fibers are light-touch receptors. (Figure 1.4; Table 1.3) [17]

1.3.2 Interneurons

Interneurons, also known as "relay neurons," connect and carry impulses between sensory or motor neurons and the CNS and connect sensory and motor neurons to form a nerve circuit. Interneurons are found in the brain and spinal cord, can communicate with each other, and establish a complex circuit. They are multipolar neurons, similar to motor neurons. After receiving signals from interneurons, motor neurons play a significant role in effector organs, such as muscles, by changing the activity of the motor units involved in the muscle.

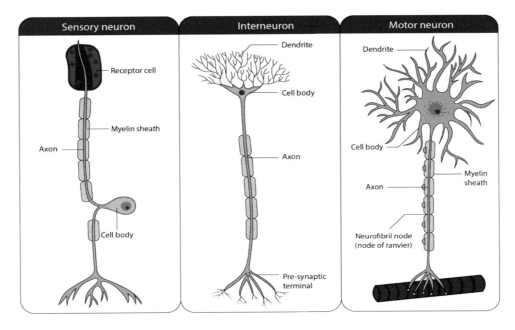

Figure 1.4 Types of neuron.

Table 1.3 Anatomical and Physiological Characteristics of Neurons

Neurons	Structure	Location	Polarity	Functions
Sensory neurons	Long dendrite and short axon	Cell body and dendrites outside the spinal cord	Unipolar	Sensory neuron transmits impulses from sensory receptors to the spinal cord and brain
Motor neuron	Short dendrite and a long axon	Cell body and dendrites within the spinal cord. Axons outside the spinal cord	Multipolar	Transmits nerve impulses from the central nervous system, spinal cord to effector structures such as skeletal muscles and glands
Interneuron	Short dendrite and short axon	Within the nervous system	Multipolar	Interconnect and transmits nerve impulses between sensory and motor neurons

1.3.3 Motor Neurons

The motor unit comprises a motor neuron in the ventral horn of the spinal cord. Motor neurons integrate neural inputs into an output signal that is suitable to regulate the strength that the effector or muscle unit engenders (Figure 1.4). Muscles are mainly regulated by a pool of motor neurons, called "motor nucleus" or "motor pool," that comprises hundreds to thousands of motor units. Small motor neurons tend to innervate muscle units with relatively long contraction times and low forces, whereas progressively large motor neurons innervate muscle units with faster contraction times and correspondingly greater forces [18].

Motor neurons descending from the cerebral cortex in the brain to the anterior horn cells of the spinal cord are called "upper motor neurons." Motor neurons traveling from the spinal

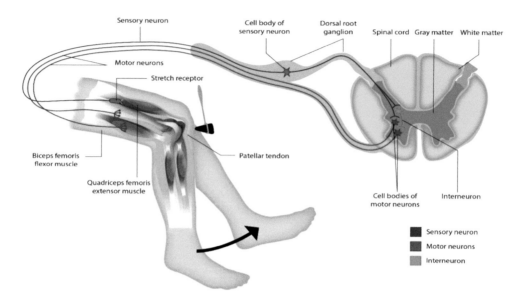

Figure 1.5 Lower motor neurons.

cord to skeletal muscles are called "lower motor neurons." Lower motor neurons play a major role in all voluntary movements. Lower motor neurons have three major types, i.e., alpha motor neurons, beta motor neurons, and gamma motor neurons (Figure 1.5).

Neuronal signals transmit through highly specialized structures called synapses [19]. The primary mechanism underlying a synaptic transmission is an action potential, an electrical signal generated by changes in the electrically excitable membrane of a neuron. Neurons are extremely specific for the swift processing and transmission of cellular signals and play a normal role.

1.4 Synapses

The term "synapse" is derived from the Greek word "synapsis," meaning union. Synapses are sites where contact between neurons or between the effector muscle and gland cells occurs. Synapses may also be defined as the site of functional communication between neurons. Here the information is transmitted not in the form of an electrical impulse but rather a chemical signal is known as neurotransmitters. The space between the axon terminal and dendritic spine or between a nerve and the muscle membrane is similar to the synaptic cleft at synapses. This entire structure is known as the neuromuscular or myoneural junction, the interconnection between a nerve ending and the muscle membrane [20]. Synapses are essential functional units that permit the processing of information in the brain.

A synapse is formed when a presynaptic knob establishes a connection with a postsynaptic cell [21]. Synapses transmit signals in only one direction, from the presynaptic cell to the postsynaptic cell. Synapses have two main types: electrical and chemical synapses. The axon terminal of presynaptic cells contains vesicles, which are occupied by specific neurotransmitters. Once an action potential in the presynaptic cell arrives in an axon terminal, the localized concentration of calcium in the cytosol is increased, which can cause some vesicles to fuse with the plasma membrane and release their substances into the synaptic cleft. Neurotransmitters diffuse across the synaptic cleft, which takes approximately 0.5 ms to bind to receptors on postsynaptic cells (Figure 1.6). The attachment of neurotransmitters changes

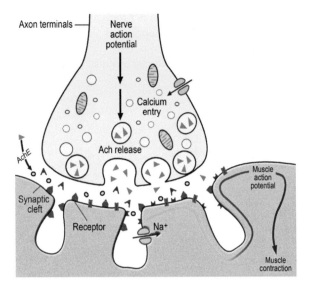

Figure 1.6 Synapse.

the permeability of the ions in the postsynaptic membrane by a structural change in membrane proteins, which in turn changes the membrane's electrical potential. The impact of this change varies depending on the type of cells. In case the postsynaptic cell is a neuron, this ionic change may induce an action potential. However, if the postsynaptic cell is a "muscle," the change in the binding of neurotransmitters causes contraction, or if it is in a "hormone," it can cause hormonal secretion [22].

1.5 Receptors

Receptors are specialized types of highly sensitive cells to certain types of changes in the external and internal environments. Receptors are for general, proprioceptive, and special senses. They detect changes in the external environment of an organism, known as the stimulus. This helps an organism pick up the sense of touch, taste, smell, etc.

The main biological function of receptors is learning about the environment, condition, position, or state of the body's external and internal environment. The body receives various stimuli of different magnitude and sources [23]. Receptors receive the stimuli, identify, assess, and produce a graded potential in a sensory neuron and respond accordingly. The graded potential causes sensory neurons to generate an action potential relayed into the CNS if the stimulus is sufficient. After detecting the stimulus, the central integration can cause a motor response to perform the required task.

The various types of stimuli, including touch, pain, temperature, smell, taste, and vision, are sensed and identified by specific types of receptors. Receptor cells are classified based on the three structural and functional criteria: cell type, position, and function. Also, they can be classified into mechanical, electrical, and chemical types [24].

Most receptors act as transducers because they convert one form of energy (light, sound, taste, touch, etc.) into another form (electrical energy). When a receptor receives a stimulus, it produces physiological alterations in its cell membrane permeability to specific ions and produces hyperpolarization of the surface membrane. This mechanism is called receptor generator potential, independent of the "all or none law." When this generator potential reaches a critical level, the receptors fire, and an action potential is produced, which travels into the nerve fibers connected to the receptor. The impulse is conducted along an afferent pathway that enters either the spinal cord or brain (Figure 1.7).

Stimulus → receptor → coordinator → effector → response (Figures 1.7 and 1.8).

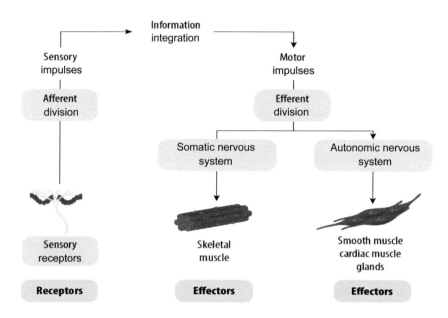

Figure 1.7 The role of receptors in receiving the stimuli, identifying, and outcomes.

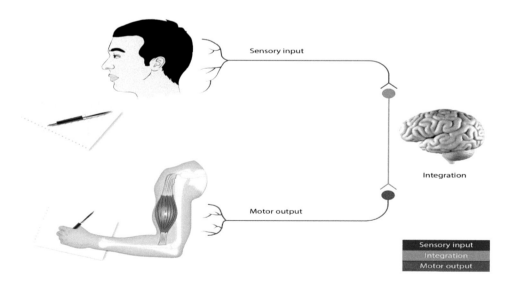

Figure 1.8 Sensory and motor receptors response.

1.6 CNS: The Brain and Spinal Cord

The nervous system is the chief controlling and coordinating system and regulates all body functions. It is classified into two main divisions: (1) the CNS, which consists of two parts, the brain and spinal cord, and (2) the PNS, which consists of cranial and spinal nerves. The brain is a highly delicate organ composed of nerve cells and tissue. The adult brain has an average weight of 1.5 kg and consists of four major parts: the cerebrum, cerebellum, pons, and medulla oblongata. These parts are anatomically and physiologically linked and work together in a well-organized and coordinated fashion.

1.6.1 Cerebrum

The cerebrum is the largest part of the brain and is located mainly in the upper skull. It consists of the left and right cerebral hemispheres, the outer or external layer called the cortex (gray matter), and the inner or internal layer called white matter. The cortex has four lobes: the frontal, parietal, temporal, and occipital lobes [25]. The cerebrum uses information from sensory organs to receive, assess, and inform all activities and respond accordingly.

The muscles on the left side of the body are controlled and regulated by the right hemisphere, and the left hemisphere controls the muscles on the right side of the body. "Each cerebral hemisphere has four lobes: the frontal, temporal, parietal, and occipital lobes." (Figure 1.9).

1.6.1.1 Frontal Lobe

The frontal lobe is the most prominent lobe, positioned in front of the cerebral hemispheres. It has momentous functions, including developing and maintaining behavior, judgment, planning, speech, speaking, writing (Broca's area), attention, and intelligence. The frontal lobe plays a significant role in prospective memory, which involves daily and future planning [26]. It is also involved in speech and language [27].

Studies have reported that disease or injury to the prefrontal cortex can cause five types of personality changes (Table 1.4), including social, behavioral, emotional, distress, and decision-making disturbances [28, 29].

1.6.1.2 Parietal Lobe

The parietal lobe is placed superior to the temporal lobe and posterior to the frontal lobe. It is divided into two physiologically important functional regions. The "anterior parietal lobe contains the primary sensory cortex, situated in the postcentral gyrus (Broadman area). The

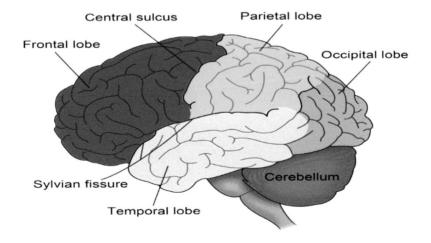

Figure 1.9 Lobes of the brain.

11

Table 1.4 Clinical Features Associated with Lesion in or Injury to Various Lobes of the Brain

Lesions	Clinical Features
Frontal lobe	• Weakness • Flaccid hemiplegia • Personality disorders • Aphasia [36]
Parietal lobe	• Aphasia • Astereognosis • Sensation impairment [31]
Temporal lobe	• Hearing impairment • Paraphasia-speech disorder • Auditory or memory • Visual hallucinations [37]
Occipital lobe	• Visual field deficits • Partial or complete blindness • Color blindness [35]

posterior parietal lobe has two regions": the superior and inferior parietal lobules [30]. The parietal lobe performs a major role in visual perception, interpretation, language, word formation, memory, planning, learning, language, recognition, and stereognosis [31].

1.6.1.3 Temporal Lobe

The temporal lobe lies in the middle cranial fossa, posterior to the frontal lobe and inferior to the parietal lobe. This lobe has two surfaces: the lateral and medial surfaces [32]. The medial temporal lobe (MTL) has a vital role in declarative memory, which relies on encoding associations between substances. Besides, it is involved in sound recognition, understanding, semantic memory, language (Wernicke's area), hearing sequencing and organization, visual perception, and facial perception [33, 34] (Figure 1.10).

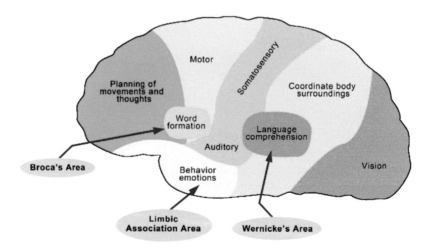

Figure 1.10 Functional areas of the brain.

12

1.6.1.4 Occipital Lobe

The occipital lobe is an essential segment of the CNS, responsible for the sense of vision [35]. It is the smallest lobe of the cerebral cortex, situated in the most posterior region of the brain, and consists of approximately 12% of the total surface area of the neocortex. It is essential for visual processing, integration, and interpretation. Lesions in the occipital lobe result in partial or complete visual impairment and/or blindness. The visual information helps determine, recognize, compare, and differentiate objects.

1.6.2 Cerebellum

The cerebellum is derived from the Latin word "cerebellum," meaning "little brain." The cerebellum is the most significant sensorimotor part of the hindbrain, having extensive biochemical and neuronal networks and linkages with the brainstem and spinal cord. It occupies approximately one-tenth of the skull cavity, comprises 10% of the total brain volume, contains 50% of all neurons, and lies behind the pons and medulla oblongata [2, 38]. The cerebellum is an oval-shaped organ with an approximate weight of 150 g, situated in the posterior cranial fossa, anterior to the fourth ventricle, pons, and medulla oblongata superiorly covered by the tentorium cerebelli and posteroinferiorly covered by the squamous occipital (Figure 1.11).

The cerebellum consists of three lobes:

- Paleocerebellum is the anterior lobe, also known as spinocerebellum.

- Neocerebellum is the posterior lobe, also known as cerebrocerebellum.

- Archicerebellum is the flocculonodular lobe, also known as vestibulocerebellum.

Nuclei of the cerebellum:

- Fastigial

- Interposed (globose and emboliform)

- Dentate

The cerebellum contains Purkinje, basket, Golgi, and stellate cells, which secrete the inhibitory neurotransmitter known as gamma-aminobutyric acid. However, the granular cells release "glutamate," which is excitatory. The stellate cells of the cerebellum release another inhibitory neurotransmitter known as "taurine." The biological outcome of deep cerebellar nuclear neurons mainly depends on inhibitory and excitatory stimulation [39].

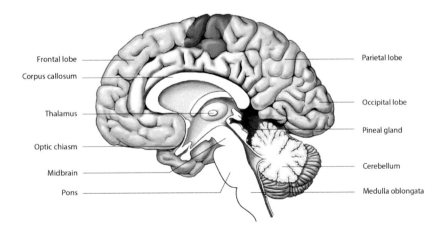

Figure 1.11 The cerebellum.

13

The cerebellum is anatomically and physiologically divided into three parts. The first part is the anterior lobe, also known as the paleocerebellum or spinocerebellum. This part of the cerebellum makes the body movements smooth and coordinated. The second part of the cerebellum, the posterior lobe, also known as the neocerebellum or cerebrocerebellum, interacts with the motor cortex in planning and programming movements. However, the third part of the cerebellum, the flocculonodular lobe, also known as the archicerebellum or vestibulocerebellum, plays a major role in maintaining balance and control of fine movements, such as eye movements (Table 1.5; Figure 1.12).

Table 1.5 Structural and Functional Divisions of Cerebellum	
Three lobes	Flocculonodular lobe Anterior lobe Posterior lobe
Three cortical layers	Molecular layer Granular layer Purkinje layer
Three Purkinje's cells afferent paths	Mossy fibers Climbing fibers Aminergic fibers
Three pairs of deep nuclei	Fastigial Interposed (globose and emboliform) Dentate
Three pairs of peduncles	Superior peduncle Middle peduncle Inferior peduncle
Three functional divisions	Paleocerebellum: Anterior lobe (spinocerebellum) Neocerebellum: Posterior lobe (cerebrocerebellum) Archicerebellum: Flocculonodular lobe (vestibulocerebellum)

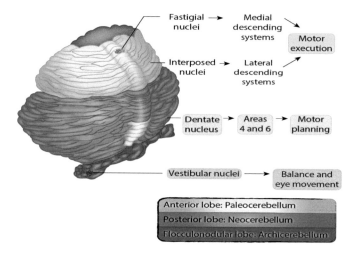

Figure 1.12 Anatomical and physiological parts of the cerebellum.

Table 1.6 Cerebellum: Disorders and Associated Features

Cerebellum Disorders	Lesions and Associated Features
Ataxia	Reeling, wide-based gait
Decomposition of movement	Patient unable to perform the fine or coordinated movements
Dysarthria	Patient unable to articulate the words correctly, the main complaints of slurred and inappropriate phrasing
Dysdiadochokinesia	Unable to perform rapid alternating movements
Dysmetria	Unable to control the range of movement
Hypotonia	Poor or reduced muscle tone
Nystagmus	Swift involuntary oscillation of the eyeballs in a horizontal, vertical, or rotary direction toward the side of the cerebellar lesion
Scanning speech	Low enunciation and hesitation in producing beginning word or syllable
Intentional tremors	A tremor in a limb as it approaches near to a targeted object

Disease or lesions in the cerebellum can cause ataxia, dysarthria, decomposition of movements, dysmetria, dysdiadochokinesia, scanning speech, hypotonia, and intentional tremors [40–42] (Table 1.6).

Cerebellar lesions or injuries mainly produce disturbances in body movements and support. Cerebellar tremors are intention tremors [43] and have three types: "static tremor (postural), kinetic tremor (the limb is held or deviate before the target and does not appropriately reach the target), and tremor in lesions of the superior cerebellar peduncle, which involve mostly the head and trunk" [44, 45]. Cerebellar diseases or lesions frequently develop ipsilaterally to the side of the lesion. However, bilateral finger tremors were reported in few cases, even though only one cerebellar hemisphere was affected [46].

After the acute or chronic cerebellar lesion, ischemia, cerebellar disease, allied clinical symptoms computed the severity of the ataxia and bradykinesia. The improvement of cerebellar symptoms depends on the nature and magnitude and etiopathology of the cerebellar lesion. The significant clinical enhancements are most probably observed during the first 15 days after the onset of the lesion; however, in some patients, the improvement continued until the third month of the lesion [47].

1.7 Brainstem

The brainstem is the most inferior part of the brain and connects the brain to the spinal cord. The human brainstem comprises three main components: the midbrain, pons, and medulla oblongata [48]. Information is relayed from the body to the cerebellum, to the cerebrum, and across the brainstem. The brainstem is an essential channel for many ascending and descending tracts and pathways, contains many cranial nerve nuclei, and is important for major integrative functions (Table 1.7). The brainstem participates in movement control, modulation of pain, autonomic reflexes, sleep, arousal, and consciousness [48], and temperature and blood pressure regulation. The brainstem has a minor interconnecting vascular network, providing collateral circulation to various parts of the brainstem; hence, it is at a higher risk for ischemia [49]. Lesions in the brainstem result in motor and cranial nerve deficits. Diseases or lesions in the

Table 1.7 Brainstem Functions
• Acts as a conduit for ascending and descending tracts
• Connects the spinal cord to various parts of higher centers in the forebrain
• Contains nuclei of cranial nerves III—XII.
• Have integrated functions, including cardiovascular and respiratory control, pain perception, temperature sensitivity, touch, proprioception, and pressure sensation
• Facilitates alertness and consciousness
• It helps in the sleep cycle

brainstem can cause various abnormal body and cranial nerve functions, leading to visual and hearing disturbances, sensation changes, muscle weakness, vertigo, coordination problems, and voice and speech problems. A recent study has specified that patients may develop transient or permanent cognitive function deficits due to brainstem lesions. Also, patients with brainstem lesions could not return to their normal daily-life activities [50].

1.7.1 Midbrain

The midbrain is the uppermost part of the brainstem, positioned beneath the cerebral cortex and above the hindbrain. The anatomical characteristics of the midbrain comprise the tectum and cerebral peduncles, with multiple nuclei and fasciculi. The caudal part, the mesencephalon, posteriorly connects the pons and rostrally joins the diencephalon, thalamus, and hypothalamus. The midbrain comprises several important nuclei and white matter tracts, most of which are involved in motor control and auditory and visual pathways. The prominent midbrain nuclei include the superior and inferior colliculus nuclei, red nucleus, substantia nigra, oculomotor nuclear complex, and trochlear nucleus. It mainly plays a role in vision, hearing, alertness, sleep and wake cycles, temperature regulation, and control of the motor control system [51].

Lesions in the midbrain may result in "distinct syndromes," often causing overlapping clinical features. The most important midbrain lesion syndrome is "Parinaud's syndrome," characterized by symptoms associated with compression of the rostral midbrain and pretectum near the level of the superior colliculus. Patients with Parinaud's syndrome may have a downgaze at rest, known as the setting sun sign. Patients may show a "pseudo-Argyll Robertson pupil"; in this condition, the pupil is poorly reactive to light but constricts with convergence. Infarction of the ventromedial midbrain results in "Weber's syndrome," characterized mainly by an ipsilateral third nerve palsy and contralateral weakness [52, 53].

1.7.2 Pons

The term "pons" is a Latin word that means "bridge." The pons is horseshoe-shaped, located in the posterior cranial fossa. The pons is anatomically linked superiorly to the midbrain, posteriorly to the fourth ventricle and cerebellum, and inferiorly to the medulla oblongata [54]. The main function of the pons is acting as a relay station between the forebrain and cerebellum and transmitting sensory information from the periphery to the thalamus. The tracts linked to the pons conduct signals from the cerebrum to the cerebellum and medulla oblongata, and tracts carry the sensory signals toward the thalamus. The pons contains many nuclei that transmit signals from the forebrain to the cerebellum and is involved in various essential physiological functions, including regulation of respiration, swallowing, sleep, eye movement, facial expressions, facial sensation, hearing, equilibrium, a sensation of taste, and body posture [48].

The nuclei of the cranial nerves, including the abducens (VI), facial nerve (VII), and vestibulocochlear nerve (VIII), and the motor part of the trigeminal nerve (V) are located in the pons. "These nuclei give rise to pontocerebellar fibers that cross the midline to form the middle cerebellar peduncle. This is particularly vulnerable to injury in malnourished individuals" [54]. The pons is directly or indirectly involved in various major body functions. Lesions in this area, which contain enormous myelinated fibers, such as the thalamus, basal ganglia, and cerebellum, result in various neurological disorders and disruption of major body physiological functions [48, 54].

1.7.3 Medulla Oblongata

The medulla oblongata, also known as the medulla, is conical in shape and is an essential component of the lower part of the brainstem. It is linked to the midbrain by the pons and continuous posteriorly to the spinal cord. The medulla oblongata consists of myelinated white matter and unmyelinated gray matter nerve fibers. It plays a significant role in conveying signals between the spinal cord and higher brain centers and controlling autonomic functions related to the cardiorespiratory system. The medulla oblongata contains the nuclei of the cranial nerves "glossopharyngeal (IX), vagus (X), accessory (XI), and hypoglossal (XII)" nerves and a part of the trigeminal (V) nerve [48]. The medulla oblongata regulates the functions of the autonomic nervous system (ANS) and connects the higher levels of the brain to the spinal cord. It also facilitates numerous basic body functions of the ANS, including cardiovascular and respiratory functions and reflex centers of coughing, sneezing, swallowing, and vomiting. Lesions or diseases affecting the middle part of the medulla may cause "medial medullary syndrome," which is characterized by partial paralysis of the opposite side of the body and loss of the touch and position sensations. However, lesions or diseases of the lateral medulla may cause "lateral medullary syndrome," which is associated with impairments of pain and temperature sensation, difficulty in swallowing, and loss of coordination.

1.8 Spinal Cord

The spinal cord is a long, delicate, cylindrical, tube-like organ and an essential part of the CNS. The spinal cord starts at the end of the brainstem and continues down to the bottom of the spine. It is 40–50 cm long and 1–1.5 cm in diameter. The spinal cord is housed in a highly protective, solid bony structure of the vertebral column, also known as the spinal column. It is surrounded by three membranes, also known as the meninges: "the outer membrane is known as the dura mater, the middle membrane is the arachnoid, and the innermost the pia mater." The spinal cord is divided into four parts—the cervical, thoracic, lumbar, and sacral spines. It comprises several segments and consists of 31 pairs of spinal nerves that carry incoming and outgoing signals or messages between the brain and body. It is a center for body reflexes, which coordinate various body reflexes. Injury to the spinal cord disturbs the pathway between the brain and body and may result in impairments in sensation, movement, autonomic function regulation, disability, and death [55].

The spinal cord consists of H-shaped gray matter and is surrounded by white matter. The gray matter contains cell bodies of motor and sensory neurons and interneurons, whereas the white matter comprises interconnecting fibers, which are primarily myelinated sensory and motor axons. The central canal is located in the center of the spinal cord, which is continuous with the ventricles and contains cerebrospinal fluid (CSF). The CSF is a clear, colorless fluid surrounding the brain and spinal cord and provides a cushion-like function to the brain and spinal cord [55].

The spinal cord act as a channel between the brain and the rest of the body. It transmits sensory information from afferent fibers to the sensory cortex and sends motor commands from the motor cortex to the muscles of the body. In some situations, the spinal cord can perform its function without signals from the brain and independently organizes reflexes through the reflex arcs. The reflex arcs permit the body to respond accordingly to the sensory information without waiting for the regulatory command from the brain and carry signals from the sensory receptor carried to the spinal cord through sensory nerve fibers, synapsed on an interneuron and carried to a motor neuron, which stimulates an effector muscle or organ [56].

The spinal nerves and segments correspond to the number of vertebrae with a minute, insignificant exception. There are 31 pairs of spinal nerves: eight pairs of cervical spinal nerves, 12 pairs of thoracic spinal nerves, five pairs of lumbar spinal nerves, five pairs of sacral spinal nerves, and one pair of coccygeal spinal nerves, parallel to a spinal segment [57] (Figure 1.13). The spinal cord consists of various bundles of nerve fibers and runs down from the brain through a canal in the center of the spine. These bones protect the spinal cord. Similar to the brain, the spinal cord is covered by the meninges and cushioned by CSF.

Spinal cord injuries frequently occur after high-energy mechanisms and can cause numerous life-threatening complications. Lesions in the spinal cord may cause tetraplegia or paraplegia, depending on whether the lesion is complete or incomplete and the level at which the damage has occurred. Tetraplegia is characterized by impaired or loss of sensory and/or motor function in the cervical segments of the spinal cord. It affects the function of the arms, legs, pelvis, and trunk. Patients with paraplegia experience loss of sensory and/or motor functions in the thoracic, lumbar, or sacral segments of the spinal cord.

Local deformation of the spine and compression of the spinal cord can cause primary spinal cord injuries. However, secondary spinal cord injuries may occur following primary damage and involve biological, biochemical, and cellular processes, including free radical damage, electrolyte disorders, inflammation, edema, and ischemia [58, 59]. These biological and biochemical disturbances can cause severe neurological effects on the human body.

A lesion or damage at the level of both sides of the cervical region of the spinal cord results in quadriplegia; if only one side is affected, it can cause hemiplegia. The C3, C4, and C5 nerve

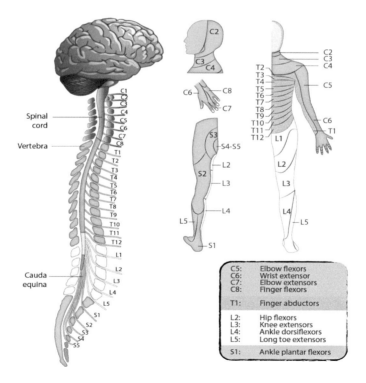

Figure 1.13 The spinal cord.

18

fibers activate the phrenic nerve, which controls the main respiratory muscle, "diaphragm"; damage to these nerve fibers, especially C4 and above, may result in a severe respiratory failure. Injury to the thoracic region limits the function of nerves related to the lower extremities and does not affect the upper extremities [60, 61]. The spinal cord communicates autonomic and somatic information; a lesion in the spinal cord can cause loss of bowel, bladder, and sexual functions and disturb the autoregulation of arterial blood pressure [60, 61].

A lesion affecting the anterior two-thirds of the spinal cord may cause a loss of motor control below the lesion with intact crude sensation; the condition is known as anterior spinal cord syndrome. This syndrome is most probably observed due to the impaired "blood supply to the anterior two-thirds of the spinal cord, damaging the corticospinal and spinothalamic tracts. This syndrome is characterized by deficits at and below the lesion, including motor loss and a loss of pain and temperature sensation. However, light touch and joint position sensation from the dorsal columns are left intact [62]. Injury to the T12–L2 segments might result in conus medullaris syndrome, whereas injury to the L3–L5 segments might lead to cauda equina syndrome. These conditions present as incomplete injuries and result in neurogenic bladder and/ or bowel, loss of sexual function, and perianal loss of sensation" [63].

1.9 Peripheral Nervous System (PNS)

PNS is divided into two main systems: the somatic nervous system (SNS) and ANS (Figure 1.14). The SNS controls the body organs, which are under voluntary control, mainly skeletal muscles. It contains one motor neuron extending from the CNS to skeletal muscles. The axons are well myelinated and rapidly conduct impulses. The SNS connects the CNS to body muscles and controls voluntary movements and reflex arcs. However, the ANS is not under voluntary control and contains a chain of two motor neurons—preganglionic and postganglionic neurons—and conducts slow impulses due to thin or unmyelinated axons. It regulates individual organs, visceral functions, and homeostasis, known as the visceral or automatic system. The ANS is divided into two main systems: the sympathetic nervous system and parasympathetic nervous system (PNS) (Figure 1.14) [5].

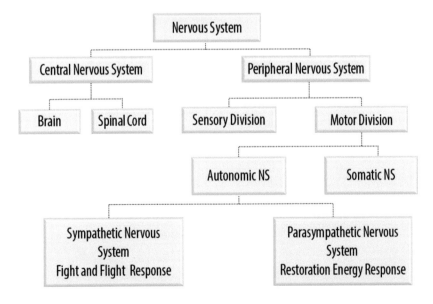

Figure 1.14 Classification of nervous system: sensory and motor divisions.

On the basis of various physiological aspects of the nervous system, the PNS is also classified into sensory and motor divisions. The cells that comprise the PNS are the nerves and ganglia. The PNS consists of neuronal cells, which lie outside of the CNS. Its main function is to connect the CNS to other body organs and systems. Ganglia act as relay positions for messages, which are transmitted through nerves. PNS nerves are classified as sensory and motor nerves.

1.10 Autonomic Nervous System (ANS)

The ANS is a highly important component of the nervous system. It controls various involuntarily body activities and is mainly an efferent system; transmits an impulse from the CNS to peripheral organ systems to regulate smooth muscles, cardiac muscles, glands, and hormonal secretion; and serves as body homeostatic function. The ANS has a significant influence on the structural and physiological integrity and electrical conductivity to timely regulate and balance various body functions according to the required situation as many body functions are not under the control of the subject [64].

The ANS consists of two main systems: the sympathetic nervous system and PNS (Figure 1.14). It consists of efferent neurons that extend from the CNS to effector structures other than the skeletal muscles. These structures include cardiac muscles; smooth muscles lining the gastrointestinal, respiratory, reproductive, and urinary tracts; and smooth muscles of the blood vessels, iris, ciliary body of the eye, and both the endocrine and exocrine glands.

The ANS is a highly dominant system that works under different environments and body conditions. The fundamental function of the ANS is to align and adjust internal functions of body organs, which are appropriate to the conditions within the body at any specific time. This system controls the physiological role of the internal organs, blood vessels, and effectors. It provides a motor control mechanism to body organs through the sympathetic, parasympathetic, and enteric nervous system (ENS) divisions (Figure 1.15).

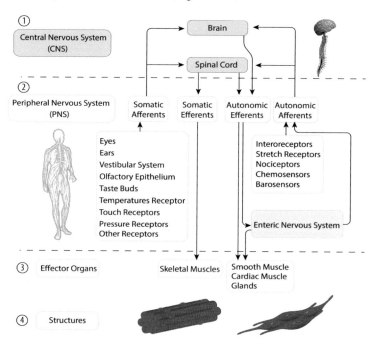

Figure 1.15 Organizational physiology of the central and peripheral nervous system.

1.10.1 Sympathetic Nervous System

The sympathetic nervous system originates from the thoracic and lumbar regions of the spinal cord (T1–T12 and L1—L3); hence, it is known as the thoracolumbar system. The sympathetic nervous system is the larger of the two parts of the ANS. It is widely distributed throughout the body, enervating different structures, including the heart, lungs, many abdominal and pelvic viscera, smooth muscles, hair follicles, and sweat glands.

The sympathetic nervous system controls metabolic functions and coordinates emotion-, exercise-, and emergency-based physiological and biological responses of the body and plays a role in life-threatening situations (i.e., the "fight or flight" response). It plays a major role during "Emergency, Embarrassment, Excitement, Exercise" situations. Hence, the sympathetic nervous system is also known as the accelerator pedal or stress response system. When the system activates, it increases the heart rate, the force of cardiac contraction, blood pressure, and respiratory rate, along with various other stress- and emotion-associated body activities.

1.10.2 Parasympathetic Nervous System

The PNS originates from the craniosacral region, including four "cranial nerves (III, VII, IX, and X) and three sacral nerves (S2, S3, and S4)." The PNS performs "rest and digest" activities; hence, this system is known to preserve and restore energy. It plays a role in the restoration of metabolic reserves and the elimination of waste products. It decreases the "heart rate, the force of contraction, stroke volume, cardiac output, and blood pressure and facilitates digestion and absorption of nutrients and consequently the excretion of waste products." The PNS is also known as the brake pedal system, as this system brings various body physiological functions toward the normal physiological response. Acetylcholine (Ach) is a chemical transmitter at pre-and postganglionic synapses in the PNS (Figure 1.16). The physiological effects caused by each system are quite anticipated during daily activities. Most changes in tissue function and organs induced by the sympathetic system work together to support strenuous physical activities, and the changes induced by the parasympathetic system are appropriate when the body is resting [65].

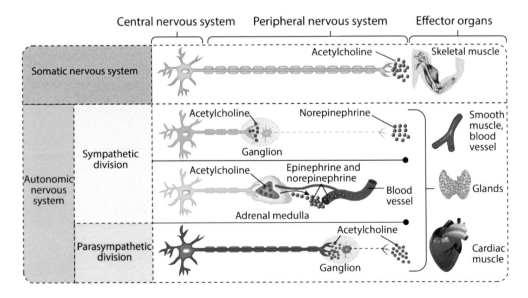

Figure 1.16 Sympathetic and parasympathetic divisions of the autonomic nervous system.

1.10.3 The Enteric Nervous System (ENS)

The ENS is the biggest part of the ANS and plays a vital role in biological and physiological functions. It consists of multiple neural circuits and controls motor functions, local blood flow, mucosal carriage, secretions, and endocrine functions. The circuits of enteric neurons are organized in networks of enteric ganglia associated with inter-ganglionic strands. They are linked with the "myenteric plexus" by some primary afferent neurons situated in the submucous plexus [66, 67].

There are biological similarities between the neurotransmitter signaling pathways of the ENS and CNS; hence, the pathophysiological processes that underlie CNS diseases often have enteric manifestations. The ENS plays a major role in coordinating the activities of vital organs. Its dysfunction is the primary cause of digestive disorders [67].

1.11 Memory

Over the past 200 years, the science of learning and memory has been under the discussion of various disciplines of science and social sciences, including philosophy, neuropsychology, and brain biology. The biological queries began during the latter part of the 20th century, as there is a great advancement in modern medical sciences and the involvement of progressive technology that made it feasible to move beyond the narrative of the exploration of mechanisms [68].

The possibility of researching its biological and physiological functions is improved by the involvement of advanced genetic, molecular, biological, and radiological techniques. Modern technology-based advancements provide comprehensive neuroanatomical information about the connectivity of memory with various regions of the brain [69,70]. The hippocampus, anterior cingulate cortex, and associated parts of the brain are essential to recall the memory, its reorganization, and consolidation during a long period after the learning [71]. Moreover, neuronal activities or structural changes in the MTL after learning are well acknowledged in the scientific literature about their role in memory [72].

The word "memory" is derived from the Latin word "memoir," defined as the ability of the brain to encode, store, retain, and recall information. Memory is a highly complex function of the brain to store and retrieve information. It provides learning abilities: acquiring knowledge processes while adapting previous experiences, which influence behaviors [73, 74]. There are two main types of memory: (1) explicit memory and (2) implicit memory. Explicit memory is the conscious recollection of previous experiences, whereas implicit memory is the past experiences that influence current behavior. Working memory is a concept that involves the amplification of thoughts, prediction, and presentation of higher intellectual functions performed by the prefrontal areas of the brain. The prefrontal area tracks much information and recollects information instantly, as it controls successive thoughts known as "working memory," associated with higher intelligence [73, 74].

The prefrontal area is divided into segments to store temporary information. By combining temporary moments of working memory, human beings could perform the following abilities:

- Predict
- Plan for the future
- Action delay until the suitable time and course of response
- Consideration of the consequences of motor actions
- Solve complicated problems
- Correlate all avenues of information and
- Control activities in accord with moral values

Memory is classified into three main types:

- Short-term memory (STM)
- Intermediate-term memory
- Long-term memory [74]

1.11.1 Short-Term Memory (STM)

STM, also known as short-term recall, storage, or primary or active memory, specifies various memory arrangements in the retention of pieces of information for a relatively short time of approximately 30s. However, long-term memory may keep and remember an amount of information for an indefinite period [75].

STM and long-term memory are closely linked, as STM works for momentary recall of limited information in the verbal domain that comes from the sensory register and is ready to be processed through attention and recognition. However, long-term memory consists of recollections of information for the performance of actions or skills, facts, rules, concepts, and events [76]. STM is physiologically linked with the concept of "working memory," which is characterized by two distinct entities. STM is a set of storage systems, whereas working memory shows cognitive operations and executive functions related to the organization of stored information [77].

STM demonstrates various representations, including illustrations, pictures, symbols, and communications, composed of a large number of features and stored information, including features representing the context in which that information was encountered [78]. STM lasts for seconds or a few minutes unless converted into intermediate- or long-term memory. STM is initiated by continuous neural activity resulting from neuronal signals traveling around a temporary memory trace in a circuit of reverberating neurons. It may be due to the presynaptic facilitation or inhibition, which occurs at synapses. The neurotransmitters secreted at terminals frequently facilitate or inhibit information storage, lasting from seconds to several minutes [74].

Subjects with parietal and temporal lobe lesions have clinical features of impaired short-term phonological capabilities but intact long-term memory [79, 80]. However, subjects with lesions in the MTL have features of impaired long-term memory but preserved STM [81]. However, some studies on human models have demonstrated mixed results.

Three are three principal processes of STM: encoding, maintenance, and retrieval. A partial or limited concentration or attention is a key characteristic of STM, thoughtful processing linked to limited attention to know the three primary cognitive events [82]. First, "encoding processes govern the transformation from perceptual representations into the cognitive/attentional focus. Second, maintenance processes keep information in the focus and protect it from interference or decay. Third, retrieval processes bring information from the past back into the cognitive focus, possibly reactivating perceptual representations" [73].

Because of STM, people forget many things, which is disturbing. There are two major explanations for forgetting: time-based decay and similarity-based interference [78]. The mechanism of forgetting happens when things, objects, or contents leave the sphere of attention and must compete with other items to regain focus (interference) or when the representation declines over time due to stochastic processes in which the chance may occur without a threshold level of dose (decay) [78]. The major argument in the scientific literature is the decay theory: with time, the information in memory erodes, making it less available for later retrieval [83].

1.11.2 Intermediate-Term Memory

Intermediate-term memory is a condition between working memory and long-term memory. This type of memory lasts for minutes, hours, or weeks and eventually disappears unless the memory traces are adequately activated to become permanent. The animal model studies have demonstrated that temporary chemical or physical changes, or both, at the presynaptic or post-synaptic membrane, may persist for a few minutes up to several weeks [84].

After encoding, an item can be preserved in an active state. The short-term synaptic plasticity (STSP) allows information recovery if the information is temporarily out of focus. STSP also initiates the mechanism of producing a stable memory, which requires longer-lasting processes, such as gene transcription. Long-term synaptic plasticity builds stable long-term memories. Whenever information in working memory switches states to a latent form, the process of encoding the information to long-term memory can be stopped in each stage, and information is lost; however, if the process continues, the memory can be brought back to the active state [85].

1.11.3 Long-Term Memory

Long-term memories are generally divided into two diverse groups: procedural and declarative memories. Procedural memories involve actions or events learned through practicing with frequent exposure to a series of motor activities, for example, driving a motorbike. However, declarative memories could be intentionally recalled and divided into two other groups: semantic memories representing distinct facts such as days, dates, word meanings, and learned notions; and episodic memories characterized by clear experiences one has, such as a birthday or wedding day. Studies have specified two groups of long-term memory: implicit and explicit. However, these terms are not used as much because they are more equivocal. Long-term memory is consolidated from STM to long-term memory primarily in the hippocampus and stored throughout the cortex [86–89]. Long-term memory can be recalled for up to years due to structural changes at the synapses, increasing or suppressing signal conduction.

Brain biology has a unique and specific function in memory development. The hippocampus plays an important role in consolidating memories from the short term to the long term. However, the amygdala enhances emotional pertinence to memories [90, 91]. "Encoding and storage of procedural memories also involve multiple brain areas. The cerebellum, basal ganglia, and association cortices contribute in learned actions as they are related to motor control and adjustment."

1.11.4 Other Types of Memory

Working memory is mainly STM used during intellectual reasoning but is terminated as the problem is resolved. Declarative memory is based on integrated thoughts, such as memory linked to important experiences, including the memory of environments, memory of time relations, and memory of experiences, and subjects can consciously remember [74]. Skill memory is mainly linked to motor activities. For example, the skills developed for playing cricket and hitting a ball toward the boundary, (1) seeing the ball, (2) speed of the ball, and (3) motions of the body on how to hit the ball. These skills are activated instantly and are based on previous learning of the cricket game and the pattern of the event [74].

1.12 Human Intelligence

Human intelligence is an ability that differentiates humans from other animal species. Human intelligence can be defined as the mental, psychological, or intellectual ability for learning, understanding, reasoning, and problem-solving [92]. It is defined as the ability to understand composite ideas, adapt professionally and competently to the environment, and engage in complex reasoning. Intelligence depends on the brain structure's normal structural and functional developments and the genetic makeup shaping brain development [93].

Neuroscience studies and cognitive theories have demonstrated that cognitive control is a core component of human intelligence. Also, larger brains and a larger volume of gray matter in various brain regions are linked to higher intelligence and cognitive functions. Existing evidence has suggested a widely distributed network of frontoparietal brain areas underlying human intelligence. Human intelligence does not reside in a single localized area of the brain [94].

The human brain is capable of the large variability in neural configurations to understand and predict variable external events. Human intelligence is linked to the integration of cognitive functions, such as perception, attention, response, thoughts, planning, and memory. The literature demonstrated that the "frontoparietal network" is interrelated to human intelligence. Also, this network has been found to trigger cognitive functions associated with perception, STM storage, and communication. However, these activities must be integrated with adapting behavior to the environment—selecting the appropriate contexts. The integration of cognitive functions and abilities depends on an individual's mental ability or "general intelligence" [95].

The human brain has outstanding intelligence and higher cognitive capabilities. Studies on brain intelligence have linked specific cognition to various brain nuclei [84, 94]. The intelligence quotient (IQ) scores have a well-established and broadly disseminated pattern of association with brain structures. The IQ scores are linked to the intracranial, cerebral, temporal lobe, hippocampal, and cerebellar volumes [96]. Intelligence is positively correlated with cortical thickness in frontal and temporal lobes areas [97, 98]. The frontal Brodmann areas [10, 45–47], parietal areas [39, 40], and temporal area [21] positively contribute to IQ scores. Intelligence cannot be credited to a specific region but is facilitated by a network of brain regions [99, 100].

The genetic and environmental variances in "mitochondrial functioning affect the integrity and efficiency of neurons and glial cells, which may contribute to individual differences in higher cognitive functioning [101, 102]. Multiple factors contribute to individual differences in intelligence, including genetics [103] and white matter tract integrity in the forceps minor, corticospinal tract, anterior thalamic radiation, rostrolateral prefrontal cortex, and inferior parietal lobe" [104–107], which play a major role in intelligence. Moreover, functional brain characteristics, the activation of frontoparietal brain networks, and connectivity related to higher-order cognitive processes [107] mediate cognitive processes, such as processing speed, attentional control, and working memory [108–111]. Furthermore, the environment affects human intelligence [110, 111].

Conclusion

This chapter provides an insight into the basic functions and structures of the various regions of the brain. It offers various facts about the biology of the brain for a conceptual knowledge of the most sensitive and highly essential structures in the brain. The structural and functional features of the brain and a basic concept of human memory and intelligence are discussed for a better understanding of the highly complex mechanisms involved in the brain. It also provides primary mechanisms that are linked to establishing notions about environmental pollution and brain biology.

References

[1] Herculano-Houzel S. The remarkable, yet not extraordinary, human brain as a scaled-up primate brain and its associated cost. *Proc. Natl. Acad. Sci. U S A.* 2012; 109: 10661–8. doi: 10.1073/pnas.1201895109.
[2] Azevedo F.A., Carvalho L.R., Grinberg L.T., Farfel J.M., Ferretti R.E., Leite R.E., Jacob Filho W., Lent R., Herculano-Houzel S. Equal numbers of neuronal and nonneuronal cells make the human brain an isometrically scaled-up primate brain. *J. Comp. Neurol.* 2009; 513(5): 532–41. doi: 10.1002/cne.21974.

[3] Pakkenberg B., Pelvig D., Marner L., Bundgaard M.J., Gundersen H.J.G., Nyengaard J.R. Aging and the human neocortex. *Exp. Gerontol.* 2003; 38: 95–9. doi: 10.1016/S0531-5565(02)00151-1.

[4] Lee B., Kang U., Chang H., Cho K.H. Th hidden control architecture of complex brain. *Netw. Sci.* 2019; 13: 154–62. doi: 10.1016/j.isci.2019.02.017.

[5] Carnagarin R., Kiuchi M.G., Ho J.K., Matthews V.B., Schlaich M.P. Sympathetic nervous system activation, and its modulation: Role in atrial fibrillation. *Front. Neurosci.* 2019; 12: 1058. doi: 10.3389/fnins.2018.01058.

[6] Splittgerber R. *Snell's Clinical Neuroanatomy. Introduction and Organization of the Nervous System*, 8th edition. Philadelphia: Wolters and Kluwer. 2019, pp. 1–2.

[7] Thau L., Reddy V., Singh P. Anatomy: Central nervous system. In: *StatPearls [Internet].* Treasure Island, FL: StatPearls Publishing. 2020. Cited date May 25, 2020.

[8] Mormann F., Koch C. Neural correlates of consciousness. *Scholarpedia.* 2007; 2(12): 1740. doi: 10.4249/scholarpedia.1740.

[9] Neuron Llinas R. *Scholarpedia.* 2008; 3(8): 1490. doi: 10.4249/scholarpedia.1490.

[10] Lodish H., Berk A., Zipursky S.L., Matsudaira P., Baltimore D., Darnell J. *Molecular Cell Biology*, 4th edition. New York: WH Freeman. 2000. ISBN-10: 0-7167-3136-3.

[11] Delandre C., Amikura R., Moore A.W. Microtubule nucleation and organization in dendrites. *Cell Cycle.* 2016; 15(13): 1685–92.

[12] Yuh-Nung J., Yeh-Jan L. Branching out mechanisms of dendritic arborization. *Nat. Rev. Neurosci.* 2010; 11(5): 316–28. doi: 10.1038/nrn2836.

[13] Craig A.M., Banker G. Neuronal polarity. *Annu. Rev. Neurosci.* 1994; 17: 267–310.

[14] Baas P.W., Deitch J.S., Black M.M., Banker G.A. Polarity orientation of microtubules in hippocampal neurons: Uniformity in the axon and nonuniformity in the dendrite. *Proc. Natl. Acad. Sci. The USA.* 1988; 85: 8335–9.

[15] Rolls M.M., Satoh D., Clyne P.J. Polarity and intracellular compartmentalization of drosophila neurons. *Neural. Dev.* 2007; 2: 1–14. https://doi.org/10.1186/1749-8104-2-7.

[16] Shepherd G.M. Dendrodendritic synapses: Past, present, and future. *Ann. N. Y. Acad. Sci.* 2009; 1170. doi: 10.1111/j.1749-6632.2009.03937.x.

[17] Abraira V.E., Ginty D.D. The sensory neurons of touch. *Neuron.* 2013; 79(4): 618–39. doi: 10.1016/j.neuron.2013.07.051.

[18] Heckman C.J., Enoka R.M. Physiology of the motor neuron and the motor unit. *Handb. Clin. Neurol.* 2004; 4: 119–47.

[19] Budday S., Steinmann P., Kuhl E. Physical biology of human brain development. *Front Cell Neurosci.* 2015; 9: 257. doi: 10.3389/fncel.2015.00257.

[20] Goyal R.K., Chaudhury A. Structure-activity relationship of synaptic and junctional neurotransmission. *Auton. Neurosci.* 2013; 176: 11–31.doi: 10.1016/j.autneu.2013.02.012.

[21] Ferrer-Ferrer M., Dityatev A. Shaping synapses by the neural extracellular matrix front. *Neuroanat.* 2018; 12: 40. doi.org/10.3389/fnana.2018.00040.

[22] Thomas C.S., Robert C. Malenka. Understanding synapses: Past, present, and future. *Neuron.* 2008; 60(3): 469–76. doi: 10.1016/j.neuron.2008.10.011.

[23] Andrzej T.S., Zmijewski Michal A., Cezary Skobowiat, Blazej Zbytek, Slominski Radomir M., Steketee Jeffery D. Sensing the environment: Regulation of local and global homeostasis by the skin neuroendocrine system. *Adv. Anat. Embryol. Cell. Biol.* 2012; 212: 115.

[24] Biga L.M., Dawson S., Harwell A., Hopkins R., Kaufmann J., LeMaster M., Matern P., Morrison-Graham K., Quick D., Runyeon J. Sensory receptors. In: *Anatomy and Physiology*, 1st edition. Corvallis, OR: Oregon State University. 2020, pp. 1–13.

[25] Khalid H.J. *Sandeep Sharma. Physiology, Cerebral Cortex Functions.* Delaware, Philadelphia: StatPearls Publishing. 2020, pp. 1–6.

[26] Neulinger K., Oram J., Tinson H., O'Gorman J., Shum D.H. Prospective memory and frontal lobe function. *Neuropsychol. Dev. Cogn. B. Aging Neuropsychol. Cogn.* 2016; 23(2): 171–83.

[27] Flinker A., Korzeniewska A., Shestyuk A.Y., Franaszczuk P.J., Dronkers N.F., Knight R.T., Crone N.E. Redefining the role of Broca's area in speech. *Proc. Natl. Acad. Sci. USA.* 2015; 112(9): 2871–5.

[28] Barrash J., Stuss D.T., Aksan N., Anderson S.W., Jones R.D., Manzel K., Tranel D. "Frontal lobe syndrome"? Subtypes of acquired personality disturbances in patients with focal brain damage. *Cortex.* 2018; 106: 65–80.

[29] Collins A., Reasoning Koechlin E., learning, and creativity: Frontal lobe function and human decision-making. *PLoS Biol.* 2012; 10(3): e1001293.

[30] Berlucchi G., Vallar G. The history of the neurophysiology and neurology of the parietal lobe. *Handb. Clin. Neurol.* 2018; 151: 3–30.

[31] Klingner C.M., Witte O.W. Somatosensory deficits. *Handb. Clin. Neurol.* 2018; 151: 185–206.

[32] Kiernan J.A. Anatomy of the temporal lobe. *Epilepsy. Res. Treat.* 2012; 2012: 176157.

[33] Conway B.R. The organization and operation of inferior temporal cortex. *Annu. Rev. Vis. Sci.* 2018; 4: 381–402.

[34] Clark R.E. Current topics regarding the function of the medial temporal lobe memory system. *Curr. Top. Behav. Neurosci.* 2018; 37: 13–42.

[35] Flores L.P. Occipital lobe morphological anatomy: Anatomical and surgical aspects. *Arq. Neuro-Psiquiatr.* 2002; 60(3A). doi.org/10.1590/S0004-282X2002000400010.

[36] Malloy P.F., Richardson E.D. Assessment of frontal lobe functions. *J. Neuropsychiatry. Clin. Neurosci.* 1994; 6(4): 399–410.

[37] Helmstaedter C., Elger C.E., Vogt V.L. Cognitive outcomes more than 5 years after temporal lobe epilepsy surgery: Remarkable functional recovery when seizures are controlled. *Seizure.* 2018; 62: 116–23.

[38] Li W.K., Hausknecht M.J., Stone P., Mauk M.D. Using a million cell simulation of the cerebellum: Network scaling and task generality. *Neural. Netw.* 2013; 47(2013): 95–102.

[39] Salman M.S., Tsai P. The role of the pediatric cerebellum in motor functions, cognition, and behavior: A clinical perspectiv. *Neuroimaging Clin. N. Am.* 2016; 26(3): 317–29.

[40] Moberget T., Ivry R.B. Cerebellar contributions to motor control and language comprehension: Searching for common computational principles. *Ann. N. Y. Acad. Sci.* 2016; 1369(1): 154–71. doi: 10.1111/nyas.13094.

[41] Yi-Cheng L., Chih-Chin H.H., Pei-Ning W., Ching-Po L., Li-Hung C. The relationship between zebrin expression and cerebellar functions: Insights from neuroimaging studies. *Front Neurol.* 2020; 11: 315. doi: 10.3389/fneur.2020.00315.

[42] AlfBrodal. Anatomical studies of cerebellar fibre connections with special reference to problems of functional localization. *Prog. Brain Res.* 1967; 25: 135–73.

[43] Bodranghien F., Bastian A., Casali C., Hallett M., Louis E.D., Manto M., Mariën P., Nowak D.A., Schmahmann J.D., Serrao M., Steiner K.M., Strupp M., Tilikete C., Timmann D., van Dun K. Consensus paper: Revisiting the symptoms and signs of cerebellar syndrome. *Cerebellum.* 2016; 15(3): 369–91.

[44] Holmes G. The symptoms of acute cerebellar injuries from gunshot wounds. *Brain.* 1917; 40: 461–535.

[45] Lawrenson C., Bares M., Kamondi A., Kovács A., Lumb B., Apps R., Filip P., Manto M. The mystery of the cerebellum: Clues from experimental and clinical observations. 2018; 29(5): 8. doi: 10.1186/s40673-018-0087-9.

[46] Cole J.D., Philip H.I., Sedgwick E.M. Stability and tremor in the fingers associated with cerebellar hemisphere and cerebellar tract lesions in man. *J. Neurol. Neurosurg. Psychiatry.* 1988; 51: 1558–68.

[47] Konczak J., Pierscianek D., Hirsiger S., Bultmann U., Schoch B., Gizewski E.R., Timmann D., Maschke M., Frings M. Recovery of upper limb function after cerebellar stroke: Lesion symptom mapping and arm kinematics. *Stroke.* 2010; 41(10): 2191–200.

[48] Robin A.H., Flashman Laura A., Chow Tiffany W., Taber Katherine H. The brainstem: Anatomy, assessment, and clinical syndromes. *J Neuropsychiatry Clin Neurosci.* 2010; 22(1): iv, 1–7. doi: 10.1176/jnp.2010.22.1.iv.

[49] Hayman L.A., Taber K.H., Dubey N. A functional atlas of brain vascular territories. In: Latchaw R.E., Kucharczyk J., Moseley M.E., editors. *Imaging of the Nervous System.* Philadelphia: Elsevier Mosby. 2005, pp. 179–97.

[50] Van Zandvoort M., de Haan E., van Gijn J. Cognitive functioning in patients with a small infarct in the brainstem. *J. Int. Neuropsychol. Soc.* 2003; 9: 490–94.

[51] Ruchalski K., Gasser M., Hathout A. Medley of midbrain maladies: A brief review of midbrain anatomy and syndromology for radiologists. *Radiol. Res. Pract.* 2012; 2012: 258524. doi: 10.1155/2012/258524.

[52] Kim J.S., Kim J. Pure midbrain infarction: Clinical, radiologic, and pathophysiologic findings. *Neurology*. 2005; 64(7): 1227–32.

[53] Derakshan I., Sabouri-Deylami M., Kaufman B. Bilateral Nothnagel syndrome. Clinical and roentgenological observations. *Stroke*. 1980; 11(2): 177–9.

[54] Kumar S., Fowler M., Gonzalez-Toldeo E., et al. Central pontine myelinolysis: An update. *Neurol. Res*. 2006; 28: 360–6.

[55] Harrow-Mortelliti M., Reddy V., Jimsheleishvili G. *Physiology, Spinal Cord in StatPearls [Internet]*. Treasure Island, FL: StatPearls Publishing. 2020.

[56] Minassian K., Hofstoetter U.S., Dzeladini F., Guertin P.A., Ijspeert A. The human central pattern generator for locomotion: Does it exist and contribute to walking? *Neuroscientist*. 2017; 23(6): 649–63.

[57] Nógrádi A., Vrbová G. Anatomy and physiology of the spinal cord. In: *Transplantation of Neural Tissue into the Spinal Cord*. Neuroscience Intelligence Unit. Boston, MA: Springer. 2006. https://doi.org/10.1007/0-387-32633-2_1

[58] Ambrozaitis K.V., Kontautas E., Spakauskas B., Vaitkaitis D. Pathophysiology of acute spinal cord injury. *Medicina (Kaunas)*. 2006; 42(3): 255–61.

[59] Kim Y.H., Ha K.Y., Kim S.I. Spinal cord injury and related clinical trials. *Clin. Orthop. Surg*. 2017; 9(1): 1–9.

[60] Belanger E., Picard C., Lacerte D., Lavallee P., Levi A.D. Subacute posttraumatic ascending myelopathy after spinal cord injury. Report of three cases. *J. Neurosurg*. 2000; 93(2 Suppl): 294–9.

[61] Al-Ghatany M., Al-Shraim M., Levi A.D., Midha R. Pathological features including apoptosis in subacute posttraumatic ascending myelopathy. Case report and review of the literature. *J. Neurosurg. Spine*. 2005 May; 2(5): 619–23.

[62] Kirshblum S.C., Burns S.P., Biering-Sorensen F., Donovan W., Graves D.E., Jha A., Johansen M., Jones L., Krassioukov A., Mulcahey M.J., Schmidt-Read M., Waring W. International standards for neurological classification of spinal cord injury (revised 2011). *J. Spinal. Cord. Med*. 2011; 34(6): 535–46.

[63] Brouwers E., van de Meent H., Curt A., Starremans B., Hosman A., Bartels R. Definitions of traumatic conus medullaris and cauda equina syndrome: A systematic literature review. *Spinal. Cord*. 2017(10): 886–90.

[64] Meenakshi R., Michael D. Gershon. The bowel and beyond: The enteric nervous system in neurological disorders. *Nat. Rev. Gastroenterol Hepatol*. 2016; 13: 517–28.

[65] McCorry L.K. Physiology of the autonomic nervous system. *Am. J. Pharm. Educ*. 2007; 71(4): 1–11. doi: 10.5688/aj710478.

[66] Keihn R.M.H-W.O., Jordan L.M., Hultborn H., Costa M., Hennig G.W., Brookes S.J.H. Intestinal peristalsis: A mammalian motor pattern controlled by enteric neural circuits. In: Keihn R.M.H-W.O., Jordan L.M., Hultborn H., editors. *Neuronal Mechanisms for Generating Locomotor Activity*. Vol. 860. New York: New York Academy of Sciences. 1998, pp. 464–6.

[67] Costa M., Brookes S J H., Hennig G.W. Anatomy and physiology of the enteric nervous system. *Gut*. 2000; 47(Suppl IV): iv15–iv19.

[68] Squire L.R. Memory and brain systems: 1969–2009. *J. Neurosci*. 2009; 29(41): 12711–6. doi: 10.1523/JNEUROSCI.3575-09.2009.

[69] Bakker A., Kirwan C.B., Miller M., Stark C.E. Pattern separation in the human hippocampal CA3 and dentate gyrus. *Sci*. 2008; 319(5870): 1640–2.

[70] Nakashiba T., Young J.Z., McHugh T.J., Buhl D.L., Tonegawa S. Transgenic inhibition of synaptic transmission reveals a role of CA3 output in hippocampal learning. *Sci*. 2008; 319(5867): 1260–4.

[71] Squire L.R. Lost forever or temporarily misplaced? The long debate about the nature of memory impairment. *Learn Mem*. 2006; 13(5): 522–9.

[72] Restivo L., Vetere G., Bontempi B., Ammassari-Teule M. The formation of recent and remote memory is associated with the time-dependent formation of dendritic spines in the hippocampus and anterior cingulate cortex. *J. Neurosci*. 2009; 29(25): 8206–14.

[73] Velez-Pardo C., Jimenez-Del-Rio M. From dark to bright to gray sides of memory: In search of its molecular basis & Alzheimer's disease. *Austin. J. Clin. Neurol*. 2015; 2(5): 1045.

[74] Guyton and Hall. *Text Book of Medical Physiology*, 13th edition. Philadelphia: Elsevier. 2016, pp. 744–50.

[75] Cascella M., Al Khalili Y. Short term memory impairment. In: *StatPearls*. Treasure Island, FL: StatPearls Publishing. 2020. PMID: 31424720.

[76] Cowan N. George Miller's magical number of immediate memory in retrospect: Observations on the faltering progression of science. *Psychol. Rev.* 2015; 122(3): 536–41.

[77] Miller E.K., Lundqvist M., Bastos A.M. Working memory 2.0. *Neuron.* 2018; 100(2): 463–75.

[78] Jonides, J., Lewis Richard L., Derek Evan Nee, Lustig Cindy A., Berman Marc G., Katherine Sledge Moore. The mind and brain of short-term memory. *Annu. Rev. Psychol.* 2008; 59: 193–224. doi: 10.1146/annurev.psych.59.103006.093615.

[79] Shallice T., Warrington E.K. Independent functioning of verbal memory stores: A neuropsychological study. *Q. J. Exp. Psychol.* 1970; 22: 261–73.

[80] Vallar G., Papagno C. Neuropsychological impairments of verbal short-term memory. In: Baddeley A.D., Kopelman M.D., Wilson B.A., editors. *The Handbook of Memory Disorders.* 2nd edition. Chichester, UK: Wiley. 2002, pp. 249–70.

[81] Scoville W.B., Milner B. Loss of recent memory after bilateral hippocampal lesions. *J. Neurol. Neurosurg. Psychiatry.* 1957; 20: 11–21.

[82] Anderson J.R. Retrieval of information from long-term memory. *Sci.* 1983; 220(4592): 25–30.

[83] Lewandowsky S., Duncan M., Brown G.D. Time does not cause forgetting in short-term serial recall. *Psychon. Bull. Rev.* 2004; 11(5): 771–90.

[84] Deary I.J., Strand S., Smith P., Fernandes C. Intelligence and educational achievement. *Intelligence.* 2007; 35: 13–21. doi: 10.1016/j.intell.2006.

[85] Kamiński J. Intermediate-term memory as a bridge between working and long-term memory. *J. Neurosci.* 2017; 37(20): 5045–7.

[86] Norris D. Short-term memory and long-term memory are still different. *Psychol. Bull.* 2017 Sep; 143(9): 992–1009.

[87] Jeneson A., Squire L.R. Working memory, long-term memory, and medial temporal lobe function. *Learn. Mem.* 2012; 19(1): 15–25.

[88] Lum J.A., Conti-Ramsden G., Page D., Ullman M.T. Working, declarative and procedural memory in specific language impairment. *Cortex.* 2012; 48(9): 1138–54.

[89] Müller N.C., Genzel L., Konrad B.N., Pawlowski M., Neville D., Fernández G., Steiger A., Dresler M. Motor skills enhance procedural memory formation and protect against age-related decline. *PLoS One.* 2016; 11(6): e0157770.

[90] Rolls E.T., Treves A. The neuronal encoding of information in the brain. *Prog. Neurobiol.* 2011; 95(3): 448–90.

[91] Haber S.N. Corticostriatal circuitry. *Dialogues Clin Neurosci.* 2016; 18(1): 7–21.

[92] Colom R., Karama S., Jung R.E., Haier R.J. Human intelligence, and brain networks dialogues. *Clin. Neurosci.* 2010; 12(4): 489–501.

[93] Tang Y.P., Shimizu E., Dube G.R. Genetic enhancement of learning and memory in mice. *Nature.* 1999; 401(6748): 63–9.

[94] Deary I.J., Penke L., Johnson W. The neuroscience of human intelligence differences. *Nat. Rev. Neurosci.* 2010; 11: 201–11. doi: 10.1038/nrn2793.

[95] Van der M.H., Dolan C.V., Grasman R.P.P.P., Wicherts J.M., Huizengan H.M., Raijmakers M.E.J. A dynamical model of general intelligence: The positive manifold of intelligence by mutualism. *Psychol. Rev.* 2007; 113: 842–61.

[96] Andreasen N.C., Flaum M., Swayze V I.I., O'Leary D.S., Alliger R., Cohen G. Intelligence and brain structure in normal individuals. *Am. J. Psychiatry.* 1993; 150: 130–4.

[97] Narr K.L., Woods R.P., Thompson P.M., Szeszko P., Robinson D., Dimtcheva T. Relationships between IQ and regional cortical gray matter thickness in healthy adults. *Cereb. Cortex.* 2007; 17: 2163–2171.

[98] Choi Y.Y., Shamosh N.A., Cho S.H., DeYoung C.G., Lee M.J., Lee J.M. Multiple bases of human intelligence revealed by cortical thickness and neural activation. *J. Neurosci.* 2008; 28: 10323–9.

[99] Karama S., Ad-Dab'bagh Y., Haier R.J., Deary I.J., Lyttelton O.C., Lepage C. Positive association between cognitive ability and cortical thickness in a representative US sample of healthy 6 to 18 year-olds. *Intelligence.* 2009; 37: 145–55.

[100] Jung R.E., Haier R.J. The parieto-frontal integration theory (PFIT) of intelligence: Converging neuroimaging evidence. *Behav. Brain. Sci.* 2007; 30: 135–54.

[101] Geary D.C. Efficiency of mitochondrial functioning as the fundamental biological mechanism of general intelligence (g). *Psychol Rev.* 2018; 125(6): 1028–50.

[102] Schubert A.L., Hagemann D. The evidence for Geary's theory on the role of mitochondrial functioning in human intelligence is not entirely convincing. *J Intell.* 2020; 8(3): 29. doi: 10.3390/jintelligence8030029.

[103] Plomin R., von Stumm S. The new genetics of intelligence. *Nat Rev Genet.* 2018 Mar; 19(3): 148–59.

[104] José Angel Pineda-Pardo, Martínez Kenia, Román Francisco J., Colom Roberto. Structural efficiency within a parieto-frontal network and cognitive differences. *Intelligence.* 2016; 54: 105–16. doi: 10.1016/j.intell.2015.12.002.

[105] Kievit R.A., Davis S.W., Griffiths J., Correia M.M., Cam-Can, Henson R.N. A watershed model of individual differences in fluid intelligence. *Neuropsychologia.* 2016; 91: 186–98.

[106] Wendelken C., Ferrer E., Ghetti S., Bailey S.K., Cutting L., Bunge S.A. Frontoparietal structural connectivity in childhood predicts development of functional connectivity and reasoning ability: A large-scale longitudinal investigation. *J Neurosci.* 2017; 37(35): 8549–58.

[107] Kirsten Hilger, Ekman Matthias, Fiebach Christian J., Basten Ulrike. Efficient hubs in the intelligent brain: Nodal efficiency of hub regions in the salience network is associated with general intelligence. *Intelligence.* 2017; 60: 10–25. doi: 10.1016/j.intell.2016.11.001.

[108] Schubert A.L., Hagemann D., Löffler C., Rummel J., Arnau S. A chronometric model of the relationship between frontal midline theta functional connectivity and human intelligence. *J. Exp. Psychol. Gen.* 2020. doi: 10.1037/xge0000865.

[109] Engle R.W. Working memory and executive attention: A revisit. *Perspect Psychol Sci.* 2018; 13(2): 190–3.

[110] Schubert Anna-Lena, Frischkorn Gidon T. Neurocognitive psychometrics of intelligence: How measurement advancements unveiled the role of mental speed in intelligence differences. *Curr. Dir. Psychol. Sci.* 2020; 29: 140–6. doi: 10.1177/0963721419896365.

[111] Cohen J.T., Bellinger D.C., Connor W.E., Shaywitz B.A. A quantitative analysis of prenatal intake of n-3 polyunsaturated fatty acids and cognitive development. *Am J Prev Med.* 2005; 294: 366–74.

Environmental Pollution

2.1 Introduction

Over the past three decades, occupational and environmental pollution has been an emerging global health problem of both developing and developed nations [1]. The swift urbanization and industrial revolution increased ecological pollution to a dangerous level. The various sources of pollution change the arrangement and composition of the environment, including air, water, and soil [1]. Environmental pollution is the combination of the biological, chemical, and physical elements, which adversely affect the everyday environment. It is an unfavorable change due to direct or indirect and human-associated activities [2].

Environmental pollutants may be "gases, dust, fumes, geochemical substances, biological organisms, chemicals, toxic metals, Radiofrequency Electromagnetic Field Radiation (RF-EMFR), radionuclides, and physical substance, heat, radiation and sound wave" [2]. The undesirable effects of these pollutants directly affect human beings or indirectly via organisms or climate change. On the basis of the pollutants' biological, physical, and chemical nature, environmental pollution is categorized into various types: air, water, soil, noise, radioactive, and thermal. All these types of pollution damage the ecosystem and are a source of multiple health hazards (Table 2.1).

The occupational and environmental diseases and the association between environment and health have a long history, observation of which led to the awareness that ill health results from a disparity between human activities and their environment [1].

After World War II, new processes, new technology, new chemicals, and novel energy sources presented more complicated problems with implications for general public health. Advancements in industry, technology, and science lead to interferences that have molded modern responses to occupational and environmental diseases. The highly advanced technology that

DOI: 10.1201/9781003212461-2

Table 2. 1 Types of Pollution
• Air pollution
• Soil pollution
• Water pollution
• Noise pollution
• Radioactive pollution
• Thermal pollution [2]

transformed the nature of the work provides many benefits and generates new public health hazards. The swift expansion of metropolitan cities and urban environments has increased the health problems related to air and water pollution. Poor sewerage clearance and disposal system further worsen the situation. Furthermore, the swift growth of mining and unplanned industries has led to polluted air and water and an excessive number of prevailing environmental and occupational diseases [3].

The environmental pollution-associated diseases are caused by dust, fumes, and gases and are influenced by the size, type, concentration, duration of exposure, biological, physical, and chemical nature of the pollutants in the breathing zone [4, 2]. Occupational and environmental diseases are caused by long-term working exposure in the industrial working environment. Air pollution is a primary type of pollution, threatening the environment, plants, animals, humans, and living organisms [5]. This is why environmental pollution is a silent killer because, as shown by the World Health Organization (WHO), about 7 million people die each year, 15.5 people per minute, due to air pollution [6] (Figure 2.1).

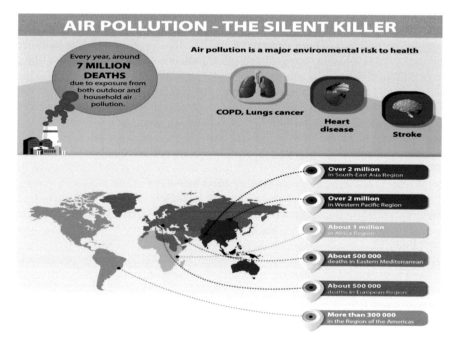

Figure 2.1 Environmental pollution and mortality (WHO Report, 2020) [6].

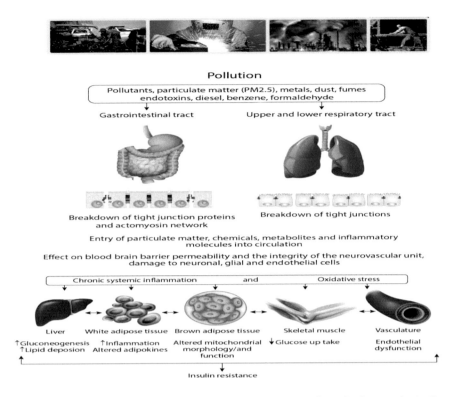

Figure 2.2 Occupational exposure to pollutants and their entry into the human body through lungs and gastrointestinal tract [14].

The occupational and environmental revelation to pollutants generated from the unplanned industrial sectors, including wood [7], welding [8], cement [9], and oil refinery [10], are evolving environmental risk factors to human health. Industrial sector employees are mostly exposed to numerous toxic substances, including coarse, fine, and ultrafine particulate matter (PM), contributing to multiple acute and chronic diseases of various body organs and systems [11].

The airborne delicate particulate matter (PM) with an aerodynamic diameter of 2.5 μm or less is one of the primary pollutants in the environment. It quickly enters the body through the respiratory and gastrointestinal tract. These PMs have a complex composition and wide distribution into the body and damage human health [12]. The "PM 2.5 μm can penetrate deep into the lungs and gastrointestinal tract (GIT) and pass through the lung and GIT barrier to enter circulating blood" and cause various negative biological and pathological impacts (Figure 2.2). The toxic effects of PM on human health mainly depend on its size, nature, biological, physical, and chemical composition [13].

PM 2.5 μm can generate from various chemical, physical, and biological sources [12]. These PM elements include "nitrogen dioxide, sulfur dioxide, nitrate, carbon mono oxide, ammonium, organic compounds, polycyclic aromatic hydrocarbons, dust, fumes, gases, and metals" [15]. The scientific literature, including experimental and epidemiologic studies, has also provided provision to the notion that exposure to work-related pollutants, dust, gases, and fumes has damaging effects on the human body physiological functions and systems, including the respiratory [9], cardiovascular [16, 17], endocrine [18], and nervous systems [19].

2.2 Sources of Pollution

Environmental pollution is the contagion of the biological, chemical, and physical elements, which adversely affect the everyday environment. Air pollution introduces gas, liquid, or solid substances that pollute the air and make the air hazardous to life and natural systems. Human-allied events and natural disasters mainly cause air pollution. The three significant sources of air pollution include: "mobile sources such as trains, trucks, cars, buses, aero-planes, sea ships; stationary sources such as power plants, oil refineries, industrial facilities, and large and small-sized factories; and area sources including agricultural areas, metropolitan cities, wood-burning fireplaces, etc." The air-polluting activities by humans such as "burning, use of fossil fuels in automobiles, toxic gas emissions from manufacturing plants, industries, and other anthropogenic activities, cause air pollution to reach a threshold that is considered noxious to human health and ecosystems" [20].

Air pollution in the form of gaseous pollutants includes "sulfur dioxide (SO_2), nitrogen oxides (NOx), ozone (O_3), carbon monoxide (CO), volatile organic compounds (VOCs), hydrogen sulfide (H_2S), hydrogen fluoride (HF), and various gaseous forms of metals" [21–23]. These elements are produced from fossil fuels, power plants, oil and gas refineries, and manufacturing industries. These toxic air pollutants are corrosive and damage ecosystems as well as humans. It is highly alarming that air pollution is higher in large metropolitan cities, where a considerable population is at risk due to crowded and poor living conditions, automobiles, unplanned industrial zones, burning in landfills, and other polluting activities.

2.3 Air Pollutants

Air pollution and pollutants are characterized into three main groups: natural, human-induced, and industrial. The arrangement and composition of air pollution depend on nature, source, emission, sunlight, and wind conditions. The most common gases that contaminate the air and its quality are "nitrogen dioxide (NO_2), nitric oxide (NO), sulfur dioxide (SO_2), ozone (O_3), and carbon monoxide (CO)". The particulate matters (PMs) are highly toxic components of air pollution, mainly consisting of carbonaceous particles, allied organic chemicals, and reactive metals [21–23]. Moreover, it also contains smoke, dust, pollens, microorganisms, waste, disposal, volatile organic compounds (VOCs), hydrocarbons, decaying plants, etc. The emissions from industries, power plants, refineries, and motor vehicles generate oxides of nitrogen (NOx) and VOCs or hydrocarbons; these substances under sunlight went into a chemical reaction and form ozone (Figure 2.3.) These pollutant particles can travel from region to region.

The most communal components of PM include "nitrates, sulfates, polycyclic aromatic hydrocarbons, endotoxin, and metals such as iron, copper, nickel, zinc, and vanadium. PM is subclassified according to particle size [i] coarse PM10 diameter <10µm; [ii] fine PM2.5, diameter <2.5µm; and [iii] ultrafine PM0.1, diameter <0.1µm" [21–23] (Figures 2.4 and 2.5).

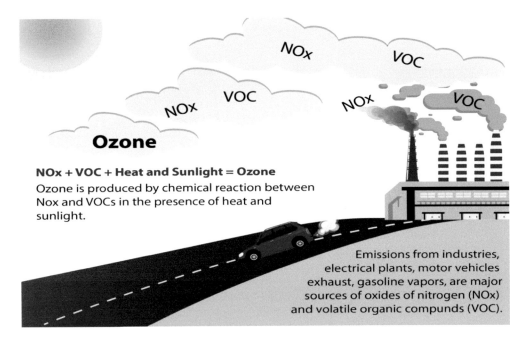

Figure 2.3 Environmental pollution and formation of ozone.

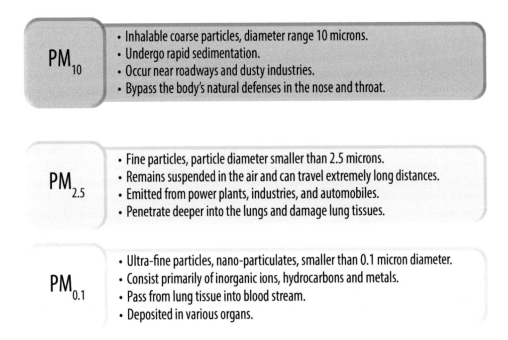

Figure 2.4 Description of particulate matter (PM).

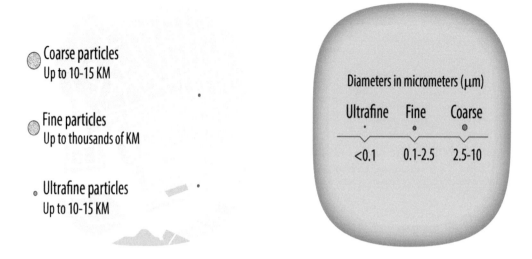

Figure 2.5 Journey of particulate matter (PM).

2.4 Ozone Formation and Pollution

The emission of pollutants generated from industrial plants, oil refineries, and motor vehicles is the primary source of ozone formation in the environment. Once "volatile organic compounds (VOCs) and Oxides of Nitrogen (NOx)" are released into the atmosphere, a chemical reaction occurs between the "volatile organic compounds and oxides of nitrogen in the presence of solar irradiation, leading to cause ozone formation" [24]. Ozone (O_3) is the three-atomic form of oxygen, a gas present in the earth's atmosphere [25]. Approximately 90% of ozone exists in the second layer of the earth's atmosphere, lower to the middle stratosphere, between the altitude of about 15 and 35 km. The stage is known as the "ozone layer," whereas about 10% are found in the lowest layer of the earth's atmosphere, known as the troposphere. An increase in ozone formation is mainly seen due to human-made air pollution. The ozone has both a positive and negative impact, which depends on its concentration of pollutants. Ozone can easily absorb the maximum percentage of biologically harmful UV-B radiation produced from the sun. The literature shows that stratospheric ozone protects life on earth against health risks such as cataracts, skin cancer, and biological and physical damages to plants and aquatic ecosystems [25].

On the contrary, high ozone levels near to earth's surface are harmful to humans and living systems. It may cause respiratory and coronary artery allied significant health problems and decrease the growth of plants and crops. In winter weather, ozone is frequently formed and transported from one place to another with large-scale meridional residual circulation. The ozone layer varies between 10 and 15 km altitude at high latitudes and 20 and 25 km in the tropics [25].

The higher concentration of NO_x and VOCs causes more ozone formation, which exerts toxic effects on the eye, nose, lungs, heart, and lung function impairment. The scientific literature has revealed that short-term and/or long-term exposures to ozone are allied with increased mortality due to respiratory, cardiovascular, and nervous system diseases. The ozone as a strong oxidant induces oxidative damage to the respiratory airways. It also causes immune-inflammatory responses within and beyond the lungs [24].

Worldwide, the high concentration of ozone is present in many residential places, mainly in and near large metropolitan cities. Increased emission of "NO_x, VOCs, and CO" fossil fuel combustion at high temperatures is highly favorable for NO_x formation. Increasing fossil-fuel-derived energies for electricity generation, vehicle exhaust, modern transportation tools, household cooking, and heating are the leading causes of increasing NO_x emissions.

Fossil-fuel combustion by-products are the world's most significant threat to public health and contribute to global inequality and environmental injustice. The emissions include many toxic air pollutants and carbon dioxide (CO_2), an essential human-produced climate-altering greenhouse gas. The primary anthropogenic sources of VOCs include vehicular exhaust, gaseous fuels, biomass, and fossil fuel combustion, and industrial solvent use. A recent study found that volatile chemicals released from pesticides, coatings, printing inks, adhesives, cleaning agents, and personal care products have emerged as a large urban source of VOCs. [26] However, the natural vegetation emissions of certain VOCs (e.g., isoprene) also contribute to ozone formation, especially at the regional levels [27–29]. All these conditions are directly or indirectly responsible for the formation of ozone and contaminate the environment.

2.5 Top Ten Highly Polluted Cities in the World

Worldwide, the larger cities, by their inherent characteristics of an overcrowded population, densely built-up areas, high-rising buildings, large numbers of automobiles, unplanned industries in the city centers, and lack of plants, are the major causes for the generation of air pollution and contamination of urban areas' environment and their weather condition [30]. Globally, the urban air pollution levels rose about 8%, mostly in metropolitan cities. A WHO 2020 [31] report showed that in approximately 80% of the large urban cities, the air pollution level is above the required safety levels. The report further identified that air pollution is murkier for cities with over 100,000 inhabitants in low- and middle-income countries, whereby 98% of the population had unhealthy air for their breathing zone.

Air pollution in urban zones significantly affects the climate and weather conditions and also markedly affects human health. The most harmful air pollutant that negatively affects human health is particulate matter with a size equal to or smaller than 2.5 μm in diameter (PM 2.5 μm). These PMs commonly exist in smoke, fumes, dust, and motor vehicle pollution [33]. PM2.5 is more hazardous, as it is fine enough to pass through the upper respiratory tract, enter into the lungs' respiratory zone, and cause long-term health problems. PM2.5 particulate matters are more hazardous when their concentration becomes more than 35.5 μg of PM2.5 per cubic meter of air in the environment [30].

The most recent world air quality report 2020 [32] based on 1,600 cities indicates that most cities or half of the world's most polluted cities were from India (Figure 2.6). This report further shows that industrial and motor vehicle exhausts were choking large parts of India with little oversight or monitoring mechanism [34]. The most polluted cities in the world are in India. These cities include Ghaziabad, Delhi, Noida, Gruguram, and Bandwan. The second highly polluted city is Hotan, China. However, Gujranwala, Faisalabad, and Raiwind are the third, fourth, and eighth highly polluted cities in Pakistan. The top 10 cities with the highest air pollution are shown in Figure 2.6 (World Air Quality Report, 2020) [32].

Rank		City, Country	Pollution in year 2019
1		Ghaziabad, India	110.2
2		Hotan, China	110.1
3		Gujranwala, Pakistan	105.3
4		Faisalabad, Pakistan	104.6
5		Delhi, India	98.6
6		Noida, India	97.7
7		Gurugram, India	93.1
8		Raiwind, Pakistan	92.2
9		Greater Noida, India	91.3
10		Bandhwari, India	90.5

Figure 2.6 Most polluted cities in the world (World Air Quality Report, 2020) [32].

2.6 Pollution and Human Health

Air pollution is a significant risk factor for several illnesses, especially allergic, respiratory, coronary artery diseases, chronic obstructive pulmonary disease (COPD), stroke, and lung cancer [35]. Moreover, air pollution impairs lung functions [8] and cognitive functions, and can cause insulin resistance and diabetes mellitus [18]. People who have bronchial asthma, COPD, and coronary artery diseases are severely affected by air pollution. Air pollutant exposures are also associated with long-term neurobehavioral conditions, cardiopulmonary diseases, metabolic syndrome, diabetes mellitus, and reproductive abnormalities [36, 18]. Air pollution causes premature death and an economic burden. The children, old aged, and immunocompromised people residing in metropolitan cities are the most vulnerable to indoor and outdoor environmental air pollution. Short-term effects include pneumonia or bronchitis, nasal, throat, eyes, and skin irritation, nausea, headaches, and dizziness [37–39]. Moreover, environmental pollution can also cause long-term damage to the nervous system, liver, kidneys, and other organs [16, 18] (Table 2.2). The toxic effect of environmental pollutants mainly depends on the pollutants' size, type, and nature. The respiratory system is highly susceptible to particulate matter (Figure 2.7; Table 2.2).

2.7 Pollution and Mortality

Environmental pollution is an unseen major threat to public health and climate. Its exposure starts from womb to infant, children, adult, and later period of the life. The children are exposed to the polluted environment in which they live, learn, play, and work (Figure 2.8). It is an invisible killer; about 7 million premature deaths every year are due to household and ambient air pollution. It shows that 13.5 people died per minute due to air pollution. The increased mortality rate is mainly due to coronary artery disease, acute respiratory infections, COPD, lung cancer, and cerebrovascular accident [37] (Figure 2.9).

Table 2.2 Air Pollution-Associated Diseases

Respiratory system	Nasal irritation, runny nose, rhinitis, cough, asthma, chronic bronchitis, COPD, pneumonia, pneumoconiosis, systemic inflammation, impaired lung function [1, 14, 16], and lung cancer
Cardiovascular system	Oxidative stress, endothelial dysfunction, atherosclerosis, thrombosis, hypertension, coronary artery disease, arrhythmia, ischemic, heart failure, and cerebrovascular accident [40, 41]
Endocrine system	Hyperglycemia, insulin resistance, type 2 diabetes mellitus [14, 18]
Reproductive system	Adverse male and female reproductive health, polycystic ovary, infertility, breast cancer [42–47]
Nervous system	Brain inflammation, impaired cognitive function, memory, low IQ, neurodegenerative diseases, white matter abnormalities, multiple sclerosis Parkinson's disease, Alzheimer's disease, autism spectrum disorders, and stroke [48]

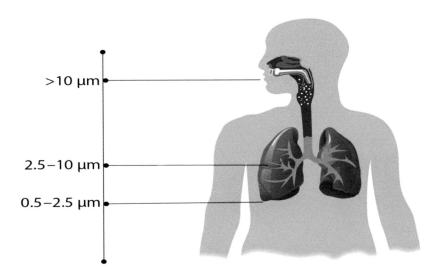

Figure 2.7 Environmental pollutants' entry into the respiratory system.

Air pollution is at dangerously high levels in many regions of the world. About 80% of people are residing in urban areas. The WHO data demonstrates that approximately "91% of the world's population lives in places where air quality exceeds the WHO guideline limits." Moreover, out of ten people, nine breathe air containing high levels of pollutants. Worldwide, one in four deaths of children under the age of 5 years is due to diseases allied to unhealthy environment and air pollution. Moreover, every year 1.7 million children under 5 years of age die due to indoor and outdoor air pollution, second-hand smoke, unsafe water, lack of sanitation, and inadequate hygiene [37].

About 4.2 million deaths every year occur due to exposure to ambient (outdoor) air pollution, and 3.8 million deaths every year are caused by household exposure to smoke from cookstoves and fuels. Air pollution causes 2.4 million deaths due to cardiovascular system diseases,

Environment Pollution and Children

Environment pollution exposure starts from womb to throughout life.

Figure 2.8 Environmental pollution exposure from fetal life to throughout life.

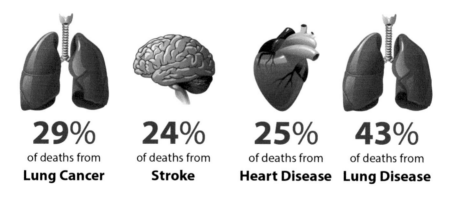

29%	**24%**	**25%**	**43%**
of deaths from	of deaths from	of deaths from	of deaths from
Lung Cancer	**Stroke**	**Heart Disease**	**Lung Disease**

Figure 2.9 Prevalence of deaths due to environmental pollution.

1.8 million deaths due to respiratory illnesses, and 1.4 million deaths due to cerebrovascular accidents (stroke) per annum. It has been reported that 24% of stroke deaths, 25% of heart disease, and 43% of all lung disease deaths are due to air pollution [37] (Table 2.3, Figure 2.9).

Worldwide, approximately 7 million deaths occurred due to household and ambient air pollution. About 94% of these deaths are from low- and middle-income countries. "The Southeast Asian and Western Pacific regions bear the burden of 2.4 and 2.2 million deaths respectively. Approximately 980000 deaths occur in Africa, 475000 in the Eastern Mediterranean region, 348000 in Europe, and 233000 in the Americas." The remaining deaths occurred in high-income

Table 2.3 Worldwide Mortality Due to Environmental Pollution	
Body system	Global deaths due to environmental pollution
Cardiovascular system	2.4 million deaths (25% deaths)
Respiratory system	1.8 million deaths (43% deaths)
Nervous system (stroke)	1.4 million deaths (24% deaths)

Source: WHO Report 2020 [37].

countries of European states (208,000), Americas (96,000), Western Pacific (83,000), and Eastern Mediterranean (18,000) [37, 38].

2.8 Pollution and Economic Burden

Environmental pollution allied diseases impose a tremendous economic burden on the nations around the world. It includes medical costs, opportunity costs due to diminished productivity of the working-age group population, health care system cost, and absence from the workplace [49]. Global organizations and research institutes on energy and clean air [50] have reported that air pollution has about $2.9 trillion in economic costs, equivalent to 3.3% of the world's GDP. In 2018, the report further estimates that the financial burden was "linked to 4.5 million deaths due to PM2.5 pollution, 1.8 billion days absence of people from the workplace, 4 million new cases of a child were diagnosed with asthma, and about 2 million preterm births" [50].

Environmental pollution impacts the global economies in various forms, including the high prevalence of illnesses, such as asthma, diabetes mellitus, COPD, and lung cancer. Environmental pollution affects every tissue of the human body [51]. All these illnesses reduce the ability of workers and lower participation rates in the labor force. The children are more vulnerable to developing frequent bronchial asthma attacks, and therefore, absent from school; this situation severely impacts their learning.

The findings further highlight that environmental pollution allied with disability from chronic diseases significantly affects the world's economies, about $200 billion in 2018, with sick leave and preterm births costing about $100 billion and $90 billion, respectively [50] (Table 2.4).

The entire annual cost of air pollution allied diseases and burden on the health care system in China was estimated at "$900 billion each year, with costs in the U.S was $600 billion annually. In 2018, the cost of air pollution equated to 6.6% of Chinese GDP, and 3% of U.S GDP" [52] (Figure 2.10).

2.9 Environmental Pollution Management

Worldwide, science and health allied organizations are trying to establish policies to minimize air pollution. These organizations are increasing awareness and education about air pollution and its hazardous effects and minimizing air pollution worldwide. At the local level, everyone should try to reduce air pollution. Encourage the community to keep the surrounding environment clean and green, grow plants, keep air-purifying indoor plants, avoid smoking, and minimize exposure to devices that generate radiofrequency electromagnetic field radiation (RF-EMFR). It is also essential to control the physical and biological agents that can adversely affect the atmosphere and cause adverse effects on human health. Environmental pollution management must implement a holistically multidisciplinary approach with combined efforts

Table 2.4 Environmental Air Pollution-Allied Economic Burden on World's Five Most Populous Cities (January 1 to June 30, 2020) [50]			
City	Population (Million)	Premature Death	Cost (Billion USD)
Tokyo	37.0	29,000	31.0
Delhi	30.0	24,000	3.5
Shanghai	26.0	27,000	13.0
Sao Paulo	22.0	3,700	3.5
Mexico City	22.0	1,100	5.5

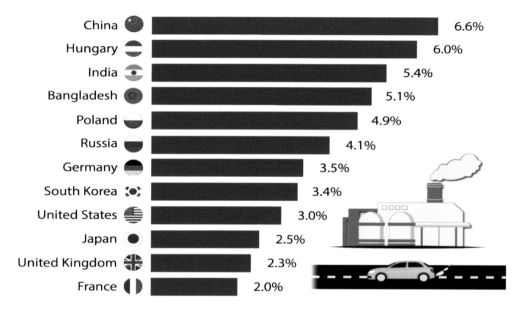

Figure 2.10 The economic burden of air pollution [50].

of public and private entities in collaboration with individual steps at a community level to minimize the environmental pollution and its associated diseases' burden.

Conclusion

The rapid unplanned urbanization and industrial revolution increased environmental pollution to a dangerous level. Environmental pollution is the contagion of the biological, chemical, and physical elements, which adversely affect the everyday environment. Environmental pollution is a major risk factor for several illnesses, mainly respiratory, coronary artery diseases, endocrine, diabetes mellitus, nervous system disorders, lung inflammation, and cancer. As per the WHO report, environmental pollution is a silent killer and causes approximately 7 million premature deaths annually, and 94% of these deaths were reported from low- and middle-income countries.

References

[1] Meo S.A., Azeem M.A., Ghori M.G., Subhan M.M. Lung function and surface electromyography of intercostal muscles in cement mill workers. *Int. J. Occup. Med. Environ. Health.* 2002; 15(3): 279–87.

[2] Rai Prabhat K. Particulate matter and its size fractionation. *Biomagnetic Monitoring of Particulate Matter.* 2016: 1–13.

[3] Corn J.K. Historical perspective. In: Harber P., Shenker Marc B., Balmes John R., editors. *Occupational and Environmental Respiratory Disease.* London: Mosby. 1996, pp. 3–4.

[4] Mengesha Y.A., Bekele A. Relative chronic effects of occupational dust on respiratory indices and health of workers in three Ethiopian factories. *Am. J. Ind. Med.* 1998; 34: 373–80.

[5] Imbus H.R. Clinical aspects of occupational medicine. In: Carl Zenz, Dickerson O. Bruce, Edward D., Horvath J.R., editors. *Occupational Medicine*, 3rd edition. St. Louis, MO: Mosby. 1994, p. 3.

[6] World Health Organization. Air pollution: The silent killer. Available at: www.who.int/airpollution/infographics/Air-pollution-INFOGRAPHICS-English-1.1200px.jpg. Cited date Jan 12, 2002.

[7] Meo S.A. Lung function in Pakistani woodworkers. *Int. J. Environ. Health Res.* 2006; 16: 193–203. doi: 10.1080/09603120600641375.

[8] Meo S.A., Azeem M.A., Subhan M.M. Lung function in Pakistani welding workers. *J. Occup. Environ. Med.* 2003; 45: 1068–73. doi: 10.1097/01.jom.0000085889.16029.6b.

[9] Meo S.A., Al-Drees A.M., Al-Masri A.A., Al-Rouq F., Azeem M.A. Effect of a duration of exposure to cement dust on respiratory function of non-smoking cement mill workers. *Int. J. Environ. Res. Public. Health.* 2013; 10: 390–899. doi: 10.3390/ijerph10010390.

[10] Meo S.A., Alrashed A.H., Almana A.A., Altheiban Y.I., Aldosari M.S., Almudarra N.F., Alwabel S.A. Lung function, and fractional exhaled nitric oxide among petroleum refinery workers. *J. Occup. Med. Toxicol.* 2015; 10: 37. doi: 10.1186/s12995-015-0080-7.

[11] Estill C.F., Slone J., Mayer A., Chen I.C., Guardia M.J. Worker exposure to flame retardants in manufacturing, construction, and service industries. *Environ. Int.* 2019; 135: 105349. doi: 10.1016/j.envint.2019.105349.

[12] Liaoa B.Q., Liue C.B., Xiec B.Q., Liud Y., Denga Y.B., Hed S.W. Effects of fine particulate matter (PM2.5) on ovarian function and embryo quality in mice. *Environ. Int.* 2019; 135: 105338. doi: 10.1016/j.envint.2019.105338.

[13] Oyewale M.M., Matlou I.M., Murembiwa S.M., Raymond P.H. health outcomes of exposure to biological and chemical components of inhalable and respirable particulate matter. *Int. J. Environ. Res. Public. Health.* 2016; 13: 10.

[14] Meo S.A., Al-Khlaiwi T., Abukhalaf A.A., Alomar A.A., Alessa O.M., Almutairi F.J., Alasbali M.M. The nexus between workplace exposure for wood, welding, motor mechanic, and oil refinery workers and the prevalence of prediabetes and type 2 diabetes mellitus. *Int. J. Environ. Res. Public. Health.* 2020; 17(11): 3992. doi: 10.3390/ijerph17113992.

[15] Srimuruganandam B., Nagendra S.M.S. Source characterization of PM10 and PM2.5 mass using a chemical mass balance model at the urban roadside. *Sci. Total. Environ.* 2012; 433: 8–19. doi: 10.1016/j.scitotenv.2012.05.082.

[16] Meo S.A., Suraya F. Effect of environmental air pollution on cardiovascular diseases. *Eur. Rev. Med. Pharmacol. Sci.* 2015; 19: 4890–7.

[17] Zhang Z., Dong B., Li S., Chen G., Yang Z., Dong Y. Exposure to ambient particulate matter air pollution, blood pressure and hypertension in children and adolescents: A national cross-sectional study in China. *Environ. Int.* 2019; 128: 103–8. doi: 10.1016/j.envint.2019.04.036.

[18] Meo S.A., Memon A.N., Sheikh S.A., Rouq F.A., Usmani A.M., Hassan A., Arian S.A. Effect of environmental air pollution on type 2 diabetes mellitus. *European. Rev. Med. Pharmacol. Sci.* 2015; 19: 123–8.

[19] Oudin A., Frondelius K., Haglund N., Källén K., Forsberg B., Gustafsson P., Malmqvist E. Prenatal exposure to air pollution as a potential risk factor for autism and ADHD. *Environ. Int.* 2019; 133: 105149. doi: 10.1016/j.envint.2019.105149.

[20] Chan N.W., Imura H., Nakamura A., Masazumi A.O. Air pollution. In: *Sustainable Urban Development Textbook*, 1st edition. Penang: Water Watch Penang and Yokohama City University. 2016, pp. 226–34.

[21] Newby D.E., Mannucci P.M., Tell G.S., Baccarelli A.A., Brook R.D., Donaldson K., et al. Expert position paper on air pollution and cardiovascular disease. *Eur. Heart. J.* 2015; 36: 83–93b. doi: 10.1093/eurheartj/ ehu458.

[22] Brook R.D., Rajagopalan S., Pope C.A. I.I.I., Brook J.R., Bhatnagar A., Diez-Roux A.V., et al. Particulate matter air pollution, and cardiovascular disease: An update to the scientific statement from the American heart association. *Circulation.* 2010; 121: 2331–78.

[23] Chin M.T. Basic mechanisms for adverse cardiovascular events associated with air pollution. *Heart.* 2015; 101: 253–6. doi: 10.1136/heartjnl-2014-306379.

[24] Zhang J., Wei Y., Fang Z. Ozone pollution: A major health hazard worldwide. *Front Immunol.* 2019; 10: 2518. doi: 10.3389/fimmu.2019.02518.

[25] Langematz U. Stratospheric ozone: Down and up through the Anthropocene. *ChemTexts.* 2019; 5(8). https://doi.org/10.1007/s40828-019-0082-7.

[26] McDonald B.C., de Gouw J.A., Gilman J.B., Jathar S.H., Akherati A., Cappa C.D. Volatile chemical products are emerging as the largest petrochemical source of urban organic emissions. *Science.* 2018; 359: 760–4.

[27] Guenther A., Geron C., Pierce T., Lamb B., Harley P., Fall R. Natural emissions of non-methane volatile organic compounds, carbon monoxide, and oxides of nitrogen from North America. *Atmosph. Environ.* 2000; 34: 2205–30. doi: 10.1016/S1352-2310(99)00465-3.

[28] Calfapietra C., Fares S., Loreto F. Volatile organic compounds from Italian vegetation and their interaction with ozone. *Environ. Pollut.* 2009; 157: 1478–86. doi: 10.1016/j.envpol.2008.09.048.

[29] Calfapietra C., Fares S., Manes F., Morani A., Sgrigna G., Loreto F. Role of biogenic volatile organic compounds (BVOC) emitted by urban trees on ozone concentration in cities: A review. *Environ. Pollut.* 2013; 183: 71–80. doi: 10.1016/j.envpol.2013.03.012.

[30] Lydia R.P. About 80% of all cities have worse air quality than what is considered healthy here are the 15 with the worst air pollution. *Business Insider.* 2016; 5: 33. Available at: www.businessinsider.com/the-cities-with-the-worlds-worst-air-pollution-who-2016-5. Cited date Jun 5, 2002.

[31] World Health Organization. Available at: www.who.int/airpollution/data/cities/en/. Cited date Feb 2020.

[32] Air Quality. World's most polluted cities 2019 (PM2.5). Available at: www.iqair.com/world-most-polluted-cities. Cited date Jun 12, 2020.

[33] Heil A., Goldammer J.G. Smoke-haze pollution: A review of the 1997 episode in Southeast Asia. *Reg. Environ. Change.* 2001; 2: 24–37.

[34] Joshi M. Half of the world's 20 most polluted cities in India Delhi. *Hindustani Times.* Available at: www.hindustantimes.com/delhi/four-out-of-top-five-polluted-cities-are-in-india-delhi-not-among-them/story-Gn2htcLbESB3BpeYJ4mY8K.html. Cited date Jun 5, 2002.

[35] Reilly J.J., Silverman E.K., Shapiro S.D. Chronic obstructive pulmonary disease. In: Longo Dan, Fauci Anthony, Kasper Dennis, Hauser Stephen, Jameson J., Loscalzo Joseph. editors. *Harrison's Principles of Internal Medicine.* 18th edition. New York: McGraw Hill. 2011, pp. 2151–9.

[36] Snow S.J., Henriquez A.R., Costa D.L., Kodavanti U.P. Neuroendocrine regulation of air pollution health effects: Emerging insights. *Toxicol Sci.* 2018; 164(1): 9–20.

[37] World Health Organization. Air pollution. Available at: http://origin.who.int/airpollution/en/. Cited date Feb 12, 2020.

[38] World Health Organization. Public health, environmental and social determinants of health (PHE). Available at: www.who.int/phe/infographics/protecting-children-from-the-environment/en/. Cited date May 12, 2020.

[39] Xu-Qin Jiang, Xiao-Dong Mei, Di Feng. Air pollution and chronic airway diseases: What should people know and do? *J Thorac Dis.* 2016; 8(1): E31–E40.

[40] Landrigan P.J., Fuller R., Acosta N.J.R., Adeyi O., Arnold R., Basu N.N., et al. The lancet commission on pollution and health. *Lancet.* 2018; 391(10119): 462–512.

[41] Hamanaka R.B., Mutlu G.M. Particulate matter air pollution: Effects on the cardiovascular system. *Front Endocrinol (Lausanne).* 2018; 9: 680.

[42] Crouse D.L., Goldberg M.S., Ross N.A., Chen H., Labrèche F. Postmenopausal breast cancer is associated with exposure to traffic-related air pollution in Montreal, Canada: A case-control study. *Environ Health Perspect.* 2010; 118: 1578–83.

[43] Hystad P., Villeneuve P.J., Goldberg M.S., Crouse D.L., Johnson K., Canadian Cancer Registries epidemiology research group exposure to traffic-related air pollution and the risk of developing breast cancer among women in eight Canadian provinces: A case-control study. *Environ. Int.* 2015; 74: 240–8.

[44] Shmuel S., White A.J., Sandler D.P. Residential exposure to vehicular traffic-related air pollution during childhood and breast cancer risk. *Environ. Res.* 2017; 159: 257–63.

[45] Carre J., Gatimel N., Moreau J., Parinaud J., Léandri R. Does air pollution play a role in infertility? A systematic review. *Environ. Health.* 2017: 16–82.

[46] Kramer U., Herder C., Sugiri D. Traffic-related air pollution and incident type 2 diabetes: Results from the SALIA cohort study. *Environ. Health. Perspect.* 2010; 118: 1273–9.

[47] Philippa D D. Overview of air pollution and endocrine disorders. *Int. J. Gen. Med.* 2018; 11: 191–20.

[48] Block M.L., Elder A., Auten R.L., Bilbo S.D., Chen H., Chen J.C. The outdoor air pollution and brain health workshop. *Neurotoxicology.* 2012; 33(5): 972–84.

[49] Landrigan P.J., Fuller R. Global health, and environmental pollution. *International Journal of Public Health.* 2015; 60: 761–2.

[50] Greenpeace. Tracking the cost of air pollution. Available at: www.greenpeace.org/international/campaign/tracking-cost-air-pollution. Cited date Jan 12, 2020.

[51] Meo S.A., Aldeghaither M., Alnaeem K.A., Alabdullatif F.S., Alzamil A.F., Alshunaifi A.I., Alfayez A.S., Almahmoud M., Meo A.S., El-Mubarak A.H. Effect of motor vehicle pollution on lung function, fractional exhaled nitric oxide and cognitive function among school adolescents. *Eur. Rev. Med. Pharmacol. Sci.* 2019 Oct; 23(19): 8678–86.

[52] World Economic Forum. This is the global economic cost of air pollution. Available at: www.weforum.org/agenda/2020/02/the-economic-burden-of-air-pollution/. Cited date July 2, 2020.

The Journey of Pollutants from Environment to Brain

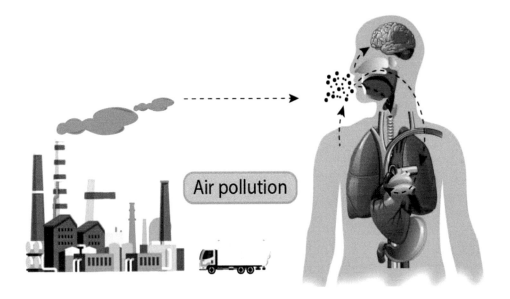

Chapter 03

Air pollution

3.1 Introduction

In the new millennium, global environmental pollution is a leading public health concern and has endangered the global health system and economies. Environmental pollutants generated from various natural, human-made, and industrial sources threaten public health by hazardous effects on multiple body systems, including respiratory [1], coronary [2], endocrine [3, 4], and the nervous system [1].

Environmental pollution consists of a complex mixture of coarse, fine, and ultrafine particles, mainly particulate matter (PM) of various sizes along with toxic substances including "dust, fumes, gases, carbon monoxide, sulfur oxides, nitrogen oxides, and other gases." It also includes VOCs, "benzene, toluene, and xylene, and metals such as lead, manganese, vanadium, iron" and biological materials [5].

Environmental pollution occurs with changes in the biological, physical, and chemical constituents of the environment. The primary sources of environmental pollution include industries, power stations, oil refineries, plastic refineries, chemical, metallurgical, and fertilizer industries. Moreover, airplanes, ships, cars, buses, trucks, trains, combustion of wood, wind-blown wildfires, and volcanoes are also sources of environmental pollution [6].

Human-made activities harm the environment by polluting the air, water, and soil. The industrial revolution was enormously successful in the development of modern technology and society. However, it also resulted in the production of massive pollutants, which contaminated the environment. Urbanization and industrialization are reaching unprecedented and upsetting proportions worldwide.

Figure 3.1 Wildfire: sources of fine and ultrafine particulate matters.

The global proportions of urbanization and industrialization have been unprecedentedly changed and have caused massive pollution. Anthropogenic air pollution accounts for about 9 million deaths per annum [7]. The developing nations are trying to achieve swift economic growth through industrialization, globalization, and regional economic integration. As part of this competition of economic growth, many countries have experienced "industrialization, urbanization, and mass consumerism." This rapid change has a significant environmental impact. People, goods, capital, technology, and information rapidly cross borders, which develop ecological problems. These problems include dust, air pollution, marine pollution, electromagnetic field radiation pollution, carbon dioxide (CO_2) emission, and food contamination, which have become common threats to the environment and human health [8].

Environmental pollutants harm the environment by exceeding normal limits. The primary pollutants are produced from these mentioned sources, and secondary pollutants are emitted as by-products of the primary ones. The contaminants can be biodegradable or non-biodegradable and of natural origin or anthropogenic [6, 9, 10].

The environmental pollution impact is not only limited to the metropolitan cities. However, it has been extended in the rural areas, jungles, and plants damaged and destroyed through human-made activities and wildfire (Figure 3.1). Once the pollutants are produced, they enter the environment and affect the region and other countries (Figure 3.2).

3.2 Environmental Pollution Journey from Country to Country

Environmental pollution is not restricted to the regional country borders; once produced, the particulate matter (PM) components stay in the environment for an extensive period, and

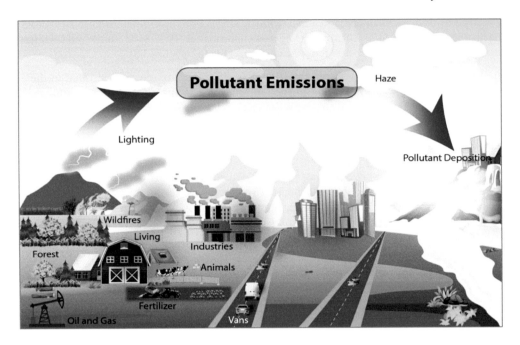

Pollutant Emissions

Haze

Lighting

Pollutant Deposition

Wildfires

Living

Industries

Forest

Animals

Fertilizer

Oil and Gas

Vans

Figure 3.2 Production and transportation of pollutants from continent to continent.

travel for a long distance, sometimes even from one country to another country and also from one continent to another (Figure 3.2). The movement of pollutants in the atmosphere is caused by transport, dispersion, and deposition. Pollutants travel with the flow of the wind, but dispersion results from local turbulence. The deposition process includes precipitation, scavenging, and sedimentation, which causes downward movement of pollutants in the atmosphere, which deposits the contaminants to the ground [11].

Pollution is often created in one place and transports through the air to another location. The particulate matter can also form a haze, an atmospheric phenomenon whereby dust, smoke, and particulate matters combine. The haze formation occurs by physical and chemical processes involving fine particulate matter interacting with water vapor under certain airflow conditions. Sometimes, when the pollutants move with the wind, a chemical reaction occurs in the atmosphere and clouds, creating a haze. It changes the biological and chemical nature of the contaminants before pollutants deposit in another place. Pollutant deposition can have adverse biological effects. Emissions in any country can affect humans and an ecosystem in countries far downwind. The climate changes and current global warming phenomena may also lead to a warmer climate and shifts in air circulation, thereby affecting the patterns of production, transport, and deposition of pollutants and pattern of health and disease [11, 12]. The effect of contaminants on human body systems depends on the type and the biological, chemical, and physical nature of the pollutants and the duration of exposures [13].

3.3 Particulate Matter

The particles in ambient air are moderately miscellaneous in their biological, chemical, and physical properties. The particulate matter (PM) component of air pollution is defined and classified by its aerodynamic diameter size, which ranges into three different forms: (1) coarse particulate matter (PM10; <10 μm); (2) fine particulate matter (PM2.5; <2.5 μm); and (3) ultrafine particulate matter (PM < 100 nm, PM <0.1 μm) [14–16] (Figure 3.3).

The "coarse particles with an aerodynamic diameter of 2.5–10 μm are deposited in the tracheobronchial (TB) region by impaction. The fine particles (< 2.5 μm) are mainly deposited in the pulmonary area by sedimentation and diffusion. However, the coarse particles with low density can be deposited in the pulmonary region by sedimentation. The soil particles, coarse with low density," enter and are deposited deep in lung alveoli.

[17]

The human respiratory system is most vulnerable to environmental pollution. Cough, sneeze reflexes, and the respiratory tract mucociliary system play a significant role in removing the pollutants for protection. Nevertheless, the deposition of particulate matter depends upon the biological, physical, and chemical nature of the matter, size of the contaminant, and duration of exposure to people [18] (Figure 3.4).

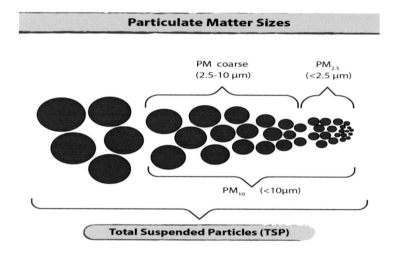

Figure 3.3 Particulate matter.

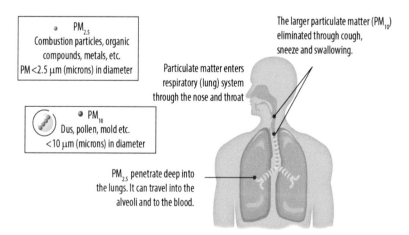

Figure 3.4 Entry of particulate matter into the respiratory system.

3.4 Deposition of Particulate Matter

The WHO highlights the concern that air pollution and its allied ailments affect all world regions. About 91% of the global population and 97% of low- and middle-income states reside in places where air pollution exceeds the recommended air quality standards [7]. Environmental pollution affects human health through multiple pathways [19, 20]. The respiratory system is the most vulnerable system for air pollutants. The entry of contaminants into the human body depends upon the upper respiratory tract, nasopharynx, larynx, trachea, conducting airway zones, and alveolar areas based on their level of exposure, aerodynamic size, and ventilation pattern [1].

The deposition of inhaled particulate matter depends on the solubility and particle size, and the deposition pattern depends on the aerodynamic diameter. The pollutants with an aerodynamic diameter of "10 μm" are deposited in the upper respiratory tract, mainly the nose and pharynx. The particles between "2.5 and 10 μm" in aerodynamic diameter can be placed throughout the tracheobronchial tree, and particles between "0.5 and 2.5 μm" are frequently deposited in the respiratory zone, mostly alveoli (Figures 3.5 and 3.6). Particles smaller than "0.1 μm" remain in the air stream and are exhaled [21]. The scientific literature demonstrates that particles equal to or smaller than "0.10 μm" enter into the general circulation and can be deposited in various tissues and human body organs, resulting in various health problems [18].

Health problems can arise as a result of the physiological failure of one or more protective mechanisms. If the subject continuously breathes dusty air, mixing between the inspired and the dead space air will cause the dust to reach the terminal airways. The dust particles tend to accumulate in the alveoli. The lungs' immune mechanism has three groupings, and these work in an integrated manner in their defensive function. There are multiple mechanisms involved at respiratory system levels, including coughing, sneezing, nostril hairs, and cilia, which collectively operate to prevent the entry of dust particles into the alveoli. The bronchial mucous contains lactoferrin, lysozyme, and antibodies. Both IgG and IgM are present, but IgA probably plays the most crucial role in the defense mechanism of the lungs. The mucous also traps small

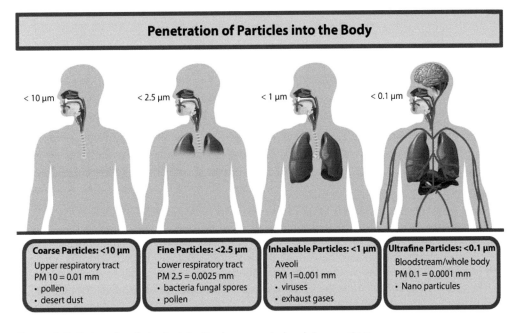

Penetration of Particles into the Body

< 10 μm   < 2.5 μm   < 1 μm   < 0.1 μm

Coarse Particles: <10 μm	Fine Particles: <2.5 μm	Inhaleable Particles: <1 μm	Ultrafine Particles: <0.1 μm
Upper respiratory tract	Lower respiratory tract	Aveoli	Bloodstream/whole body
PM 10 = 0.01 mm	PM 2.5 = 0.0025 mm	PM 1=0.001 mm	PM 0.1 = 0.0001 mm
• pollen	• bacteria fungal spores	• viruses	• Nano particules
• desert dust	• pollen	• exhaust gases	

Figure 3.5 Entry of pollutants into the lungs and circulatory system.

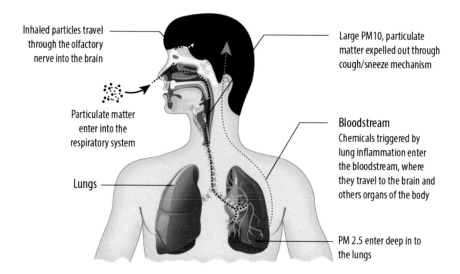

Inhaled particles travel through the olfactory nerve into the brain

Particulate matter enter into the respiratory system

Lungs

Large PM10, particulate matter expelled out through cough/sneeze mechanism

Bloodstream
Chemicals triggered by lung inflammation enter the bloodstream, where they travel to the brain and others organs of the body

PM 2.5 enter deep in to the lungs

Figure 3.6 Entry of pollutant into the lungs and brain.

particles out of inspired air and prevents them from reaching the alveoli. The mucociliary system does not expel out the bacteria and small particles that reach the alveoli. In this situation, the phagocytosis activates by pulmonary alveolar macrophages and kills the bacteria [22].

It is vital to understand how nanoparticles and ultrafine particulate matter enter the brain and cause neuronal damage. These particles enter the brain through the lungs and blood circulation, while another mechanism responsible is through the nose and olfactory system. Exposure to ambient air pollution increases morbidity and mortality and contributes to the global disease burden. The particulate matters with the size of "PM2.5 μm" are the most consistent and robust predictor of mortality in long-term exposure studies [23].

3.5 Translocation of Pollutants from Lungs to the Brain

The translocation of particles from the lungs has been highlighted in the modern scientific literature [24]. It has resulted in an increased awareness of researchers to understand non-pulmonary targets of particulate air pollutants. The nanoparticles leave the lungs and are deposited in the extrapulmonary tissues [24, 25]. "The ultrafine (nano-size particles) and fine particles are the most notorious air pollution components, penetrating lung tissue compartments to reach the capillaries and circulating cells" [24, 25].

The deposition of these particulate matters in the brain is a growing concern for the science community and the public. Literature shows that ultrafine particles swiftly translocate from the lungs into the cells, tissues, and blood and enter into the systemic circulation [26]. About 50% of inhaled ultrafine PM are deposited in the alveolar regions, pass through the alveolar-capillary barrier, and enter the pulmonary interstitium. The particles can cross the endothelial cells, move into the blood circulation, and are deposited in other organs, including the heart and brain, to produce more severe health consequences [27, 28].

The entry of pollutants into the brain is directly through the nasal olfactory pathway or indirectly through the lungs, gastrointestinal, and circulatory system. The pollutants can cause inflammation in the respiratory system and release "pro-inflammatory cytokines into the circulation. Moreover, they activate the hypothalamus–pituitary–adrenal (HPA) axis and increase

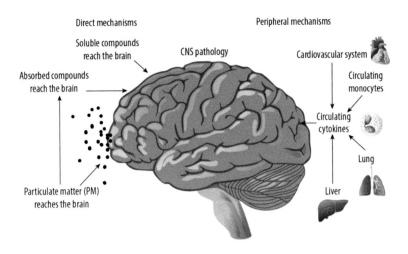

Figure 3.7 Entry of pollutants from nose and lungs to the brain.

the blood levels of stress hormones, including cortisol" [29–32]. The pollutants provide stimulus to release inflammatory cytokines that reach the brain (Figure 3.7), cause neuroinflammation, and impair various neuronal functions [29–32]. These pollutants can cause Alzheimer's disease, Parkinson's disease, neurodevelopmental disorders, and cerebrovascular accident [32].

3.6 Translocation of Pollutants from Nose to Brain

The ultrafine particulate matter (PM) is considered the highly reactive toxic air pollution element and can penetrate the brain directly through the olfactory nerve transport mechanism [22]. The upper respiratory tract "nasal cavity" provides another transport pathway for the direct entry of some inhaled materials into the brain through the "olfactory neurons and olfactory bulbs." Literature reveals that the movement of PM and ultrafine particles through the nasal cavity enter the brain across the olfactory epithelium [27, 28].

The deposition of particles can occur in the "piriform cortex, olfactory tubercule, amygdala and entorhinal cortex" [33] (Figure 3.8). The mechanism demonstrates the connections with the orbitofrontal cortex, thalamus, hypothalamus, and hippocampal regions [34]. The particulate matter can also cross synapses within the olfactory pathway and travel via secondary and tertiary neurons to various areas in the brain [35]. It has been reported that metal compounds such as "manganese, iron, cadmium, thallium, mercury, cobalt, zinc, and carbon particles move into the brain following inhalation or intranasal, tracheal instillation exposures" [34]. The suggested mechanism through which transport of these particles occurs is through the olfactory nerve transport system (Figure 3.8).

3.7 Pollutants and Blood–Brain Barrier

The ultrafine particulate matters enter the brain through the respiratory system and general circulation. Another pathway is direct through the olfactory pathway into the brain [29]. Either way, once the ultrafine particulate matters enter the brain, they cross the blood–brain barrier (BBB) (Figure 3.9).

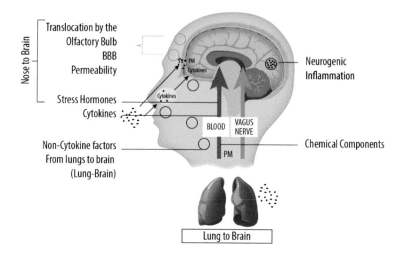

Figure 3.8 Pollutants enter from nose to brain and lungs to brain.

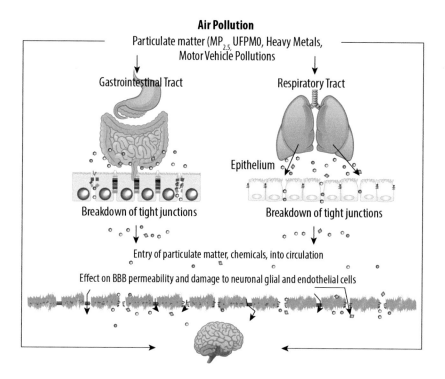

Figure 3.9 Mechanism of entry of pollutants into the blood and nervous system.

The BBB comprises of numerous cells such as "endothelia, pericytes, astrocytes, and blood vessels within the brain parenchyma. The BBB is a chemical and physical barrier comprised of multiple cell types, metabolizing enzymes, and transporter proteins that protect the brain from external insult." This transporter system provides an impermeable barrier covering most of the regions of the brain. The BBB efficiently guards the CNS interstitial against toxic substances.

It has been demonstrated that air pollutants swiftly change the biology and physiology of BBB; hence, this change alters the function. The nanoparticles can injure endothelial cells, damage the BBB, and enter the brain [36]. Human exposure to air pollution shows endothelial cell damage in the cerebral vasculature [37], damage to the BBB, and entry into the brain, especially in the higher centers.

More recently, Shou et al. (2020) [38] found that exposure to PM2.5 causes reduced integrity of tight junction proteins in the BBB. The authors further reported that chronic exposure to "PM2.5 induced a neuroinflammatory reaction, cognitive impairment, and inflammation in the hypothalamus and olfactory bulb. The CNS dysfunction may be due to neuroinflammatory reactions; reduced integrity of the BBB allowed pulmonary inflammation to neuronal alterations" [38].

Air pollution causes "CNS oxidative stress, neuroinflammation, neuron damage, enhancement of abnormal filamentous proteins (Aβ and α synuclein), BBB changes, and cerebrovascular damage, linking the pathways through which air pollution impacts the CNS disease pathology" [29]. These findings show that environmental pollutants, mainly PM, can cause BBB damage and impairs brain functions (Table 3.1; Figure 3.10).

Table 3.1 Environmental Pollution and Blood–Brain Barrier (BBB) Breakdown

- Environmental pollution
- Disruption or breaks down of the blood–brain barrier (BBB)
- Leakage of blood-derived toxic products and pathogens enter the brain
- Increased BBB permeability through the endothelial cells
- Increased migration of inflammatory molecules into the brain
- Activation of microglial cells
- Increase reactive oxygen species (ROS)
- Increase neuroinflammatory processes
- Damage in the myelin sheath
- Neuronal damage

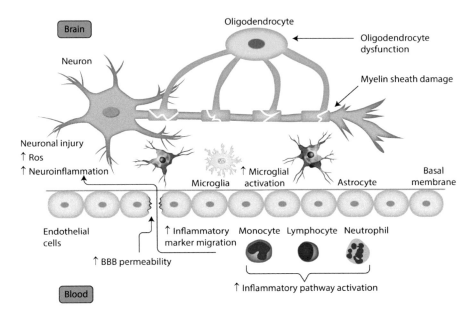

Figure 3.10 Environmental pollution and injury to the blood–brain barrier.

3.8 Pollution and Brain

Environmental pollution has a significant impact on the CNS. The scientific investigations revealed various links between air pollution and brain diseases and behavioral impairments, neuroinflammation, and neurodegeneration [33]. The recent epidemiological literature has raised apprehensions about the health hazards of air pollution on brain biological outcomes. These results on the brain include "brain inflammation and white matter abnormalities, leading to a rising risk for autism spectrum disorders, lower intelligence quotient, neurodegenerative diseases Parkinson's disease, Alzheimer's disease, and stroke" [39].

Literature has also acknowledged that people living or working in urban areas with higher air pollution are more likely to have impaired cognitive function and brain pathology [40, 41].

Calderón-Garcidueñas et al. (2008) [42] reported that exposure to air pollution causes neuroinflammation, alters brain innate immune responses, and causes the accumulation of alpha-synuclein in childhood. Air pollution is considered a risk factor for Alzheimer's and Parkinson's diseases among people living in a polluted environment. Similarly, Li et al. (2019) [43] demonstrated that PM2.5 could cause injuries in multiple organs, including the lung, brain, heart, testis, and intestine. The authors established evidence that air pollution can cause significant pathological and functional alterations in multiple organs, including the brain.

Zhu et al. (2020) [44] conducted a study and reported that PM2.5 exposure might lead to oxidative stress and neuroinflammation. These responses disturb cellular and molecular mechanisms and cause mitochondrial dysfunction, synaptic damage, neuronal apoptosis, and neurodegenerative diseases. The authors established a mechanism through which PM2.5 causes neurodegenerative changes in the brain. The neurodegenerative changes in the brain further impair behavioral and cognitive functions.

Air pollution is a multifactorial, complex environmental condition assaulting all the body systems, including the brain and its various centers. Air pollution affects the brain through different cellular, molecular, and inflammatory pathways that directly damage brain structures or lead to neurological diseases. There is also a strong link in which air pollution can cause ischemia, stroke, multiple sclerosis, and various neurological disorders [45, 46]. Animal and human studies demonstrated that air pollution causes brain oxidative stress, neuroinflammation, and neuron damage. Air pollution enhances the abnormal filamentous proteins (Aβ and α synuclein), BBB changes, and cerebrovascular injury. These findings link the pathways through which air pollution impacts the nervous system disease pathology [29].

Air pollution affects brain biology by producing pro-inflammatory mediators [47]. The biochemical arrangement of air pollution shows both spatial and temporal variations that reflect local sources, including motor vehicle pollutants, industry, and natural biologic processes. The outdoor and indoor pollutants can penetrate the body and reach the brain, resulting in adverse impacts on the nervous system [23, 26] (Figure 3.11).

The health hazards depend upon the number of pollutants, size, surface area, chemical nature, emission source, physical properties, and deposition of contaminants. The effects of contaminants on human health may be acute or chronic. The acute effects are manifested shortly after exposure in hours or days. However, the chronic effects usually take months and years [48, 49].

The brain is vulnerable to ultrafine ambient PM. The nano-size particles could cross the BBB and enter the brain [50]. The biological components of PMs can cause neurodevelopmental disorders, including "schizophrenia, autism, mental retardation, and neurodegenerative diseases" [51]. Exposure to lead and manganese have substantial implications for neuronal damage and brain disease. The clinical manifestation of lead neurotoxicity includes reduced cognitive function, attention, encephalopathy, and convulsions [52]. The scientific literature has also revealed that exposure to air pollutants can cause damage to the sensory neurons present in the olfactory

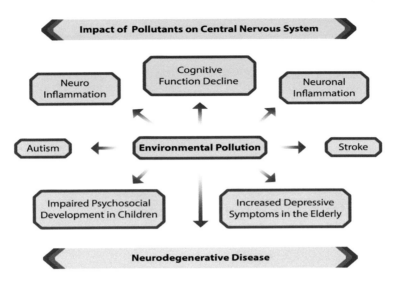

Figure 3.11 Effect of pollutants on brain.

epithelium and associated functional changes [53]. Air pollution mixtures such as ozone exposure cause cerebral edema [54], neurodegeneration in the hippocampus, striatum, and substantia nigra [55], altering human behavior and attitude.

The most probable mechanism that mediates "air pollution-induced adverse effects on the human brain" is oxidative stress. The particles escape through the respiratory tract and enter the circulatory system and ultimately reach the brain. Moreover, the ultrafine PM can enter the brain through the olfactory system [27, 28]. The current literature reveals that the brain tissue of people residing in highly polluted areas shows an "increase in CD-68, CD-163, and HLA-DR positive cells. These indicate infiltrating monocytes or resident microglia activation, elevated pro-inflammatory markers, Interleukin-1β, IL1-β; cyclo-oxygenase 2, COX2." There is an increase in Aβ42 deposition, a hallmark of Alzheimer's disease, BBB damage, endothelial cell activation [56], and brain lesions in the prefrontal lobe [57]. Moreover, "pro-inflammatory markers such as COX2 and IL1-β, as well as the CD-14 marker for innate immune cells, were localized in the frontal cortex, substantia nigra, and vagus nerves [58]. Furthermore, animal studies have also shown that air pollution causes cytokine production" [59, 60]. These studies significantly support the hypothesis and show that air pollution has notable adverse impacts on the nervous systems (Figure 3.12).

Air pollution allied effects on the nervous system are primarily molecular, neuroinflammatory, and neurotoxic, which implicate the various brain diseases. The literature demonstrates the neurological effects in adults and children after extended-term exposure to air pollutants. Air pollution components increase the risk for neurodevelopmental and neurodegenerative disease, and cognitive function impairment. In human adults, the biological and molecular markers of systemic inflammation (IL-6 and fibrinogen) were raised. This instant response on the IL-6 level is possibly due to the production of acute-phase proteins. The progression of oxidative stress and atherosclerosis seems to be the mechanisms involved in the neurological disturbances caused by long-term air pollution. Moreover, a consequence of oxidative stress is neuroinflammation, which occurs in the brain and is involved in the impairment of developmental maturation and various neurodegenerative disorders [60].

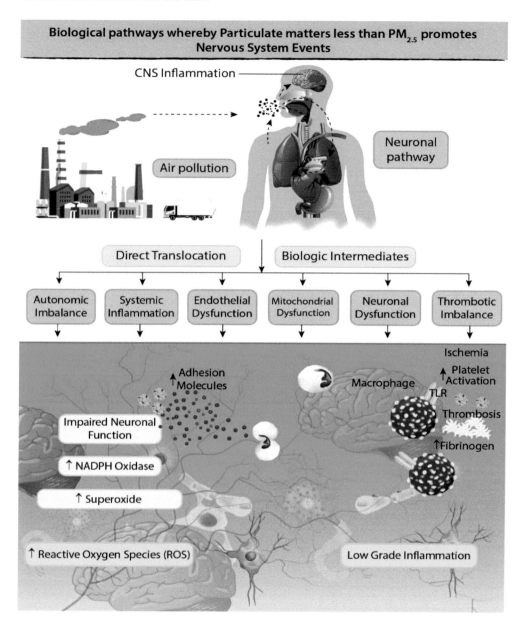

Figure 3.12 Pollution-induced adverse effects on the human brain.

Conclusions

Air pollution is a complex mixture of environmental toxicants. The air pollutants, fine particulate matter (PM 2.5 μm), and ultrafine particulate matter (PM 100 nm, PM 0.1 μm) travel from the environment to the brain via the respiratory, olfactory, and gastrointestinal system into the blood circulation. These environmental toxicants assault the nervous system through several cellular and molecular pathways to cause neuronal damage. Environmental pollution can cause oxidative stress, neuroinflammation, neurodegeneration, cerebral vascular damage, white matter abnormalities, declined cognitive functions, Alzheimer's disease, Parkinson's disease, autism spectrum disorders (ASDs), and stroke. Worldwide, people face elevating global environmental issues, and the developing countries are experiencing complex, severe, and fast-growing pollution problems. Environmental pollution is more than just a health issue; it is a social and economic issue that has had devastating consequences on public health, economy, and brain biology. Health officials must take strict preventive measures and implement strategies to ensure that policies are effective at regional and international levels to minimize pollution worldwide.

References

[1] Meo S.A., Aldeghaither M., Alnaeem K.A., Alabdullatif F.S., Alzamil A.F., Alshunaifi A.I., Alfayez A.S., Almahmoud M., Meo A.S., El-Mubarak A.H. Effect of motor vehicle pollution on lung function, fractional exhaled nitric oxide and cognitive function among school adolescents. *Eur. Rev. Med. Pharmacol. Sci.* 2019; 23(19): 8678–86.

[2] Meo S.A., Suraya F. Effect of environmental air pollution on cardiovascular diseases. *Eur. Rev. Med. Pharmacol. Sci.* 2015 Dec; 19(24): 4890–7.

[3] Meo S.A., AlMutairi F.J., Alasbali M.M., Alqahtani T.B., AlMutairi S.S., Albuhayjan R.A., Al Rouq F., Ahmed N. Men's health in industries: Plastic plant pollution and prevalence of pre-diabetes and type 2 diabetes mellitus. *Am. J. Mens. Health.* 2018; 12(6): 2167–72.

[4] Meo S.A., Memon A.N., Sheikh S.A., Rouq F.A., Usmani A.M., Hassan A., Arian S.A. Effect of environmental air pollution on type 2 diabetes mellitus. *Eur. Rev. Med. Pharmacol. Sci.* 2015 Jan; 19(1): 123–8.

[5] Phalen R.F., Mendez L.B., Oldham M.J. New developments in aerosol dosimetry. *Inhal. Toxicol.* 2010; 22(Suppl 2): 6–14.

[6] Manisalidis I., Stavropoulou E., Stavropoulos A., Eugenia Bezirtzoglou. Environmental and health impacts of air pollution: A review. *Front. Public. Health.* 2020; 8: 14. doi: 10.3389/fpubh.2020.00014.

[7] World Health Organization (WHO). Air pollution. Available at: www.who.int/newsroom/air-pollution. Cited date Dec 2018.

[8] Otsuka K. Shift in China's commitment to regional environmental governance in Northeast Asia? *Journal Journal of Contemporary East Asia Studies.* 2018: 16–34. https://doi.org/10.1080/24761028.2018.1504643.

[9] Colbeck I., Lazaridis M. Aerosols and environmental pollution. *Sci. Nat.* 2009; 97: 117–31. doi: 10.1007/s00114-009-0594-x.

[10] Incecik S., Gertler A., Kassomenos P. Aerosols and air quality. *Sci. Total. Env.* 2014; 355: 488–9. doi: 10.1016/j.scitotenv.2014.04.012.

[11] Feng X.U., Nan Xiang, Yoshiro Higano. How to reach haze control targets by air pollutants emission reduction in the Beijing-Tianjin-Hebei region of China? *PLoS One.* 2017; 12(3): e0173612. Published online 2017 Mar 10. doi: 10.1371/journal.pone.0173612.

[12] Bottenheim J.W., Dastoor A., Sun-Ling G., Kaz Higuchi, YI-Fanli. Long-range transport of air pollution to the arctic. 2004: 1–15. doi: 10.1007/b94522.

[13] Meo S.A., Al-Drees A.M., Al Masri A.A., Al Rouq F., Azeem M.A. Effect of the duration of exposure to cement dust on respiratory function of non-smoking cement mill workers. *Int. J. Environ. Res. Public. Health.* 2013 Jan 16; 10(1): 390–8. doi: 10.3390/ijerph10010390.

[14] Oberdorster G., Ferin J., Lehnert B.E. Correlation between particle size, in vivo particle persistence, and lung injury. *Environ Health Perspect.* 1994; 102(Suppl 5): 173–9.

[15] Li N Sioutas C., Cho A., Schmitz D., Misra C., Sempf J. Ultrafine particulate pollutants induce oxidative stress and mitochondrial damage. *Environ Health Perspect.* 2003; 111: 455–60.

[16] Kumar P., Morawska L., Birmili W., Paasonen P., Hu M., Kulmala M., et al. Ultrafine particles in cities. *Environ. Int.* 2014; 66: 1–10.

[17] Deng Q., Deng L., Miao Y., Guo X., Li Y. Particle deposition in the human lung: Health implications of particulate matter from different sources. *Environ. Res.* 2019; 169: 237–45. doi: 10.1016/j.envres.2018.11.014.

[18] Meo S.A., Azeem M.A., Ghori M.G., Subhan M.M. Lung function and surface electromyography of intercostal muscles in cement mill workers. *Int. J. Occup. Med. Environ. Health.* 2002; 15(3): 279–87.

[19] Oberdörster G., Utell M.J. Ultrafine particles in the urban air: To the respiratory tract-and beyond? *Environ Health Perspect.* 2002; 110(8): A440–1.

[20] Oberdörster G1., Sharp Z., Atudorei V., Elder A., Gelein R., Kreyling W., Cox C. Translocation of inhaled ultrafine particles to the brain. *Inhal. Toxicol.* 2004; 16(6–7): 437–45.

[21] Sheppard D., Hughson W.G., Shellito J. Occupational lung diseases. In: Joseph La Dou. editors. *Occupational Medicine.* Appleton: Lange. 1990, pp. 221–36. Dayman H. The expiratory spiro-gram. *Am. Rev. Resp. Dis.* 1960; 83: 842–55.

[22] Walter J.B., Israel M.S. The body's defense against infection. In: *General Pathology*, 6th edition. Edinburgh: Churchill Livingstone. 1987, pp. 102–3.

[23] Qian Li, Congbo Song, Hongjun Mao. Particulate matter and public health. *Encyclopedia of Environ. Health.* 2019; 31–5. doi.org/10.1016/B978-0-12-409548-9.10988-1.

[24] Kreyling W.G., Semmler-Behnke M., Moller W. Ultrafine particle-lung interactions: Does size matter? *J. Aerosol. Med.* 2006; 19: 74–83.

[25] Kreyling W.G., Semmler M., Erbe F., Mayer P., Takenaka S., Schulz H., Oberdorster G., Ziesenis A. Translocation of ultrafine insoluble iridium particles from lung epithelium to extrapulmonary organs is size dependent but very low. *J. Toxicol. Environ. Health. A.* 2002; 65: 1513–30.

[26] Gurgueira S.A., Lawrence J., Coull B., Murthy G.G., Gonzalez-Flecha B. Rapid increases in the steady-state concentration of reactive oxygen species in the lungs and heart after particulate air pollution inhalation. *Environ Health Perspect.* 2002; 110: 749–55.

[27] Lewis J., Bench G., Myers O., Tinner B., Staines W., Barr E., et al. Trigeminal uptake and clearance of inhaled manganese chloride in rats and mice. *Neurotoxicology.* 2005; 26(1): 113–23.

[28] Elder A., Gelein R., Silva V., Feikert T., Opanashuk L., Carter J. Translocation of inhaled ultrafine manganese oxide particles to the central nervous system. *Environ Health Perspect.* 2006; 114(8): 1172–80.

[29] Block M.L., Calderón-Garcidueñas L. Air pollution: Mechanisms of neuroinflammation and CNS disease. *Trends. Neurosci.* 2009; 32: 506–16.

[30] Genc S., Zadeoglulari Z., Fuss S.H., Genc K. The adverse effects of air pollution on the nervous system. *J. Toxicol.* 2012; 2012: 782462. doi: 10.1155/2012/782462.

[31] Maher B.A., Ahmed I.A., Karloukovski V., MacLaren D.A., Foulds P.G., Allsop D., Mann D.M., Torres-Jardón A., Calderon-Garciduenas L. Magnetite pollution nanoparticles in the human brain. *Proc. Natl. Acad. Sci. USA.* 2016; 113: 10797–801.

[32] Chen C., Nakagawa S. Planetary health and the future of human capacity: The increasing impact of planetary distress on the human brain. *Challenges.* 2018; 9(2): 41; https://doi.org/10.3390/challe9020041.

[33] Lucchini R.G., Dorman D.C., Elder A., Veronesi B. Neurological impacts from inhalation of pollutants and the nose-brain connection. *Neurotoxicology.* 2012; 33(4): 838–41. doi: 10.1016/j.neuro.2011.12.001.

[34] Sunderman F.W., Jr. Nasal toxicity, carcinogenicity, and olfactory uptake of metals. *Ann. Clin. Lab. Sci.* 2001; 31: 3–24.

[35] Leavens T.L., Rao D., Andersen M.E., Dorman D.C. Evaluating transport of manganese from olfactory mucosa to striatum by pharmacokinetic modeling. *Toxicol. Sci.* 2007; 97: 265–78.

[36] Chen L., Yokel R.A., Hennig B., Toborek M. Manufactured aluminum oxide nanoparticles decrease expression of tight junction proteins in brain vasculature. *J. Neuroimmune. Pharmacol.* 2008; 3: 286–95.

[37] Calderon-Garciduenas L., Solt A.C., Henriquez-Roldan C., Torres-Jardon R., Nuse B., Herritt L., et al. Long-term air pollution exposure is associated with neuroinflammation, an altered innate immune response, disruption of the blood–brain barrier, ultrafine particulate deposition, and accumulation of amyloid beta-42 and alpha-synuclein in children and young adults. *Toxicol. Pathol.* 2008b; 36: 289–310.

[38] Shou Y., Zhu X., Zhu D., Yin H., Shi Y., Chen M., Lu L., Qian Q., Zhao D., Hu Y., Wang H. Ambient PM2.5 chronic exposure leads to cognitive decline in mice: From pulmonary to neuronal inflammation. *Toxicol Lett.* 2020; 331: 208–17. doi: 10.1016/j.toxlet.2020.06.014.

[39] Allen J.L., Klocke C., Morris-Schaffer K., Conrad K., Sobolewski M., Cory-Slechta D.A. Cognitive effects of air pollution exposures and potential mechanistic underpinnings. *Curr. Environ. Health. Rep.* 2017; 4(2): 180–91.

[40] Weuve J., Puett R.C., Schwartz J., Yanosky J.D., Laden F., Grodstein F. Exposure to particulate air pollution and cognitive decline in older women. *Arch. Intern. Med.* 2012; 172: 219–27.

[41] Finkelstein M.M., Jerrett M. A study of the relationships between Parkinson's disease and markers of traffic-derived and environmental manganese air pollution in two Canadian cities. *Environ. Res.* 2007; 104: 420–32.

[42] Calderón-Garcidueñas L., Solt A.C., Henríquez-Roldán C., Torres-Jardón R., Nuse B., Herritt L., Villarreal-Calderón R., Osnaya N., Stone I., García R., Brooks D.M., González-Maciel A., Reynoso-Robles R., Delgado-Chávez R., Reed W. Long-term air pollution exposure is associated with neuroinflammation, an altered innate immune response, disruption of the blood–brain barrier, ultrafine particulate deposition, and accumulation of amyloid beta-42 and alpha-synuclein in children and young adults. *Toxicol Pathol.* 2008; 36(2): 289–310. doi: 10.1177/0192623307313011.

[43] Li D., Zhang R., Cui L., Chu C., Zhang H., Sun H., Luo J., Zhou L., Chen L., Cui J., Chen S., Mai B., Chen S., Yu J., Cai Z., Zhang J., Jiang Y., Aschner M., Chen R., Zheng Y., Chen W. Multiple organ injury in male C57BL/6J mice exposed to ambient particulate matter in a real-ambient PM exposure system in Shijiazhuang, China. *Environ Pollut.* 2019; 248: 874–87. doi: 10.1016/j.envpol.2019.02.097.

[44] Zhu X., Ji X., Shou Y., Huang Y., Hu Y., Wang H. Recent advances in understanding the mechanisms of PM2.5-mediated neurodegenerative diseases. *Toxicol Lett.* 2020; 329: 31–7. doi: 10.1016/j.toxlet.2020.04.017.

[45] Genc S., Zadeoglulari Z., Fuss S.H., Genc K. The adverse effects of air pollution on the nervous system. *J Toxicol.* 2012; 2012: 782462. doi: 10.1155/2012/782462.

[46] Mateen F.J., Brook R.D. Air pollution as an emerging global risk factor for stroke. *JAMA.* 2011; 305(12): 1240–1.

[47] Block M.L., Elder A., Auten R.L., Bilbo S.D., Chen H., Chen J.C. The outdoor air pollution and brain health workshop. *Neurotoxicology.* 2012; 33(5): 972–84.

[48] Meo S.A., Al-Drees A.M., Rasheed S., Meo I.M., Khan M.M., Al-Saadi M.M., Alkandari J.R. Effect of the duration of exposure to polluted air environment on lung function in subjects exposed to crude oil spill into seawater. *Int. J. Occup. Med. Environ. Health.* 2009; 22(1): 35–41.

[49] Meo S.A., Bashir S., Almubarak Z., Alsubaie Y., Almutawa H. Shisha smoking: Impact on cognitive functions impairments in healthy adults. *Eur. Rev. Med. Pharmacol. Sci.* 2017; 21(22): 5217–22.

[50] Lockman P.R., Koziara J.M., Mumper R.J., Allen D.D. Nanoparticle surface charges alter blood–brain barrier integrity and permeability. *J. Drug. Target.* 2004; 12: 635–41.

[51] Meyer U., Feldon J., Fatemi S.H. In-vivo rodent models for the experimental investigation of prenatal immune activation effects in neurodevelopmental brain disorders. *Neuroscience and biobehavioral reviews.* 2009; 33: 1061–79.

[52] Mendola P., Selevan S.G., Gutter S., Rice D. Environmental factors associated with a spectrum of neurodevelopmental deficits. *Ment. Retard. Dev. Disabil. Res. Rev.* 2002; 8: 188–97.

[53] Tonelli L.H., Postolache T.T. Airborne inflammatory factors: "From the nose to the brain". *Front Biosci (Schol Ed).* 2010; 2: 135–52.

[54] Cretu D.I., Sovrea A., Ignat R.M., Filip A., Bidian C., Cretu A. Morpho-pathological and physiological changes of the brain and liver after ozone exposure. *Rom. J. Morphol. Embryol.* 2010; 51: 701–6.

[55] Pereyra-Munoz N., Rugerio-Vargas C., Angoa-Perez M., Borgonio-Perez G., Rivas-Arancibia S. Oxidative damage in substantia nigra and striatum of rats chronically exposed to ozone. *J. Chem. Neuroanat.* 2006; 31: 114–23.

[56] Calderón-Garcidueñas L., Azzarelli B., Acuna H., Garcia R., Gambling T.M., Osnaya N., Monroy S., D.E.L Tizapantzi M.R., Carson J.L., Villarreal-Calderon A., Rewcastle B. Air pollution and brain damage. *Toxicol. Pathol.* 2002; 30(3): 373–89.

[57] Calderón-Garcidueñas L., Solt A.C., Henríquez-Roldán C., Torres-Jardón R., Nuse B., Herritt L., Villarreal-Calderón R., Osnaya N., Stone I., García R., Brooks D.M., González-Maciel A., Reynoso-Robles R., Delgado-Chávez R., Reed W. Long-term air pollution exposure is associated with neuroinflammation, an altered innate immune response, disruption of the blood–brain barrier, ultrafine particulate deposition, and accumulation of amyloid beta-42 and alpha-synuclein in children and young adults. *Toxicol. Pathol.* 2008; 36(2): 289–310.

[58] Calderón-Garcidueñas L., Mora-Tiscareño A., Ontiveros E., Gómez-Garza G., Barragán-Mejía G., Broadway J., et al. Air pollution, cognitive deficits and brain abnormalities: A pilot study with children and dogs. *Brain. Cogn.* 2008; 68(2): 117–27.

[59] Campbell A., Oldham M., Becaria A., Bondy S.C., Meacher D., Sioutas C., Misra C., Mendez L.B., Kleinman M. particulate matter in polluted air may increase biomarkers of inflammation in mouse brain. *Neurotoxicology.* 2005; 26(1): 133–40.

[60] Kleinman M.T., Araujo J.A., Nel A., Sioutas C., Campbell A., Cong P.Q., Li H., Bondy S.C. Inhaled ultrafine particulate matter affects CNS inflammatory processes and may act via MAP kinase signaling pathways. *Toxicol. Lett.* 2008; 178(2): 127–30.

Environmental Pollution and Brain Development

4.1 Introduction to Brain Development

In the past three decades, significant progress has been made in neuroscience and understanding the brain's biological and molecular mechanisms and its development. The literature elucidates the neurobiology of brain development as a neural organization via cellular and molecular events. The biology of brain development consists of multifaceted sequences of various events, which emerge from the growing neurobiological processes that underlie neuronal, cognitive, social, and psychological development [1, 2].

The process of brain development is very long, starting at the beginning of the third gestational week with the variation of neural progenitor cells and continuing through adolescence and later life. The mechanisms involved in normal brain development range from the molecular events of gene expression to those influenced by environmental factors [3, 4]. The interruption of either can alter the anatomical, physiological, and biological outcomes [5]. Brain development is categorized into multiple processes that operate in a highly organized, genetically controlled, and constantly changing context. These natural processes facilitate the development of the complex and dynamic structures of the human brain [6].

The biological processes and environmental elements play a significant role in translating the cellular mechanisms into the complex morphology of the human brain. The particular imprint of neurulation, neuronal differentiation, migration, and assembling during cortical folding remains complex to understand [7, 8]. The brain is highly sensitive to learning key events. According to the body's needs and different environmental

DOI: 10.1201/9781003212461-4

63

situations, brain cells can change throughout life [9], and cognitive development is variable at different stages of life. The brain is an extremely accommodating organ that controls and coordinates various functions according to the situation, environment, and requirements [9].

4.2 Embryology of Brain Development

Fertilization is a highly sequenced, coordinated, hormonal, and molecular event that involves merging a sperm cell with an egg cell. Human conception mainly occurs during the second week after the menstrual cycle. The sperm and egg cells unite to form a diploid cell, called a zygote [10].

After fertilization, during the first week of conception, or the third week of pregnancy, the zygote moves down the fallopian tube toward the uterus, where it divides to form a cluster of cells, known as a "morula." During the fourth week of pregnancy, swiftly dividing cells called the "blastocyst" begin to burrow into the uterine lining or endometrium by a process referred to as implantation [11]. In the blastocyst, the inner group of cells becomes the embryo, and the outer layer rises to develop the placenta. In the third week of conception, or the fifth week of pregnancy, the embryo has three layers: the "endoderm," which gives rise to columnar cells and internal organs, the "mesoderm," which gives rise to muscle cells and connective tissue, and the "ectoderm," which gives rise to skin cells and the CNS and PNS [3, 12].

From the embryonic period, through birth and infancy, to kindergarten age, the brain undergoes critical stages of development characterized by rapid structural and biological changes [13, 14]. During this period, neuronal cells are born that migrate to their final destination, and neuronal networks are established [15, 16]. From a biological perspective, three different stages exist. The first stage involves neural tube formation, neuronal division, and migration, the second stage is neuronal connectivity, and the third stage consists of the formation of synapses and synaptic connections [17]. The literature classifies these stages of brain development into neurulation, proliferation, migration, differentiation, synaptogenesis, and myelination [15, 16].

4.2.1 Neurulation

The first step in brain development is the formation of the neural tube, by the process known as primary neurulation, which requires cell signaling and hereditary coordination by an assortment of genes. The process starts with an open neural plate and ends with the neural plate bending in precise and diverse steps. These steps eventually lead to the neural plate closing to form a neural tube that serves as the embryonic brain and spinal cord [3].

During the early second week of development, the evolving embryo is transformed into a three-layered spherical structure. In one area of the sphere, the cells thicken to form a neural plate. On the 21st day of development, a "trench" in the neural plate creates a neural groove. This plate folds over and forms a tube, which slowly closes, first at the bottom and then at the top, in a zipper-like pattern. The inner cells form the CNS, brain, and spinal cord, while the outer cells form the ANS [16]. The "cells that border the neural and non-neuronal ectoderm that migrate and form layers are called neural crest cells (Figure 4.1). The neural crest develops into the peripheral nervous system spinal and cranial nerves" [3, 16].

At the closure of a neural tube, it develops three primary vesicles and subsequently five secondary vesicle structures. The anterior part of the tube becomes the forebrain, further transforming into the telencephalon, which "includes the cerebral hemispheres, the

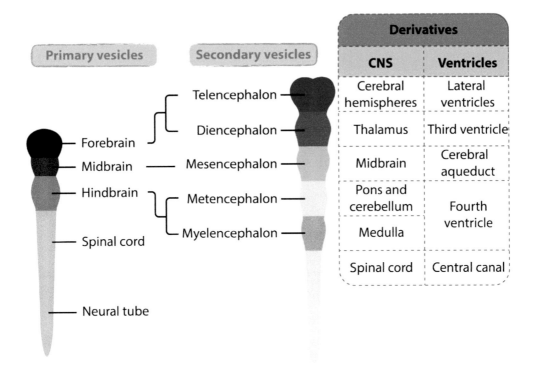

		Derivatives	
Primary vesicles	Secondary vesicles	CNS	Ventricles
Forebrain	Telencephalon	Cerebral hemispheres	Lateral ventricles
	Diencephalon	Thalamus	Third ventricle
Midbrain	Mesencephalon	Midbrain	Cerebral aqueduct
Hindbrain	Metencephalon	Pons and cerebellum	Fourth ventricle
	Myelencephalon	Medulla	
Spinal cord		Spinal cord	Central canal
Neural tube			

Figure 4.1 Brain development: primary and secondary vesicles.

diencephalon (thalamus and hypothalamus), and the basal ganglia [18]. The cells located around the middle vesicle become the midbrain, which connects the diencephalon to the hindbrain." The rearmost part of the tube gives rise to the hindbrain, consisting of the pons, cerebellum, and medulla oblongata. Finally, the remaining cells give rise to the spinal cord (Figure 4.2) [16].

4.2.2 Neuronal Proliferation

Once the overall neural tube structures are organized, the cells that line the tube's innermost part, known as the ventricular zone, proliferate at logarithmic growth. These cells swiftly multiply, form a second zone, the marginal zone containing axons and dendrites. The proliferative stage starts from gestation week 4 and continues to week 12, with the significance that the newborn brain exhibits a larger number of neurons than the adult brain. Eventually, the excessive formation of neurons is balanced by apoptosis or programmed cell death [16]. This stage of brain development extends during the first half of the gestational period and is categorized by the formation of approximately 4,166 neurons per second, or 250,000 neurons per minute [19]. The neuronal division and relocation are linked in thickness and surface area with cortical growth [20]. However, until the mid-gestational period, the growth-induced cortical stress is too trivial to induce cortical folding, and the cortical surface remains smooth [21].

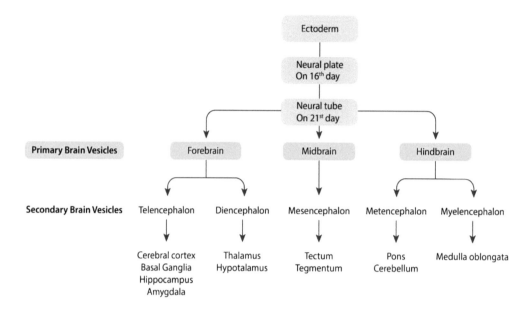

Figure 4.2 Embryological stages of brain development.

4.2.3 Neuronal Migration

Neuronal cell migration is a vital process in the development of the normal brain. The proliferative germinal zones exhibit two primary functions, i.e., to produce the appropriate number of cells for the particular brain region and to produce the precise class of cells that need to migrate in the right direction and position [22]. The neurons travel to their final destinations to achieve the normal anatomical and biological processes. The anatomical and physiological development of the brain depends on neurogenesis, neuronal migration, neuronal circuit formation, differentiation, and positioning of neurons [22]. Neuronal cell migration brings the cells into an appropriate spatial position with a biological linkage with other cells [23]. The newborn neurons form the neuroepithelium, a proliferative layer of the neural tube. These cells migrate from the germinal zone and disperse throughout the nervous system to reach their final destination to become part of an appropriate neuronal circuit [24]. Neuronal migration can be divided into three phases: (1) extension of the leading process, (2) nucleokinesis, and (3) retraction of the trailing process [25, 26].

"Neuronal migration follows a radial or tangential migratory pathway, depending on the part of the developing nervous system in which the neurons originate. In radial migration, neurons follow a path perpendicular to the neuroepithelial surface and proceed alongside radial glial fibers [27]. The tangential migration of neurons is parallel to the pial surface [28, 29]. A dual-phase neuronal migration referred to as a switching migration also is present, which is a combination of tangential and radial migration" [30]. For an appropriate destination, the migrating neurons select a pathway using physical substrates and chemical cues of either a diffusible or non-diffusible nature [29].

Functioning neuronal circuits are based on neuronal migration in an appropriate spatiotemporal pattern. Any defect in neuronal migration may cause a neurological disorder. The neuronal migration defects (NMDs) are a heterogeneous group of developmental defects that can cause devastating brain diseases. The clinical manifestations of NMDs include schizophrenia, autism, ataxia, epilepsy, cognitive abnormalities, and cerebellar degeneration [31–34].

4.2.4 Neuronal Differentiation

Neuronal cell differentiation is the process of maturation of the neurons, in which they change from one cell type to another and become more specialized to perform various functions. This process occurs numerous times during development to evolve from a simple zygote to complex tissues and specific neuronal cell types. The neuronal cells closely coordinate between growth and differentiation programs [35]. When a neuron migrates to its targeted destination, it can differentiate into a mature neuron with axons and dendrites. The development of axons is facilitated by growth cones, which promote growth toward the targets and away from other cells. Molecular signals control these progressions. Dendrite development, however, occurs by a different process. The dendrites appear as thick strands, with a few spine-like structures extending from the cell body. As they mature, the number and density of spines surges, which increases the chances of dendrites establishing contact with a neighboring axon. This connection between axons and dendrites is the basis for synaptic connections, which are highly essential for brain function [35].

4.2.5 Synaptogenesis

After neuronal proliferation, migration, and differentiation, a developing neuron reaches its final destination in the nervous system. Synaptogenesis is the formation and maturation of synaptic contacts in the development of the CNS and PNS [36]. The synapses are generally observed around the 20th week of gestation. Synapse formation and stabilization in the brain is an essential process and requires bidirectional communication between presynaptic and postsynaptic partners (Figure 4.3) [37].

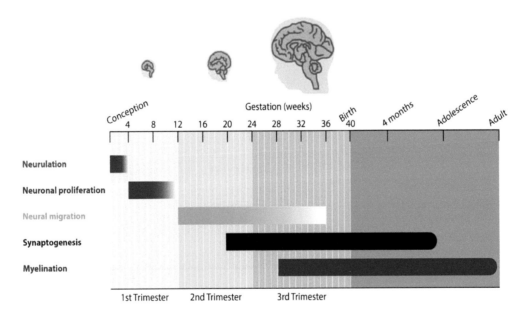

Figure 4.3 Stages of early brain development.

In neurons, the over-formation of synapses is followed by a gradual reduction. Various regions in the brain attain peak synapse production at different levels. For example, the peak is reached somewhere between the 4th and 8th postnatal months in the visual cortex, but the prefrontal cortex does not reach its peak until the 15th postnatal month. This time variance in the height of synapse production is vital, as the later the peak of synapse formation, the longer the region remains plastic [16].

4.2.6 Myelination

Myelination, the last step in brain development [38], starts during the 28th week of pregnancy (Figure 4.3). The developmental peak occurs during the first year of life and continues into young adulthood [39]. This process comprises wide-ranging variations in the oligodendrocyte and membrane architecture. During this progression, the axons wrapped in fatty cells facilitate neuronal activity and communication. The myelinated axons transmit electrical impulses faster than the unmyelinated axons. The timing of myelination is variable depending on the area of the brain involved. Some sensory and motor cells myelinate earlier during the preschool-age period; however, in the regions involved in higher cognitive function, such as the prefrontal cortex, the process continues until adolescence or early adulthood (Figure 4.4) [40]. In addition to these processes, the brain is constantly changing due to environment and experience and responds according to the situation. The developmental and learning processes of the brain continue throughout life [9]. The neuronal connections for different brain functions start at birth and continue until adulthood (Figure 4.4).

4.3 Fundamental Features of the Brain

In an average, healthy, and young adult, the brain exhibits an approximate volume of 1,350 cm^3, a surface area of 1,820 cm^2, and a cortical thickness of approximately 2.7 mm. The brain is comprised of about 100 billion neurons, of which 20 billion are located in the cerebral cortex [41] (Table 4.1). Each cortical neuron has approximately 7,000 synaptic networks to other neurons, resulting in a total of 0.15 quadrillion synapses and around 150,000 km of myelinated nerve fibers [42]. The folding of the cortical surface is essential as a mechanism to maximize the number

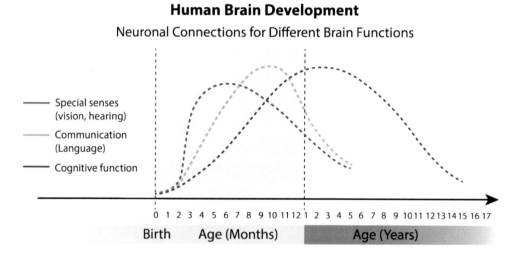

Figure 4.4 Neuronal connection for different brain functions.

Table 4.1 Major Biological Features of the Human Brain [41–45]	
Major parts	Cerebrum, cerebellum, and brainstem
Hemispheres	Right and left
Lobes	Each hemisphere exhibits four lobes: frontal, temporal, parietal, and occipital
Cells	Brains contain two types of cells: nerve cells or neurons and glial cells
Volume	About 1,260 cm^3 in the adult male and 1,130 cm^3 in the adult female
Surface area	1,820 cm^2
Weight	Average 1,200–1,600 g, 2% of body weight, male 1,180–1,620 g, female 1,030–1,400 g
Cortical thickness	2.7 mm
Neurons	About 100 billion
Neurons in the cerebralcortex	About 20 billion (18—20%)
Synaptic connections	Each cortical neuron has 7,000 synaptic networks with other neurons
Myelinated nerve fibers	50,000 km
Cranial nerves	12 pairs
Spinal nerves	31 pairs: 8 cervical, 12 thoracic, 5 lumbar, 5 sacral, and 1 coccygeal

of cortical neurons and minimize the total fiber length within the limited space inside the skull [43]. Environmental elements and physical forces are well-acknowledged in facilitating a significant role in regulatory pattern, selection, and brain surface morphogenesis [44, 45].

4.4 Brain Development in Early Life

The development of the brain is a long process starting a few days after conception and continuing through adolescence and later life. The brain experiences the most dramatic developmental changes during the first few years of life. The early years of a child's life are highly essential for the later years of learning, gaining knowledge, skills, behavior, and overall brain development [46].

The brain changes in shape and size in response to the surrounding environment and events encountered in the early years of life. The anatomical development continues through the embryological period and early infant life. Biological and cognitive development continue and change during infancy, childhood, and adulthood (Figure 4.4) [3, 4].

The precise, physiological processes of brain development start before birth and continue throughout childhood to reach their full potentials. It is the role and responsibility of parents and family to provide a safe environment to protect growing children so that their brains are healthy during the initial and later development phases [47]. Brain development is affected by infant's and children's exposures and personal experiences with people and the surrounding environment. There are components to the brain's adaptation to the environment beyond neurons and synapses, and preparing new synapses is associated with learning. Brain development

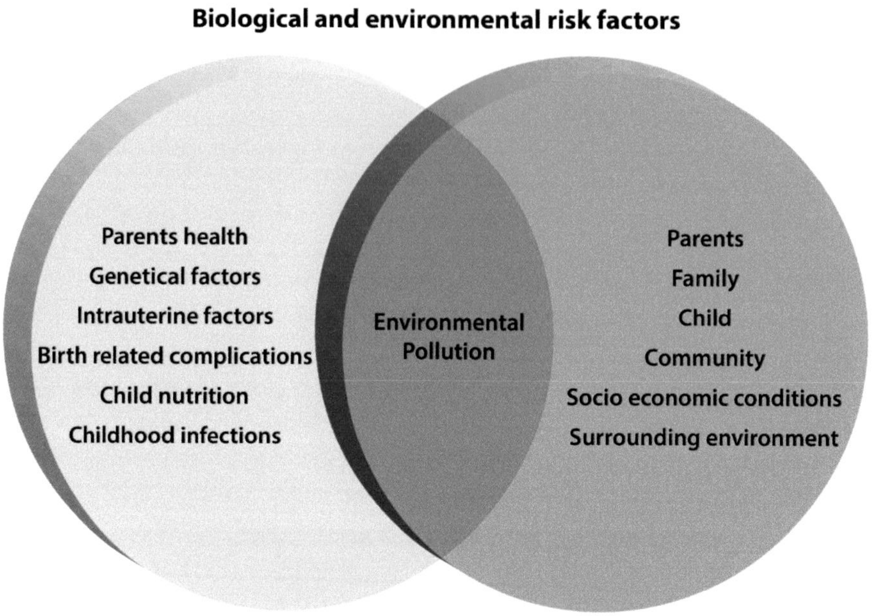

Biological and environmental risk factors

Figure 4.5 Biological and environmental factors affecting early brain development.

and learning capabilities are increased in a safe environment protected from stress, injuries, and environmental pollution [42].

There are multiple factors that directly and/or indirectly affect brain development, including parent's health, pregnancy period, normal sleep, nutrition, smoking, alcohol, stress, exposure to toxins or infections, surrounding environments, and environmental pollution (Figure 4.5) [42, 48–51].

4.5 Effect of Environmental Pollution on Brain Development

Like other basic needs, fresh and clean air is a requirement for normal human evolution and well-being. Rapid urbanization and industrialization are prominent causes of environmental pollution. Surroundings may now no longer meet the criteria for standard air quality due to the presence of toxic levels of air pollutants (Figure 4.6), which cause inhabitants to face acute and long-term detrimental effects concerning biological and neurobehavioral development [52]. Neurodevelopmental disorders also increased among children over the last three decades [53]. Considering how changes in the human genome occur slowly, this suggests that non-genetic factors are the driving forces behind this rapid increase. Many reports, including epidemiological studies, show in-utero or childhood exposure to environmental pollution and its effects on brain development [54].

Environmental pollution is an emerging public health threat and a leading contributor to the burden of disease. The current estimate demonstrates that approximately 2 billion children are exposed to environmental pollution worldwide [55]. The WHO reported marked urbanization globally, with 98% of large cities located in low- and middle-income countries that do not meet safety standards. Similarly, about 56% of large cities are over the contamination limit in developed areas, and 80% of people are overexposed to pollution [56, 57].

Figure 4.6 Sources of environmental pollution.

The most common components of air pollutants are "particulate matter (PM), ozone (O3), carbon monoxide (CO), sulfur dioxide (SO2), nitrogen oxide (NO), and lead (Pb)." These pollutants are hazardous to human health. Air pollution particles with a PM diameter of 2.5 and 1.0 μm, and ultrafine particles with a diameter < 100 nm predominantly affect the brain [58, 59]. The PM2.5 and PM1.0 are mainly produced from emissions of motor vehicle pollutants, fuel combustion, wildfires, oil refineries, power plants, and metal processing industries. Moreover, tobacco smoke, ovens, flames, and pesticides are also significant sources of detrimental substances [59, 60].

The environmental pollutants PM 0.1 μm can be translocated and found in the brain within 4–24 h after inhalation. Nasal PM 0.1 μm can travel through the olfactory nerves to the brain. Researchers also reported that exposures to aerosols of PM 0.1 μm exhibit the maximum brain uptake in the olfactory bulb, even 7 days after exposure. Up to 20% of the PM0.1 deposited on the olfactory mucosa move to the olfactory bulb [61]. This pathway may avoid the BBB and more directly affect the human brain [62]. The PM0.1 translocate and directly damages the neuronal tissue and also affects ANS functions [63].

Exposure to environmental pollutants, lead, and arsenic exhibits long-lasting chronic consequences for neurodevelopment. The children and adolescents exposed to air pollution impaired neurodevelopmental functions, attention deficit-hyperactivity disorder, ASDs, and cognitive function impairment [64]. Among school-aged children, prenatal air pollution exposure can cause damage to brain structure [31] (Figure 4.7). The hazardous effects of air pollution on brain functions include impairment of sensory processing, cognitive development, emotion, retention, memory, and academic performance [55]. Thus, air pollution negatively affects brain functions and development by neural, behavioral, and cognitive impairment.

The health hazards of air pollution are more detrimental to children and adolescents who live in large metropolitan cities. Air pollutants enter the body due to the failure of common barriers intended to ward against the entry of toxic particles to the nasal, gastrointestinal, and lung epithelial tissues and the breakdown of the BBB. The long-term exposure to air pollution causes extensive neuroinflammation, neuronal loss, and ultimately neurodegenerative changes and cognitive function impairment.

Overpopulation, poor living environment, and environmental pollution can result in hazardous effects on the brain (Figure 4.8). The damaging effects depend on the type, size, chemical, biological, and physical nature of the pollutants, duration of exposure, and the overall health status of those exposed. A wide range of effects are due to free radical formation, oxidative stress, neuronal damage, and early hallmarks of Alzheimer's and Parkinson's diseases [65, 66]. Motor vehicle pollutants constitute a significant source of urban air pollution associated with neuroinflammation and Alzheimer's and Parkinson's diseases [65]. The motor

Figure: 4.7 Exposure of children to outdoor air pollution.

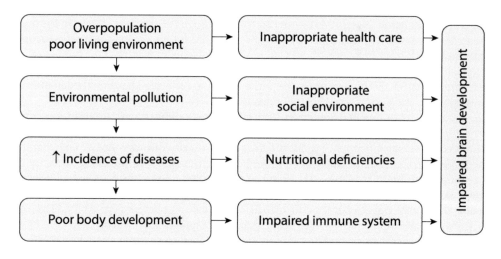

Figure 4.8 Factors affecting brain development.

vehicle air pollutants promote increased "deprivation of tight junction proteins in the cerebral microvasculature, resulting in altered blood–brain barrier permeability and expression of neuroinflammatory markers" [66].

Oxidative stress results in dysregulation of inflammatory responses, progressive neurodegeneration, altered brain restoration, and brain plasticity changes similar to those seen in Alzheimer's disease [34]. Cigarette smoking is a leading cause of oxidative stress and is responsible for decreased expression of presynaptic function, axonal transport, and neurodegenerative changes [67]. Prenatal exposure to pollutants causes permanent changes in neurotransmitters and alters brain development, producing deficits in brain biology and memory impairment [68–69].

Several types of environmental pollution have been identified as being responsible for causing oxidative stress, which is the leading cause of various inflammatory and degenerative processes in the nervous system, both at the cellular and molecular levels (Figure 4.9). Once the inflammatory and neuronal degenerative processes begin, various brain disorders, including cognitive function impairment, impaired memory, dementia, depression, and other inflammatory response disorders, follow.

Berg et al. (2020) [70] demonstrated that when children during their early life reside close to highly trafficked roadways, exposure to pollutants alters neurodevelopment, delaying growth and development of psychomotor reflexes. Similarly, Patten et al. (2020) [71] conducted a study on exposure to traffic-related air pollution and its association with increased risk for various neurodevelopmental disorders. The authors found that exposure to traffic-related air pollution during the gestational and early postnatal period increased the risk of neurodevelopmental disorders.

Nghiem et al. (2019) [72] assessed the impact of maternal exposure to dioxin on fetal brain development outcomes in early childhood. The results showed that exposure to prenatal dioxin

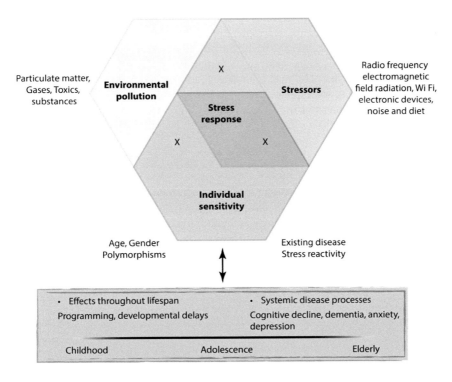

Figure 4.9 Effect of environmental pollution on brain development.

affects neuronal activity, functional connectivity, and language development. Similarly, Cho et al. (2020) [73] reported that polycyclic aromatic hydrocarbon (PAH) exposure was linked to cortical thinning, impaired verbal learning, and memory loss in healthy adults. The authors also suggest that PAH is an environmental risk factor for neurodegeneration.

Power et al. (2018) [74] measured the linkage between brain MRI findings and PM exposures. They demonstrated that long-term exposure to PM was not related to markers of cerebrovascular disease. Cserbik et al. (2020) [75] examined the associations between PM2.5 exposure on brain morphometry and cognition among children. The authors found that residential PM2.5 exposure was related to "hemispheric specific differences in the gray matter across cortical regions of the frontal, parietal, temporal, and occipital lobes, subcortical areas, and the cerebellum. Thus, PM2.5 exposure may affect brain development and cognitive and emotional" functions. Both animal and human studies show that environmental pollution exhibits detrimental developmental effects and causes brain function impairment.

4.6 Vulnerability of Children to Environmental Pollution

The scientific literature demonstrates that infants and children are highly susceptible to adverse effects of environmental pollution, mainly due to their higher respiratory rate and poorly developed natural barriers in the lungs and brain against the inhaled particles [76] (Table 4.2). Epidemiological studies indicate that these harmful consequences are related to prenatal and postnatal exposure at all ages. The air pollutant impact occurs due to the failure of natural barriers against the entry of toxic PM into the brain, promoting neuroinflammation and neurodegenerative processes [55].

The development of usual barriers, including the "blood–brain barrier, nasal, gut, and lung epitheliums," is vital for the overall development of human health. These barriers are more compromised in young urbanites exposed to air pollution, thus reducing the brain's capability to protect against potentially dangerous toxic, fine, and ultrafine particles.

When the toxic pollutants enter the body, the physiological and immune responses are activated in the blood and CSF. The cytokines, "interleukin-1β (IL-1β), interleukin-6 (IL-6), and tumor necrosis factor-alpha (TNF-α)", are released and promote the swelling of tissue. These mechanisms contribute to widespread neuroinflammation in the brain, leading to damage to neural tissue in various areas of the brain. The most targeted and affected parts include the prefrontal area, frontal cortices, olfactory bulbs, and hippocampus [77].

Environmental pollution exposure results in high concentrations of inflammatory mediators in the brain. Moreover, white matter hyperintensity and neuronal demyelination occur

Table 4.2 Factors Affecting the Vulnerability of Children to Environmental Pollution

- High respiratory rate
- High exposures to air pollution
- More time playing on the ground
- High hand-to-mouth behavior
- Immature barriers that protect against the entry of toxic particles
- Sensitivity to air pollutants and toxic substances
- Immature metabolic pathways
- Inability to metabolize toxic chemicals
- Lack of the enzymes to remove toxic substances
- Easily disrupted early developmental processes
- Immature immune systems
- Immature body protective systems

due to the decreased blood flow [78]. These effects impair the neuronal functional capabilities and result in cognitive deficits [79, 80] (Table 4.3). These findings support the hypothesis that air pollution exhibits a harmful impact on myelination and the function of neurons. Impaired blood flow negatively impacts the circulatory architecture of the brain and damages various regions in the brain. Furthermore, exposure to environmental pollution leads to the development of characteristic Alzheimer's disease pathology [80].

The literature also provides evidence that air pollution can cause environmental stress and an early brain function discrepancy in children and adolescents (Figure 4.10). The neuroinflammatory "changes in endothelial barriers, blood vessel arrangement, and blood–brain barrier breakdown contribute to cognitive impairment" and neurodegenerative states. These events negatively impact the cortical structures, and subsequent brain biology, behavior, and cognitive performance development are also affected [81]. Environmental pollution may pose a serious threat to healthy brain development, with biological, physiological, and anatomical changes

---

Table 4.3 Summary: Impact of Environmental Pollution on Brain Development

- Entry of pollutant into the brain through lungs, olfactory nerve, and GIT
- Chronic oxidative stress
- Neuronal inflammation
- Decreased blood flow
- Hypoxia
- Neuronal demyelination
- Neuronal death
- Impairment of brain development

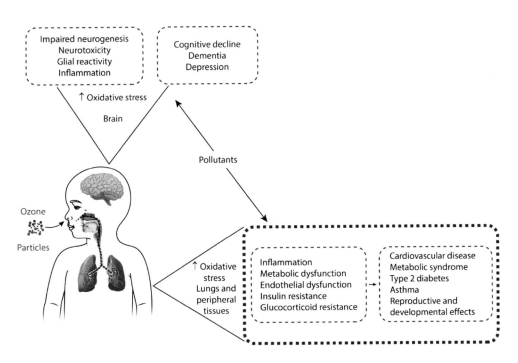

Figure 4.10 Environmental pollution exposures during early childhood and complications developed later in life.

aligned with neurodegeneration. The air pollution dose–response and threshold approaches are the leading causes of chronic urban exposure.

The WHO warned that in children under 5 years of age, one in every four deaths is related to environmental pollution [82]. Environmental pollution thus poses an immense threat to children's health, mainly brain biology. During the last two decades, in addition to the risks of environmental pollution on brain development worldwide, people are facing another challenging issue of excessive use of mobile phones. These electronic devices emit RF-EMFR.

4.7 Effects of Radiofrequency Electromagnetic Field Radiation on the Brain

In recent years, RF-EMFR generated from mobile phones, mobile phone base-station towers (MPBSTs), and other electronic devices have been considered a part of environmental pollution and named "electromagnetic smog." Increasing exposure led to a growing concern about the possible health hazards of RF-EMFR. Mobile phones, base-station towers, and their associated devices increased human exposure to RF-EMFR. Recent studies indicate an association between RF-EMFR emitted by cell phones with various hazardous effects on brain biology [83–88]. However, other studies showed no significant impact of RF-EMFR on the brain and no threat to brain functions [89–92]. However, considerable literature exists to highlight concerns related to RF-EMFR and children's brain development [86].

When comparing the effects of pollution on a child and an adult, it must be noted that major structural variances are present between the head of a child and an adult. Children exhibit thinner skulls compared to adults; thus, infants' and children's skull and brain tissues are more vulnerable to RF-EMFR. Moreover, the myelin sheath in a child's brain is also developing, and after infancy, the myelination process slows down but continues till adulthood. The immature myelin sheath and the unprotected axons can be easily damaged by environmental pollution and RF-EMFR, which can lead to neuronal degeneration and impaired brain biology [93–94]. The literature indicates that a child's brain is vulnerable to environmental pollutants and RF-EMFR, which may therefore impair brain development.

4.7.1 Radiofrequency Electromagnetic Field Radiation and the Skull

An infant's and child's head varies anatomically from the head of an adult and of an aged person. These anatomical differences are essential to understand the existing vulnerability and durability of the skull bone to protect brain tissues from trauma, pollution, and radiation during various periods of life. The average thickness of the skull bone of an adult is 6–8 mm. The skull thickness increases as children grow [95], and similarly, the cranial capacity also changes to house the growing and developing brain of the child. The fetus's skull consists of cartilage, which gradually ossifies until birth. A few areas, such as fontanels, are not ossified for long, even after the birth. The biological advantage of such incomplete ossification is to ease the descent of the fetus through the birth canal at the time of delivery. However, the skull of an adult is entirely ossified and strengthened [96–97]. The anatomical characteristics of the skull bones of infants and children are interesting, since the bone's remodeling and transformation are due to stress on the bone.

The remodeling of bony tissue in the cranium continues until the age of 18 years. The thickness of the skull of an infant under the age of 2 is about 3–4 mm, and the thickness of the skull of a child between the ages of 5 and 10 years is 4–5 mm, while in an adult of 20 years, it is about 6–8 mm [98]. The thickness of the skull rapidly increases during the first year and slows down during the second year, while skull density increases swiftly during the first three years of life. Both skull thickness and density continue to increase up to adulthood [98, 99]. Among infants under the age of 2 years, skull bone density is about 750–850 mg/cm^3, whereas the density of children age 5–10 years is 850–1,000 mg/cm^3, and in adults 20 years old, it is about 1,000–1,100 mg/cm^3 [Figure 4.11]. The skull bone density increases as subjects become older; thus, when

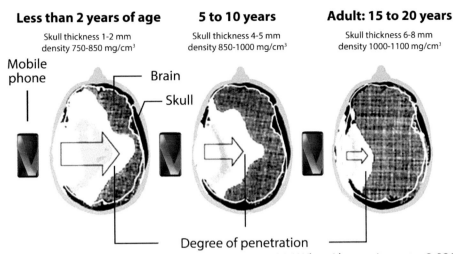

Less than 2 years of age
Skull thickness 1-2 mm
density 750-850 mg/cm³

5 to 10 years
Skull thickness 4-5 mm
density 850-1000 mg/cm³

Adult: 15 to 20 years
Skull thickness 6-8 mm
density 1000-1100 mg/cm³

Mobile phone

Brain

Skull

Degree of penetration

Absorption rate: 4.50 W/kg   Absorption rate: 3.20 W/kg   Absorption rate: 2.930 W/kg

Figure 4.11 Skull bone thickness, density, and penetration of RF-EMFR.

a child is exposed to RF-EMFR, the compact skull bone, which is the brain's protective covering, is not fully thickened, making children's brains more vulnerable to any harmful substances, including environmental pollutants and RF-EMFR. Modern literature acknowledges that children exposed to the mobile phone, base-station towers, and exposure to RF-EMFR could account for various health problems [99–100].

The characteristics of RF-EMFR, the frequency generated from cell phones, the intensity of the RF waves, absorption rates, and degree of energy transfer from the electromagnetic field at a particular point in the absorber are important to understand [101]. The frequency of RF-EMFR emitted by mobile phone devices is about 900–1,800 MHz and 900–2,400 MHz [102]. The intensity of radiation is restricted to specific limits in certain tissues. The capability of RF-EMFR to enter body tissues and organs depends on "the conductivity and permittivity of the tissue and the radiation wavelength, which is inversely related to the frequency" (Figure 4.12). At lower frequencies, the penetration of RF-EMFR is higher, i.e., devices operating at the 900 MHz range can irradiate the body more, approximately 25% compared with 20% penetration at 1,800 MHz. Understanding that penetration of RF-EMFR into specific tissues and organs may be more obvious than others is also essential, possibly since some organs are less protected and more vulnerable than others. Penetration by RF-EMFR depends on various physical, biological, and physiological factors such as age, gender, position, and location of the tissues, organs, and systems of the human body (Figure 4.12).

4.8 Radiofrequency Electromagnetic Field Radiation and the Brain

The development of advanced telecommunication systems and smartphone technology dramatically increased the extent and magnitude of RF-EMFR exposure. The RF-EMFR generated from cell phones, base-station towers, and other electronic devices exert thermal and non-thermal effects. Short-term and long-term exposure to RF-EMFR may exhibit adverse effects on human health, mainly in children, as their tissues are under the developing phase [86–88]. However, literature also reveals that RF-EMFR exhibits no significant harmful effects on the human brain.

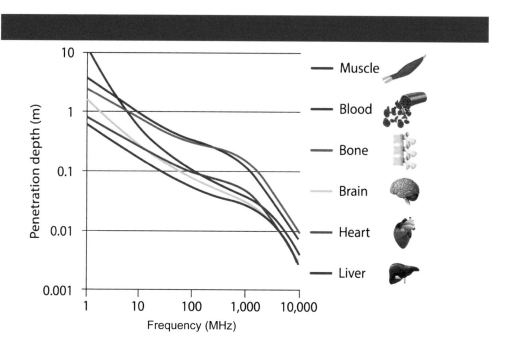

Figure 4.12 Penetration depth of RF-EMFR in body tissue [101–105].

Chang-Ta Chiu et al. (2015) [106] reported that mobile phone usage was associated with headaches and migraines among children. Children who regularly use mobile phones were reported to exhibit a worse health status compared to their health a year before, according to Van Rongen et al. (2009) [107]. Oftedal et al. (2000) [108] demonstrated that long-standing and unnecessary use of mobile phones could cause increased hotness around or in the ear and cause headaches and fatigue. In addition, the literature also demonstrates more health symptoms among mobile phone users, possibly due to "electromagnetic hypersensitive" [109–110]. Zarei et al. (2019) [111] demonstrated an association between offspring speech problems and maternal exposure to the use of cordless phones before and during pregnancy.

Meo et al. (2019) [87] investigated the effect of RF-EMFR generated from mobile phone base station towers, which are mainly placed in crowded, commercial, and residential areas, including near schools. The towers were fixed within 200 m from the school buildings. In school 1, RF-EMFR was 2.010 µW/cm2 with a frequency of 925 MHz, and in school 2, RF-EMFR was 10.021 µW/cm2 with a frequency of 925 MHz. The students were exposed to RF-EMFR for six hours a day, five days a week, for a total period of two years. The authors identified substantial damage in motor screening tasks and spatial working memory tasks among school children exposed to higher RF-EMFR produced by MPBSTs. Researchers concluded that increased exposure to RF-EMFR was associated with decreased gross motor skills, spatial working memory, and attention in school adolescents compared to students who were exposed to low RF-EMFR.

Raika et al. (2017) [112] found a relationship between cell phone use and warming around the ear, headache, thinking and attention difficulties, and sleep disturbances. However, Röösli and Hug (2001) [113] did not show any associations between RF-EMFR exposure and non-specific health symptoms. Researchers also identified that long-term exposure to RF-EMFR could cause a dramatic decrease in neuronal growth and function [114]. The development of neurons, axonal growth, and formation of synapses depend on myelination by the oligodendrocytes. The lack of myelination impairs the action potential and causes neuronal degeneration [93].

Alkis et al. (2019) [115] demonstrated that 900–1,800 and 2,100-MHz RF-EMRF generated from cell phones could cause "oxidative damage, increase lipid peroxidation, and oxidative DNA damage formation in the frontal lobe of the brain tissues." In contradiction to these findings, Jeong et al. (2018) [116] reported that RF-EMRF exposure did not change the levels of oxidative stress, DNA damage, apoptosis, astrocytes, or microglia markers in the brains of mice. Similarly, Durdik et al. (2019) [92] conducted a study on exposure to radiation produced by cell phones on various "biochemical markers, reactive oxygen species, DNA single- and double-strand breaks." Their findings demonstrate that exposure to pulsed radiation increases reactive oxygen species (ROS) but did not result in sustained DNA damage and apoptosis.

Earlier experimental studies reported adverse effects of RF-EMFR generated from mobile phones and other devices on the development of the CNS during gestation [117–120], as well as memory [121], lack of attention, and concentration [122, 123]. However, other studies exist that do not show any significant effect of RF-EMFR on neuronal structure, brain development, function [124, 125], and memory [126].

Conclusion

The development of the brain is a complex process that includes embryological, genetic, biological, and environmental events. Brain development is a lifelong process, starting at the beginning of the third gestational week. Some processes are completed at birth, whereas others continue to develop throughout life. The events contributing to brain development are based on a series of molecular mechanisms of genetic expressions to environmental input, the disruption of which can alter the development and biological outcomes. Environmental pollution during the gestational period and/or later exhibits detrimental effects and impairs brain biology's developmental and cognitive processes. The literature on environmental pollution and its adverse impact on brain biology is well documented; however, inconsistent evidence exists about the effects of RF-EMFR on brain functions. At this point, further research underlying the impacts of RF-EMFR generated from various electronic devices on neurodevelopment is needed to reach a more definitive conclusion. The local, regional, and international public health authorities should update the community and provide adequate precautionary guidance for the public to minimize environmental pollution and educate them on the potential health risks. For safety, the use of mobile phones by school-age children can and may be minimized. The MPBSTs should be placed at a distance from thickly residential areas, homes, daycare centers, schools, hospitals, and places frequented by pregnant women to minimize exposure to RF-EMFR and associated possible hazardous effects.

References

[1] Keunen K., van der Burgh H.K., de Reus M.A., Moeskops P., Schmidt R., Stolwijk L.J., et al. Early human brain development: Insights into macroscale connectome wiring. *Pediatr Res.* 2018; 84(6): 829–36. https://doi.org/10.1038/s41390-018-0138-1.

[2] Keunen K., Counsell S.J., Benders M.J.N.L The emergence of functional architecture during early brain development. *NeuroImage.* 2017; 160: 2–14.

[3] Nikolopoulou E., Galea G.L., Rolo A., Greene N.D., Copp A.J. Neural tube closure: Cellular, molecular and biomechanical mechanisms. *Development.* 2017; 144(4): 552–66. https://doi.org/10.1242/dev.145904, PMID: 28196803.

[4] Lei J., Calvo P., Vigh R., Burd I. Journey to the center of the fetal brain: Environmental exposures and autophagy. *Front Cell Neurosci Neurosci.* 2018.2018; 12: 118. https://doi.org/10.3389/fncel.2018.00118.

[5] de Prado Bert P., Mercader E.M.H., Pujol J., Sunyer J., Mortamais M. The effects of air pollution on the brain: A review of studies interfacing environmental epidemiology and neuroimaging. *Curr Environ Health Rep.* 2018; 5(3): 351–64.

[6] Douet V., Chang L., Cloak C., Ernst T. Genetic influences on brain developmental trajectories on neuroimaging studies: From infancy to young adulthood. *Brain Imaging Behav.* 2014; 8(2): 234–50. https://doi.org/10.1007/s11682-013-9260-1.

[7] Stiles J., Jernigan T.L. The basics of brain development. *Neuropsychol Rev.* 2010; 20(4): 327–48. https://doi.org/10.1007/s11065-010-9148-4.

[8] Budday S., Steinmann P., Kuhl E. Physical biology of human brain development. *Front Cell Neurosci.* 2015; 9: 257. https://doi.org/10.3389/fncel.2015.00257.

[9] Keuroghlian A.S., Knudsen E.I. Adaptive auditory plasticity in developing and adult animals. *Prog Neurobiol.* 2007; 82(3): 109–21.

[10] Georgadaki K., Khoury N., Spandidos D.A., Zoumpourlis V. The molecular basis of fertilization (Review). *Int J Mol Med.* 2016; 38(4): 979–86. https://doi.org/10.3892/ijmm.2016.2723.

[11] Darnell D., Gilbert S.F. Neuroembryology. *Wiley Interdiscip Rev Dev Biol.* 2017; 6(1). https://doi.org/10.1002/wdev.215.

[12] Wilde J.J., Petersen J.R., Niswander L Genetic, epigenetic, and environmental contributions to neural tube closure. *Annu Rev Genet.* 2014; 48: 583–611.

[13] Vasung L., Abaci Turk E., Ferradal S.L., Sutin J., Stout J.N., Ahtam B., et al. Exploring early human brain development with structural and physiological neuroimaging. *Neuroimage.* 2019; 187: 226–54. https://doi.org/10.1016/j.neuroimage.2018.07.041.

[14] Silbereis J.C., Pochareddy S., Zhu Y., Li M., Sestan N. The cellular and molecular landscapes of the developing human central nervous system. *Neuron.* 2016; 89(2): 248–68.

[15] Bystron I., Blakemore C., Rakic P. Development of the human cerebral cortex: Boulder committee revisited. *Nat Rev Neurosci.* 2008; 9(2): 110–22.

[16] Tierney A.L., Nelson C.A. Brain development and the role of experience in the early years. *Zero Three.* 2009; 30(2): 9–13.

[17] Raybaud C., Ahmad T., Rastegar N., Shroff M., Al Nassar M. The premature brain: Developmental and lesional anatomy. *Neuroradiology.* 2013; 55(Suppl 2): 23–40. https://doi.org/10.1007/s00234-013-1231-0.

[18] Ishikawa Y., Yamamoto N., Yoshimoto M., Ito H. The primary brain vesicles revisited: Are the three primary vesicles (forebrain/midbrain/hindbrain) universal invertebrates? *Brain Behav Evol.* 2012; 79(2): 75–83. https://doi.org/10.1159/000334842.

[19] Blows W.T. Child brain development. *Nurs Times.* 2003; 99: 28–31.

[20] Sun T., Hevner R.F. Growth and folding of the mammalian cerebral cortex: From molecules to malformations. *Nat Rev Neurosci.* 2014; 15(4): 217–32. https://doi.org/10.1038/nrn3707.

[21] Budday S., Steinmann P., Kuhl E. The role of mechanics during brain development. *J Mech Phys Solids.* 2014; 72: 75–92. https://doi.org/10.1016/j.jmps.2014.07.010.

[22] Rahimi-Balaei M., Bergen H., Kong J., Marzban H. Neuronal migration During the development of the cerebellum. *Front Cell Neurosci.* 2018; 12: 484. https://doi.org/10.3389/fncel.2018.00484.

[23] Marín O., Valiente M., Ge X., Tsai L.H. Guiding neuronal cell migrations. *Cold Spring Harb Perspect Biol.* 2010; 2(2): a001834. https://doi.org/10.1101/cshperspect.a001834.

[24] Cooper J.A. Cell biology in neuroscience: Mechanisms of cell migration in the nervous system. *J Cell Biol.* 2013; 202(5): 725–34. https://doi.org/10.1083/jcb.201305021.

[25] Tsai L.H., Gleeson J.G. Nucleokinesis in neuronal migration. *Neuron.* 2005; 46(3): 383–8. https://doi.org/10.1016/j.neuron.2005.04.013.

[26] Tsai H.H., Niu J., Munji R., Davalos D., Chang J., Zhang H., et al. Oligodendrocyte precursors migrate along the vasculature in the developing nervous system. *Science.* 2016; 351(6271): 379–84. https://doi.org/10.1126/science.aad3839.

[27] Marín O., Rubenstein J.L. A long, remarkable journey: Tangential migration in the telencephalon. *Nat Rev Neurosci.* 2001; 2(11): 780–90. https://doi.org/10.1038/35097509.

[28] Nadarajah B., Brunstrom J.E., Grutzendler J., Wong R.O., Pearlman A.L. Two modes of radial migration in the early development of the cerebral cortex. *Nat Neurosci.* 2001; 4(2): 143–50. https://doi.org/10.1038/83967.

[29] Hatanaka Y., Zhu Y., Torigoe M., Kita Y., Murakami F. From migration to settlement: The pathways, migration modes and dynamics of neurons in the developing brain. *Proc Jpn Acad Ser B Phys Biol Sci.* 2016; 92(1): 1–19. https://doi.org/10.2183/pjab.92.1.

[30] Kawaji K., Umeshima H., Eiraku M., Hirano T., Kengaku M. Dual phases of migration of cerebellar granule cells guided by axonal and dendritic leading processes. *Mol Cell Neurosci.* 2004; 25(2): 228–40. https://doi.org/10.1016/j.mcn.2003.10.006.

[31] Deutsch S.I., Burket J.A., Katz E. Does subtle disturbance of neuronal migration contribute to schizophrenia and other neurodevelopmental disorders? Potential genetic mechanisms with possible treatment implications. *Eur Neuropsychopharmacol.* 2010; 20(5): 281–7. https://doi.org/10.1016/j.euroneuro.2010.02.005.

[32] Guerrini R., Parrini E. Neuronal migration disorders. *Neurobiol Dis.* 2010; 38(2): 154–66. https://doi.org/10.1016/j.nbd.2009.02.008.

[33] Marzban H., Del Bigio M.R., Alizadeh J., Ghavami S., Zachariah R.M., Rastegar M. Cellular commitment in the developing cerebellum. *Front Cell Neurosci.* 2014; 8: 450. https://doi.org/10.3389/fncel.2014.00450.

[34] Qin R., Cao S., Lyu T., Qi C., Zhang W., Wang Y. CDYL deficiency disrupts neuronal migration and increases susceptibility to epilepsy. *Cell Rep.* 2017; 18(2): 380–90. https://doi.org/10.1016/j.celrep.2016.12.043.

[35] Léopold P. Neuronal differentiation: TOR and insulin receptor pathways set the tempo. *Cell.* 2004; 119(1): 4–5.

[36] Chikawa M. Synaptogenesis. In: Binder M.D., Hirokawa N., Windhorst U., editors. *Encyclopedia of Neuroscience.* Berlin, Heidelberg: Springer. 2009. https://doi.org/10.1007/978-3-540-29678-2_5829.

[37] Cohen-Cory S. The developing synapse: Construction and modulation of synaptic structures and circuits. *Science.* 2002; 298(5594): 770–6. https://doi.org/10.1126/science.1075510, PMID: 12399577.

[38] Snaidero N., Simons M. Myelination at a glance. *J Cell Sci.* 2014; 127(14): 2999–3004. https://doi.org/10.1242/jcs.151043.

[39] Fields R.D. White matter in learning, cognition, and psychiatric disorders. *Trends Neurosci.* 2008; 31(7): 361–70. https://doi.org/10.1016/j.tins.2008.04.001.

[40] Nelson C.A., de Haan M., Thomas K.M. Neural bases of cognitive development. In: Damon W., Lerner R., Kuhn D., Siegler R., editors. *Handbook of Child Psychology*, 2nd edition. NJ: John Wiley & Sons. 2006, pp. 2–58.

[41] Herculano-Houzel S. The human brain in numbers: A linearly scaled-up primate brain. *Front Hum Neurosci.* 2009; 3(31): 31. https://doi.org/10.3389/neuro.09.031.2009.

[42] Pakkenberg B., Pelvig D., Marner L., Bundgaard M.J., Gundersen H.J.G., Nyengaard J.R., et al. Aging and the human neocortex. *Exp Gerontol.* 2003; 38(1–2): 95–99.

[43] Zilles K., Palomero-Gallagher N., Amunts K. Development of cortical folding during evolution and ontogeny. *Trends Neurosci.* 2013; 36(5): 275–84. https://doi.org/10.1016/j.tins.2013.01.006.

[44] Bayly P.V., Okamoto R.J., Xu G., Shi Y., Taber L.A. A cortical folding model incorporating stress-dependent growth explains Gyral wavelengths and stress patterns in the developing brain. *Phy Zs Biol.* 2013; 10(1): 016005. https://doi.org/10.1088/1478-3975/10/1/016005.

[45] Franze K. The mechanical control of nervous system development. *Development.* 2013; 140(15): 3069–77. https://doi.org/10.1242/dev.079145.

[46] Shonkoff J.P., Phillips D.A. National Research Council (US) and Institute of Medicine (US) Committee on integrating the science of early childhood development. In: Shonkoff J.P., Phillips D.A., editors. *Developing Brain, from Neurons to Neighborhoods: The Science of Early Childhood Development.* Washington, DC: National Academies Press. 2000. Available at: www.ncbi.nlm.nih.gov/books/NBK225562/. Cited date Oct 30, 2020.

[47] Centers for Disease, Control and Prevention (CDC). Early brain development and health. Available at: www.cdc.gov/ncbddd/childdevelopment/early-brain-development.html. Cited date Oct 30, 2020.

[48] Jiang F., Jiang F. Sleep and early brain development. *Ann Nutr Metab.* 2019; 75(Suppl 1): 44–54. https://doi.org/10.1159/000508055.

[49] Sailer S., Sebastiani G., Andreu-Férnández V., García-Algar O. Impact of nicotine replacement and electronic nicotine delivery systems on fetal brain development. *Int J Environ Res Public Health.* 2019; 16(24): 5113. https://doi.org/10.3390/ijerph16245113.

[50] Cisneros-Franco J.M., Voss P., Thomas M.E., de Villers-Sidani E. Critical periods of brain development. *Handb Clin Neurol.* 2020; 173: 75–88. Https://doi.org/10.1016/B978-0-444-64150-2.00009-5.

[51] Peeples L. News feature: How air pollution threatens brain health. *Proc Natl Acad Sci U S A*. 2020; 117(25): 13856–60. https://doi.org/10.1073/pnas.2008940117.

[52] Suades-González E., Gascon M., Guxens M., Sunyer J. Air pollution and neuropsychological development: A review of the latest evidence. *Endocrinology*. 2015; 156(10): 3473–82.

[53] Weintraub K. The prevalence puzzle: Autism counts. *Nature*. 2011; 479(7371): 22–4.

[54] Maekawa F., Nakamura K., Nakayama S.F. Editorial: Chemicals in the environment and brain development: The importance of neuroendocrinological approaches. *Front Neurosci*. 2017; 11: 133. https://doi.org/10.3389/fnins.2017.00133.

[55] D'Angiulli A. Severe urban outdoor air pollution and children's structural and functional brain development, from evidence to precautionary strategic action. *Front Public Health*. 2018; 6(6: 95): 95. https://doi.org/10.3389/fpubh.2018.00095.

[56] United Nations Children's Funds (UNICEF). *Clean Air for Children. The Impact of Air Pollution on Children*. New York. Available at: www.unicef.org/publications/files/UNICEF_Clear_the_Air_for_Children_30_Oct_2016.pdf. Cited date Jun 2, 2020.

[57] Van Donkelaar A., Martin R.V., Brauer M., Hsu N.C., Kahn R.A., Levy R.C., et al. Global estimates of fine particulate matter using a combined geophysical-statistical method with information from satellites, models, and monitors. *Environ Sci Technol*. 2016; 50(7): 3762–72.

[58] Schraufnagel D.E. The health effects of ultrafine particles. *Exp Mol Med*. 2020; 52(3): 311–7. https://doi.org/10.1038/s12276-020-0403-3.

[59] Di Domenico M., de Menezes Benevenuto S.G., Tomasini P.P., Yariwake V.Y., de Oliveira Alves N., Rahmeier F.L., et al. Concentrated ambient fine particulate matter (PM2.5) exposure induce brain damage in pre and postnatal exposed mice. *Neurotoxicology*. 2020; 79: 127–41. https://doi.org/10.1016/j.neuro.2020.05.004.

[60] Kingsley S.L., Eliot M.N., Carlson L., Finn J., Macintosh D.L., Suh H.H., et al Proximity of US schools to major roadways: A nationwide assessment. *J Expo Sci Environ Epidemiol*. 2014; 24(3): 253–9.

[61] Oberdörster G., Sharp Z., Atudorei V., Elder A., Gelein R., Kreyling W., et al. Translocation of inhaled ultrafine particles to the brain. *Inhal Toxicol*. 2004; 16(6–7): 437–45. https://doi.org/10.1080/08958370490439597.

[62] Tian L., Shang Y., Chen R., Bai R., Chen C., Inthavong K., et al. Correlation of regional deposition dosage for inhaled nanoparticles in human and rat olfactory. *Part Fibre Toxicol*. 2019; 16(1): 6. https://doi.org/10.1186/s12989-019-0290-8.

[63] Heusser K., Tank J., Holz O., May M., Brinkmann J., Engeli S., et al. Ultrafine particles and ozone perturb norepinephrine clearance rather than centrally generated sympathetic activity in humans. *Sci Rep*. 2019; 9(1): 3641. https://doi.org/10.1038/s41598-019-40343-w.

[64] Sunyer J., Dadvand P. Prenatal brain development as a target for urban air pollution. *Basic Clin Pharmacol Toxicol*. 2019; 125(Suppl 3): 81–8. https://doi.org/10.1111/bcpt.13226.

[65] Levesque S., Taetzsch T., Lull M.E., Kodavanti U., Stadler K., Wagner A., et al. Diesel exhaust activates and primes microglia: Air pollution, neuroinflammation, and regulation of dopaminergic neurotoxicity. *Environ Health Perspect*. 2011; 119(8): 1149–55.

[66] Oppenheim H.A., Lucero J., Guyot A.C., Herbert L.M., McDonald J.D., Mabondzo A., et al. Exposure to vehicle emissions results in altered blood–brain barrier permeability and expression of matrix metalloproteinases and tight junction proteins in mice. *Part Fibre Toxicol*. 2013; 10: 62.

[67] Ho Y.S., Yang X., Yeung S.C., Chiu K., Lau C.F., Tsang A.W., et al. Cigarette smoking accelerated brain aging and induced pre-Alzheimer-like neuropathology in rats. *PLoS One*. 2012; 7(5): e36752.

[68] Umezawa M., Tainaka H., Kawashima N., Shimizu M., Takeda K. Effect of fetal exposure to titanium dioxide nanoparticle on brain development-brain region information. *J Toxicol Sci*. 2012; 37(6): 1247–52.

[69] Schröder N., Figueiredo L.S., de Lima M.N. Role of brain iron accumulation in cognitive dysfunction: Evidence from animal models and human studies. *J Alzheimers Dis*. 2013; 34(4): 797–812.

[70] Berg E.L., Pedersen L.R., Pride M.C., Petkova S.P., Patten K.T., Valenzuela A.E., et al. Developmental exposure to near roadway pollution produces behavioral phenotypes relevant to neurodevelopmental disorders in juvenile rats. *Transl Psychiatry*. 2020; 10(1): 289. https://doi.org/10.1038/s41398-020-00978-0.

[71] Patten K.T., González E.A., Valenzuela A., Berg E., Wallis C., Garbow J.R., et al. Effects of early life exposure to traffic-related air pollution on brain development in juvenile Sprague-Dawley rats. *Transl Psychiatry*. 2020 May 27; 10(1): 166. https://doi.org/10.1038/s41398-020-0845-3.

[72] Nghiem G.T., Nishijo M., Pham T.N., Ito M., Pham T.T., Tran A.H., et al. Adverse effects of maternal dioxin exposure on fetal brain development before birth assessed by neonatal electroencephalography (EEG) leading to poor neurodevelopment; a 2-year follow-up study. *Sci Total Environ.* 2019; 667: 718–29. https://doi.org/10.1016/j.scitotenv.2019.02.395.

[73] Cho J., Sohn J., Noh J., Jang H., Kim W., Cho S.K., et al. Association between exposure to polycyclic aromatic hydrocarbons and brain cortical thinning: The environmental pollution-induced neurological effects (EPINEF) study. *Sci Total Environ.* 2020; 737: 140097. https://doi.org/10.1016/j.scitotenv.2020.140097.

[74] Power M.C, Lamichhane A.P., Liao D., Xu X., Jack C.R., Gottesman R.F., et al. The association of long-term exposure to particulate matter air pollution with brain MRI findings: The ARIC study. *Environ Health Perspect.* 2018; 126(2): 027009. HTTPS://doi.org/10.1289/EHP2152.

[75] Cserbik D., Chen J.C., McConnell R., Berhane K., Sowell E.R., Schwartz J., et al. Fine particulate matter exposure during childhood relates to hemispheric-specific differences in brain structure. *Environ Int.* 2020; 143: 105933. https://doi.org/10.1016/j.envint.2020.105933.

[76] Vanos J.K. Children's health and vulnerability in outdoor microclimates: A comprehensive review. *Environ Int.* 2015; 76: 1–15.

[77] Calderón-Garcidueñas L., Franco-Lira M., Mora-Tiscareño A., Medina-Cortina H., Torres-Jardón R., Kavanaugh M. Early Alzheimer's and Parkinson's disease pathology in urban children: Friend versus foe responses-it is time to face the evidence. *BioMed Res Int.* 2013; 2013: 161687.

[78] Calderón-Garcidueñas L., Mora-Tiscareño A., Styner M., Gómez-Garza G., Zhu H., Torres-Jardón R., et al. White matter hyperintensities, systemic inflammation, brain growth, and cognitive functions in children exposed to air pollution. *J Alzheimers Dis.* 2012; 31(1): 183–91.

[79] Casado Naranjo I., Portilla Cuenca J.C., Duque De San Juan B., García A.F., Sevilla R.R., Serrano Cabrera A., et al Association of vascular factors and amnestic mild cognitive impairment: A comprehensive approach. *J Alzheimers Dis.* 2015; 44(2): 695–704.

[80] Braak H., DelTredici K. The pathological process underlying Alzheimer's disease in individuals under thirty. *Acta Neuropathol.* 2011; 121(2): 171–81.

[81] Brockmeyer S., D'Angiulli A. How air pollution alters brain development: The role of neuroinflammation. *Transl Neurosci.* 2016; 7(1): 24–30.

[82] World Health Organization. Air pollution and child health: Prescribing clean air. Available at: www.who.int/ceh/publications/air-pollution-child-health/en/. Cited date Jan 2, 2019.

[83] Schoeni A., Roser K., Röösli M. Symptoms and the use of wireless communication devices: A prospective cohort study in Swiss adolescents. *Environ Res.* 2017; 154: 275–83.

[84] Kalafatakis F., Bekiaridis-Moschou D., Gkioka E., Tsolaki M. Mobile phone use for 5 minutes can cause significant memory impairment in humans. *Hell J Nucl Med.* 2017; 20(Suppl): 146–54.

[85] Foerster M., Thielens A., Joseph W., Eeftens M., Röösli M. A prospective cohort study of adolescents' memory performance and individual brain dose of microwave radiation from wireless communication. *Environ Health Perspect.* 2018; 126(7): 077007.

[86] Sage C., Burgio E. Electromagnetic fields, pulsed radiofrequency radiation, and epigenetics: How wireless technologies May affect childhood development. *Child Dev.* 2018; 89(1): 129–36. https://doi.org/10.1111/cdev.12824.

[87] Meo S.A., Almahmoud M., Alsultan Q., Alotaibi N., Alnajashi I., Hajjar W.M. Mobile phone base station tower settings adjacent to school buildings: Impact on students' cognitive health. *Am J Mens Health.* 2019; 13(1): 1557988318816914. https://doi.org/10.1177/1557988318816914.

[88] Hong S., Huang H., Yang M., Wu H., Wang L. Enriched environment decreases cognitive impairment in elderly rats with prenatal mobile phone exposure. *Front Aging Neurosci.* 2020; 12: 162. https://doi.org/10.3389/fnagi.2020.00162.

[89] Haarala C., Bergman M., Laine M., Revonsuo A., Koivisto M., Hämäläinen H. Electromagnetic field emitted by 902 MHz mobile phones shows no effects on children's cognitive function. *Bioelectromagnetics.* 2005 (Suppl 7): S144–50. https://doi.org/10.1002/bem.20142.

[90] Besset A., Espa F., Dauvilliers Y., Billiard M., de Seze R. No effect on cognitive function from daily mobile phone use. *Bioelectromagnetics.* 2005; 26(2): 102–8.

[91] Curcio G. Exposure to mobile phone-emitted electromagnetic fields and human attention: No evidence of a causal relationship. *Front. Public.* Health. 2018; 6: 42. doi: 10.3389/fpubh.2018.00042.

[92] Durdik M., Kosik P., Markova E., Somsedikova A., Gajdosechova B., Nikitina E., Horvathova E., Kozics K., Davis D., Belyaev I. Microwaves from mobile phone induce reactive oxygen species but not DNA damage, preleukemic fusion genes and apoptosis in hematopoietic stem/progenitor cells. *Sci. Rep.* 2019; 9(1): 16182.

[93] Skaist A. The effects of RF-EMF on the child brain. *The Sci. J. Lander College of Arts and Sci.* 2019; 12(2): 40–5.

[94] Boda E., Rigamonti A.E., Bollati V. Understanding the effects of air pollution on neurogenesis and gliogenesis in the growing and adult brain. *Curr. Opin. Pharmacol.* 2020; 50: 61–6. doi: 10.1016/j.coph.2019.12.003.

[95] Kirk S., David P., Gregory R., Tracy S.N., Charles H., Chelsea M., Don T., Fred P., Sergei T., Linda J.L. Automated measurement of pediatric cranial bone thickness and density from clinical computed tomography. *Conf. Proc. IEEE. Eng. Med. Biol. Soc.* 2012; 2012: 4462–5. doi: 10.1109/EMBC. 2012.6346957.

[96] Tirpude A., Fulpatil M.P., Kole S. Norms for size and closure time of anterior fontanel: Study on babies in Nagpur region. *National. J. Clin.* Anat. 2016; 5(2): 78–85.

[97] Jin S.W., Sim K.B., Kim S.D. Development and growth of the normal cranial vault: An embryologic review. *J. Korean. Neurosurg. Soc.* 2016; 59(3): 192–6. doi: 10.3340/jkns.2016.59.3.192.

[98] Delye Hans, Clijmans Tim, Mommaerts Maurice Yves, Sloten Jos Vnder, Goffin Jan. Creating a normative database of age-specific 3D geometrical data, bone density, and bone thickness of the developing skull: A pilot study. *J. Neurosurg. Pediatr.* 2015; 16(6): 687–702.

[99] Rathee S., Duhan M. Comparative study of mobile phone emitted EM waves on the human brain at different charging levels. *Int. J. Biomed. Eng. Technol.* 2016; 22(2): 178–88.

[100] Michelozzi Paola, Capon Alessandra, Kirchmayer Ursula, Forastiere Francesco, Biggeri Annibale, Barca Alessandra, Perucci Carlo A. Adult and childhood leukemia near a high-power radio station in Rome, Italy. *Am. J. Epidemiol.* 2002; 155(12): 1096–103.

[101] Houston B.J., Nixon B., King B.V., De Iuliis G.N., Aitken R.J. The effects of radiofrequency electromagnetic radiation on sperm function. *Reproduction.* 2016; 152(6): R263–76. doi: 10.1530/ REP-16-0126.

[102] Bolte J.F., Eikelboom T. Personal radiofrequency electromagnetic field measurements in The Netherlands: Exposure level and variability for everyday activities, times of day and types of area. *Environment. International.* 2012; 48: 133–42 doi: 10.1016/j.envint.2012.07.006.

[103] Panagopoulos D.J., Johansson O., Carlo G.L. Polarization: A key difference between man-made and natural electromagnetic fields, in regard to biological activity. *Sci. Rep.* 2015; 514914. doi: 10.1038/srep14914.

[104] Liu C., Duan W., Xu S., Chen C., He M., Zhang L., Yu Z., Zhou Z. Exposure to 1800 MHz radiofrequency electromagnetic radiation induces oxidative DNA base damage in a mouse spermatocyte-derived cell line. *Toxicol. Lett.* 2013; 218: 2–9. doi: 10.1016/j.toxlet.2013.01.003.

[105] Gabriel S., Lau R.W., Gabriel C. The dielectric properties of biological tissues: III. Parametric models for the dielectric spectrum of tissues. *Phys. Med. Biol.* 1996; 41: 2271–93. doi: 10.1088/ 0031- 9155/41/11/003.

[106] Chiu Chang-Ta, Chang Ya-Hui, Chen Chu-Chieh, Ming-Chung K.O, Chung-Yi L.I. Mobile phone use and health symptoms in children. *J. Formos. Med. Assoc.* 2015; 114(7): 598–604. doi: 10.1016/j. jfma.2014.07.002.

[107] Van Rongen E., Croft R., Juutilainen J., Lagroye I., Miyakoshi J., Saunders R. Effects of radiofrequency electromagnetic fields on the human nervous system. *J. Toxicol. Environ. Health B. Crit. Rev.* 2009; 12: 572–97.

[108] Oftedal G., Wilén J., Sandström M., Hansson K. Mild symptoms experienced in connection with mobile phone use. *Occup. Med (Lond).* 2000; 50: 237–45.

[109] Hillert L., Berglind N., Arnetz B.B., Bellander T. Prevalence of self-reported hypersensitivity to electric or magnetic fields in a population-based questionnaire survey. *Scand. J. Work. Environ. Health.* 2002; 28: 33–41.

[110] Schröttner J., Leitgeb N. Sensitivity to electricity—Temporal changes in Austria. *BMC. Public. Health.* 2008; 8: 310.

[111] Zarei S., Vahab M., Oryadi-Zanjani M.M., Alighanbari N., Mj Mortazavi S. Mother's Exposure to electromagnetic fields before and during pregnancy is associated with risk of speech problems in offspring. *J. Biomed. Phys. Eng.* 2019; 9(1): 61–8.

[112] Durusoy Raika, Hassoy Hür, Özkurt Ahmet, Ali Osman Karababa. Mobile phone use, school electromagnetic field levels, and related symptoms: A cross-sectional survey among 2150 high school students in Izmir. *Environ. Health.* 2017; 16(1): 51. doi: 10.1186/s12940-017-0257-x.

[113] Martin Röösli L., Kerstin Hug. Wireless communication fields and non-specific symptoms of ill health: A literature review. *Wien. Med. Wochenschr.* 2011; 161: 9–10: 240–50. doi: 10.1007/s10354-011-0883-9.

[114] Eghlidospour M., Ghanbari A., Mohammad Javad Mortazavi S., Azari H. Effects of radiofrequency exposure emitted from a GSM mobile phone on proliferation, differentiation, and apoptosis of neural stem cells. *Anat. Cell. Biol.* 2017; 50(2): 115–23.

[115] Alkis M.E., Bilgin H.M., Akpolat V., Dasdag S., Yegin K., Yavas M.C., Akdag M.Z. Effect of 900-1800-, and 2100-MHz radiofrequency radiation on DNA and oxidative stress in the brain. *Electromagn. Biol. Med.* 2019; 38(1): 32–47. doi: 10.1080/15368378.2019.1567526.

[116] Jeong Y.J., Son Y., Han N.K., Choi H.D., Pack J.K., Kim N., Lee Y.S., Lee H.J. Impact of long-term RF-EMF on oxidative stress and neuroinflammation in aging brains of C57BL/6 mice. *Int. J. Mol. Sci.* 2018 Jul 19; 19(7): 2103. doi: 10.3390/ijms19072103.

[117] Salford L.G., Brun A.E., Eberhardt J.L., Malmgren L., Persson B.R. Nerve cell damage in mammalian brain after exposure to microwaves from GSM mobile phones. *Environ. Health. Perspect.* 2003; 111(7): 881–8.

[118] Odaci E., Kaplan S., Sahin B., Bas O., Gevrek F., Aygun D., et al. Effects of low-dose oxcarbazepine administration on developing cerebellum in the newborn rat: A stereological study. *Neurosci. Res. Commun.* 2004; 34: 28–36.

[119] Odaci E., Bas O., Kaplan S. Effects of prenatal exposure to a 900 MHz electromagnetic field on the dentate gyrus of rats: A stereological and histopathological study. *Brain. Res.* 2008; 1238: 224–32.

[120] Aldad T.S., Gan G., Gao X.B., Taylor HS Fetal radiofrequency radiation exposure from 800–1900 mhz-rated cellular telephones affects neurodevelopment and behavior in mice. *Sci. Rep.* 2012; 2: 312.

[121] Ntzouni M.P., Stamatakis A., Stylianopoulou F., Margaritis L.H. Short-term memory in mice is affected by mobile phone radiation. *Pathophysiology.* 2011; 18(3): 193–9.

[122] Deniz O., Kaplan G., Selçuk S., Terzi M.B., Altun M., Yurt G., Aslan K.K., Davis K.D. Effects of short and long-term electromagnetic field exposure on the human hippocampus. *J. Microsc. Ultrastructure.* 2017; 5(4): 191–7.

[123] Ntzouni M.P., Skouroliakou A., Kostomitsopoulos N., Margaritis L.H. Transient and cumulative memory impairments induced by GSM 1.8 GHz cell phone signal in a mouse model. *Electromagn. Biol. Med.* 2013; 32(1): 95–120.

[124] Tsurita G., Nagawa H., Ueno S., Watanabe S., Taki M. Biological and morphological effects on the brain after exposure of rats to a 1439 MHz TDMA field. *Bioelectromagnetics.* 2000; 21(5): 364–71.

[125] Cosquer B., Kuster N., Cassel J.C. Whole-body exposure to 2.45 GHz electromagnetic fields does not alter 12-arm radial-maze with reduced access to spatial cues in rats. *Behav. Brain. Res.* 2005; 161(2): 331–5.

[126] Sienkiewicz Z.J., Blackwell R.P., Haylock R.G., Saunders R.D., Cobb B.L. Low-level exposure to pulsed 900 MHz microwave radiation does not cause deficits in the performance of a spatial learning task in mice. *Bioelectromagnetics.* 2000; 21(3): 151–9.

Biology of Memory and Electromagnetic Field Radiation

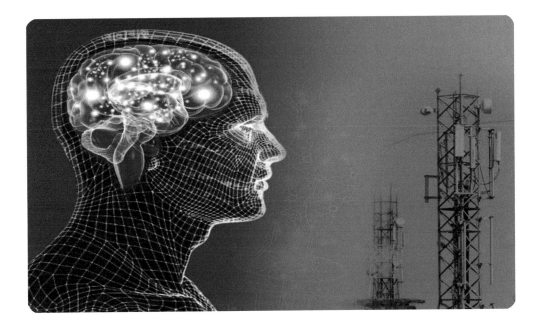

5.1 Biology of Memory: Introduction

Learning and memory are some of the most desirable, captivating capabilities of the human mind. It is these very abilities that distinguish human beings from all other animals. Learning is a biological process of gaining new knowledge, and memory allows one to recall and restructure it. Most of the knowledge and skills are not intrinsic but are learned [1]. The brain has a vital role in receiving, analyzing, and filtering the information learned, remembered, and forgotten in a specific sequence. A molecular, cellular, and circuit mechanism triggers how memories are created, stored, retrieved, and lost [1]. Healthy people can recall up to 10,000 visual images [2].

The word "memory" originated from the English word "memorie," Anglo-French "memoire or memorie," and the Latin word "memoria" and "memor," meaning "mindful" or "remembering." The role of the brain is to process, store, and remember information. It can recall previous events and recall formerly absorbed facts, incidents, impressions, skills, and practices [3]. While providing an ability to learning, it also helps gain knowledge processes and adapt to past experiences that then influence present behavior [3]. Memory is an integral part of daily routine activities and allows one to effectively interact with the environment and maintain communication with other individuals. Memory involves encoding, storage, consolidation, and retrieval of information [4, 5].

5.2 Historical Background

Since the last two millennia, researchers have been trying to understand memory through various models. First, it was assumed that human memory was of two types: "natural memory" and "artificial memory." However, this theory was gradually reformed. David Hartley, a famous English philosopher, developed another theory that the CNS is responsible for encoding memories through the motions [6, 7]. After that, William James from America and Wilhelm Wundt from Germany researched human memory functions. In the 1870s and 1880s, the notion of neural plasticity was hypothesized. However, in the mid-1880s, Herman Ebbinghaus, a young German philosopher, developed a method of studying memory. He conducted series of experiments using various syllables and established a link with meaningful words, and his findings of concepts of learning and forgetting were highlighted in the literature. He classified memory into three types and this classification was more acceptable to the scientific community, which was as follows, sensory memory, STM, and long-term memory [6, 7] (Figure 5.1).

Furthermore, in 1904, Richard Semon, a German biologist, suggested that an experience leaves behind physical traces on particular neuron webs in the brain. Sir Frederick Bartlett, a British psychologist, considered the founding father of cognitive psychology, established how the brain stores memories. In the 1940s, the field of neuropsychology emerged; Karl Lashley found that memory traces in the brain were widely localized and distributed throughout the cortex [6, 7].

In 1949, a Canadian neuropsychologist, Donald Olding Hebb, highlighted how the neurons contribute to psychological processes. He established the concept that the neurons "which fire together, wire together," suggesting that memory encoding occurs as connections between the neurons. This concept was referred to as "Hebb's Rule" and is sustained by discovering the mechanisms of memory consolidation, long-term potentiation (LTP), and neural plasticity [6, 7]. Besides, Eric Kandel identified the molecular changes during learning and the role of neurotransmitters [8, 9].

In the 1950s and 1960s, memory literature became acknowledged as the "cognitive revolution," leading to many new biological theories on how memory works. In 1956, George Miller published an article on STM. The basic concepts in cognitive science are working memory,

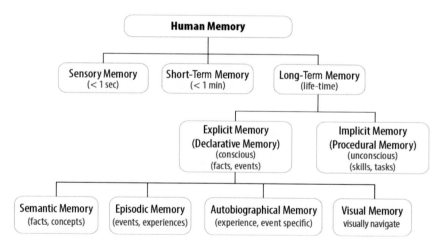

Figure 5.1 Classification of memory.

often called STM or immediate memory, which involves the availability of recent data, events, and thoughts [10, 11].

In 1968, Richard Atkinson and Richard Shiffrin described the "multi-store" model of memory, which involves a "sensory memory, a short-term memory, and a long-term memory" (Figure 5.3). It emerged as a widely accepted model of memory. They reported that information transfer from store to store linearly, information transfer like a computer with an input, process, and output. The information is detected by the sense organs and enters into the sensory memory and STM [12, 13]. (Figure 5.2; 5.3).

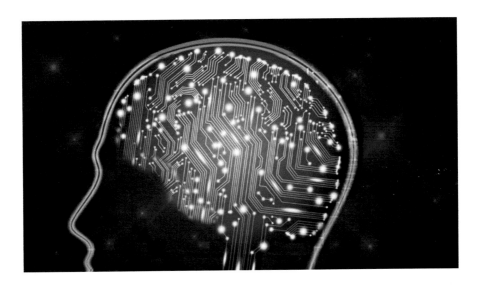

Figure 5.2 Cognitive functions imagination.

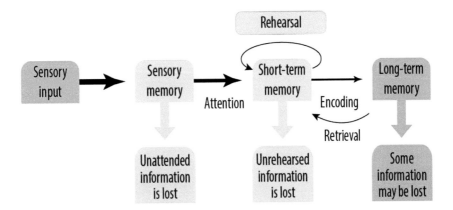

Figure 5.3 Atkinson and Shiffrin "multi-store" model of memory.

Craik and Lockhart presented another model, known as the "levels-of-processing model" in 1972 (Figure 5.4). This processing model concentrates on how memory involves complex processing and predicts how in-depth information is processed, how evidence is programmed and encoded, and how nice it is remembered. The deeper the level of processing, the easier it is to recall the information [14]. Craik and Tulving (1975) studied the levels of processing and retention framework, which claims that the best-recalled information is that which has been processed by meaning. They also hypothesized that deeper processing would take longer than shallow processing [15] (Table 5.1). The sensory memory receives the environmental input through various senses such as visual, audio, gustatory, and touch, whose contents are transferred to STM and then to long-term memory (Figure 5.4; Table 5.2).

Rose et al. (2015) [16] described how the brain maintains to-be-remembered information in working memory when the brain begins processing other information. Working memory cognitive models that recall include retrieval from long-term memory when things are displaced from focal attention. The behavioral processes and neural substrates support long-term memory and recall items after being displaced from the focal attention [16].

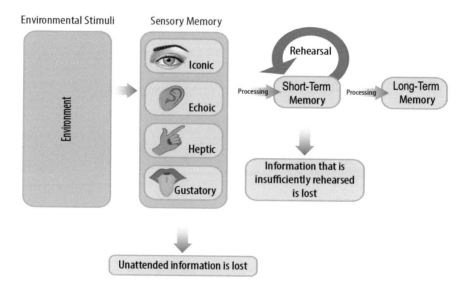

Figure 5.4 Craik and Lockhart "levels-of-processing model" of memory.

Table 5.1 The Levels of Processing (Processing Model)
1. **Shallow processing:** It includes maintenance, rehearsal, and repetition to retain STM contents.

   i. **Structural processing (looks like):** It encodes the physical characteristics of contents. For example, the teacher asks the students about what the alphabets, letters "A," "B," "D," look like.

   ii. **Phonemic processing (sound like):** It encodes sound characteristics.

2. **Deep processing:** It involves rehearsal, a meaningful analysis such as thinking and associations of data, leading to better recall. For example, a teacher asks the student to tell the meaning of words and link the word with previous knowledge.

   i. **Semantic processing:** It encodes what a word means and links it to like words and similar meanings [14].

There are different types of memory and learning, and various anatomical areas in the nervous system subserve most of them. Most forms of memory include "short-term" and long-term. Generally, repetitive/reverberating action potential sustains brief forms of STM, while persisting molecular and cellular modifications sustain long-term forms of memory. Memory can be divided mainly into non-declarative and declarative types. The memory events consist of three components: learning, storage, and recall.

5.3 Brain and Memory

The CNS is the chief controlling and coordinating system of the human body. The major components that have been identified and interlinked to control the memory are the frontal, temporal, parietal, and occipital lobes (Figure 5.5). Moreover, the three specific parts are the cerebrum, cerebellum, and brainstem. The brainstem consists of midbrain, pons, and medulla oblongata (Figure 5.6). The cerebellum plays a vital role in balance, fine and gross motor skills, and cognitive functions such as attention, language, emotion, and the processing of procedural memories [17, 18].

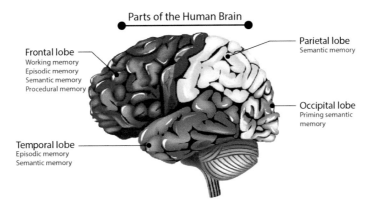

Figure 5.5 The role of various lobes of the brain in memory.

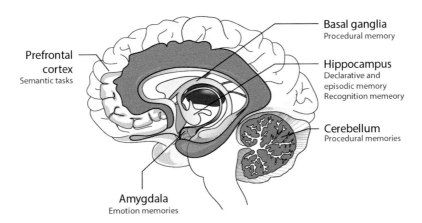

Figure 5.6 The role of various regions of the brain in memory.

The cerebrum has two left and right hemispheres, and both hemispheres are linked by corpus callosum and commissures. It plays a role in memory, attention, awareness, thoughts, consciousness, and language [19].

The frontal lobe (prefrontal cortex) is involved in handling and computing "short-term memories and storing long-term memories. The temporal lobe ornate with the senses of smell and sound, the processing of semantics in both speech and vision," and has a role in forming long-term memory. The hippocampus transfers the information from short-term to long-term memory and control of spatial memory and behavior. Another system vital to memory is the basal ganglia; its part called "neostriatum" is crucial for the synthesis and retrieval of procedural memory [20] (Figure 5.7; Figure 5.8).

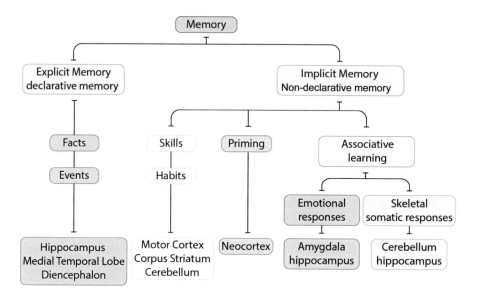

Figure 5.7 Explicit and implicit memory.

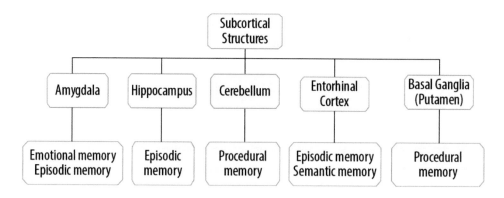

Figure 5.8 Structural classification of memory systems based on their anatomy.

Source: Modified from Squire, 2009; Van Strien, 2009 [35–36].

5.4 Sensory Memory

Sensory memory is the shortest memory to store the imitations of sensory information after completing original stimuli. It plays a role through the special senses, including vision, hearing, taste, smell, and touch, which are retained correctly for the shortest duration. Sensory memory is an ultra-short-term memory; it stores information for 100 milliseconds and swiftly declines within 100 milliseconds after perceiving an item [21].

Sensory memory is a system wherein information is stored briefly with great detail and low resistance to visual interference. The contents of sensory memory are usually transferred into a more durable memory known as STM, storing items for a few seconds. The STM capacity is severely limited; hence, some substances from STM are transferred into long-term memory, where contents can stay for a long time or a lifetime period [22].

It has been demonstrated that in sensory memory, the retaining period depends on the nature of the task that sensory memory is called to subserve [23]. The sensory memory can be affected by the delivery of spatial attention. The "sensory memories, consciousness would be attention-free, while working memory/access consciousness would require attention. The information is transferred from the sensory memory into short-term memory via the process of attention, which effectively filters only those stimuli which are of interest" [22].

5.5 Short-Term Memory

The STM lasts for seconds or a minute unless converted into long-term memory. This memory is initiated by continuous neural activity resulting from neuronal signals traveling around a temporary memory trace in a circuit of reverberating neurons. It may be due to the presynaptic facilitation or inhibition, which occurs at synapses. The "neurotransmitter chemicals secreted at terminals frequently cause facilitation or inhibition, lasting for seconds up to several minutes" [3, 24].

STM lasts for about 1–30 seconds or a minute. "Immediate memory" is also an alternate term for this concept. STM is theorized as the temporary holding of the information before becoming either a long-term or permanent one [24].

STM, such as remembering the name of any student who attends class for the first time, can easily be forgotten unless stored in long-term memory. This is known as consolidation, after which it can be kept permanently. However, the capability to retrieve it from a long-term store rests on elements, including how frequent direct communication is with that student. Short-term or working memory regulates the new information assembled in an impermanent knowledge domain [25] while being processed into new knowledge. Children with impaired STM demonstrate poor performance in the instant reproduction of verbal sequences and/or visuospatial patterns. They also have problems learning verbal information, including relations between theoretical concepts such as those essential in biology, chemistry, new vocabulary, and definitions [26].

Moreover, they face difficulties in calculation and reasoning, in comprehensions and mathematics. This shortage of information can be in either visual or verbal information or

Table 5.2 Types of Sensory Memory

- **Iconic memory:** Memory for visual perception
- **Echoic memory:** Memory for auditory perception
- **Olfactory memory:** Memory for smell perception
- **Haptic memory:** Memory for touch perception
- **Gustatory memory:** Memory for taste perception

both. However, completing primary and secondary education can be achieved despite having a STM deficit.

The STM problems are mainly due to lesions in the inferior parietal lobule and inferior frontal gyrus, mainly seen in a cerebrovascular accident with the involvement of the left middle cerebral artery. In children, it occurs due to traumatic brain injury. STM disorders are a typical result of traumatic brain injury and mainly affect episodic memory, which causes cognitive impairment [27, 28]. Children with dyslexia usually demonstrate poor verbal STM and working memory spans. The children with "Down syndrome, trisomy 21," "Williams syndrome, 7q11.23," "Turner syndrome, and Klinefelter syndrome" have visual STM [29, 30].

Biologically, memory is the most straightforward set of encoded neural connections in the brain, being the re-recreation or reformation of present and previous experiences by the synchronous firing of neurons included in that particular event. Memory is associated with the various forms of learning, the process in which an individual may get worldly knowledge and alter it in a specific behavior. During the process of learning, the neurons fire together to produce a particular experience [29, 30]. For example, by studying a specific language, we gradually learn it, using our memory to retrieve just learned words. Hence, memory is dependent upon learning for storing and retrieving information. However, learning also depends upon memory to a degree where the knowledge is kept. Memory provides the setting in which connotations and inference connect novel information.

STM is an impermanent memory managed at any point in time and referred to as "the brain's Post-it stick note." It retains a small amount of information with limited capacity to restore the information in an active, readily available state for a short time, 10–15 seconds or maximum for a minute [10]. It is also suggested that the human working memory can retain 5–9 objects at once [7, 2]. The STM capacity and duration may be enhanced if a subject can articulate the words or digits aloud instead of being read sub-vocally [29, 30].

STM is not a comprehensive concept of information. The information may disappear speedily unless the subject makes a conscious effort to retain it. STM is a crucial stage toward long-term memory. The transfer of information to long-term memory establishes more permanent storage of the information, which can be improved by repetition of the information by connotation and associating with previously acquired knowledge [31].

The subjects with parietal and temporal lobe lesions have clinical features of impaired short-term phonological capabilities but intact long-term memory [31–32]. However, subjects with a lesion in the MTL have features of impaired long-term memory but preserved STM [33]. However, some human model studies also demonstrate mixed results. Forgetting STM is an annoying problem; sometimes, people fail to retrieve or recall the information encoded just a few seconds earlier. There are two primary explanations for forgetting; one is a time-based decay, and the second is a similarity-based interference. In these conditions, the subject fails to retrieve or recall the information [34].

5.6 Working Memory

Working memory is frequently used, and it is easily interchangeable with STM. It denotes the theoretical context of structures and processes used for the provisional packaging of information. The central executive area of the prefrontal cortex plays a primary function in STM and working memory. It assists as an impermanent assembly for STM, where information is reserved for reasoning processes [25]. The central executive controls two neural loops: visual information (visual cortex) and language (Broca's area). These two areas momentarily embrace the information until it has been obliterated.

The STM decays after 10–14 seconds; it may be stored for a minute, depending on the content. The STM can be extended by repetition or rehearsal; hence, the data, record, and knowledge re-enter the short-term storage and are recollected for a further period. In routine, the contents in the STM are apprehended for short-term. The content progressively pushes out the older content unless the older content is actively protected against interference by rehearsal. However, the outside intervention tends to cause disturbances in STM retention [13]. Short-term memories may convert into long-term memory through consolidation, involving rehearsal and meaningful association.

5.7 Long-Term Memory

The long-term memory can be recalled up to years. Due to structural changes at the synapses, these changes increase or suppress signal conduction [3, 21]. The long-term memory lasts for minutes, days, weeks, or years and is eventually lost unless the memory traces are activated enough to become more permanent. The chemical or physical changes or both at presynaptic or postsynaptic membrane can persist for a few minutes up to several weeks and years [3].

The long-term memory assembled an unlimited amount of information for a long-time period and encoded the information semantically. Long-term memory involves a physiological process in the neurons; the mechanism is known as LTP. While the subject learned information, the neural circuits communicate through synapses and transference of neurotransmitters across the synapse to receptors. The repeated use of the exact contents of information enhances the efficiency of synapse connections, and their biological potentials facilitate the nerve impulses along with the specific neural circuits. It may include networks to the visual, auditory cortex, and associated cortical areas [2, 20, 21].

This biological process is entirely different from "short-term memory." The "short-term memory" is facilitated through temporary procedures of neuronal communication in the "frontal, prefrontal, and parietal lobes" of the brain. Long-term memory is sustained by permanent changes in neural connections in the brain. The hippocampus acts as a transit point for long-term memory. The forgetting of items may occur in long-term memory when the synaptic networks and/or neural network weaken; or when the activation of a new network is superimposed over an older one, it may interfere with the old memory.

5.7.1 Declarative (Explicit) Memory

Declarative memory, also called explicit memory, is the memory of facts and events that can be consciously "recalled or declared" and can store such information for years [37]. The declarative memory is sub-divided into "episodic memory" and "semantic memory." Declarative memory is encoded, consolidated, and stored mainly in the hippocampus, entorhinal, and perirhinal regions of the MTL [38] (Table 5.3). The "working memory capacity and declarative knowledge do not predict learning; however, working memory capacity and declarative knowledge do predict performance during practice" [39].

Table 5.3 Types of Long-Term Memory

Declarative (explicit) memory
Episodic
Semantic
Retrospective
Prospective
Procedural (implicit) memory

5.7.1.1 Episodic (Semantic) Memory

The declarative memory is further subdivided into episodic memory and semantic memory.

The autobiographical memories of previous proceedings and thoughts of forthcoming circumstances encompass both episodic and semantic contents correlating the "internal" (episodic) and "external" (semantic) events [40, 41].

Semantic memory is frequently applied as a basis of knowledge and notions learning for the cognitive system. Episodic memory recollects personal experiences and events during daily activities. It demonstrates autobiographical events, including the time, place, and emotions. The arrangements of events are stored in episodic memory and build associations in semantic memory [42].

Semantic memory is a highly organized record of facts, concepts, meanings, and knowledge about the outside world. The episodic and semantic memory requires a similar encoding process. However, semantic memory mainly involves activating the frontal and temporal cortexes, while episodic memory actions focus on the hippocampus. Once administered in the hippocampus, episodic memories are consolidated and stored in the neocortex [41, 42].

5.7.1.2 Retrospective Memory

In retrospective memory, the content may be remembered in the past. It includes semantic, episodic, and autobiographical memory. The cognitive factors are highly dominant than behavioral factors in shaping retrospective memory performance. The prospective memory components involve recognition processes, whereas the retrospective memory component requires recall of the substances [43].

5.7.1.3 Prospective Memory

The prospective memory refers to remembering or making a "mental note" of performing an action or event in the future. The prospective memory requires one to recall and remind themselves to execute the intended action or given task at the appropriate time. The central aspect of this prospective memory component is what also separates it from traditional retrospective memory; the participant must recall himself without being prompted by an experimenter [44].

Individuals have different methods to manage prospective memory tasks compared to those they use for retrospective memory tasks, including constructing plans for when and how the intended activity will be accomplished. In this memory, recovery of items, actions, commitments "appears to be accomplished through cognitive processes. These processes range from self-initiated retrieval or monitoring, resource-demanding, to relatively automatic retrieval processes stimulated by environmental cues associated with the intended action" [44].

Prospective memory allows the subjects to make plans, store them, retain them, and bring them back to one's consciousness at the right place and time to execute them. In prospective memory, the content to be remembered is in the future. It may be either event-based or time-based, often activated by a time signal such as delivering a lecture at 8 am and dropping a child at school. It allows establishing plans, retain, and bring them to the consciousness at the right time and place [45–47]. A hindrance to the role of prospective memory manifests mainly in Alzheimer's disease, which may provide an early prognostic indicator of cognitive decline [48].

5.7.2 Procedural (Implicit) Memory

Procedural memory is also called implicit memory. It is the unconscious memory of skills and how the skills were used, such as riding on a bike. The memory is acquired through repetition and practice. In this memory, previous experiences facilitate the performance of a task [37]. The procedural memories are encoded and stored by the "putamen, caudate nucleus and the motor cortex" [49, 50]. The earlier learned skills such as riding a bicycle are stored in the putamen. However, the inborn actions such as the sucking reflex of infants are stored in the caudate nucleus. Moreover, the cerebellum plays a role in the timing and coordination of actions and skills. Procedural or implicit or memory is less accessible to conscious awareness and enables gradual learning of habits and skills [51].

5.8 Memory Encoding

Encoding is a biological process that demonstrates the early processes, actions, or achievement of the information. It involves capturing information by sensory systems and its conversion by neuronal coding beyond simple perception [52]. The memory process starts with consideration in which a memorable event causes neurons to fire more occasionally, making the experience more intense, and the event is encoded as a memory. Encoding starts at various stages. The initial stage is creating STM from the ultra-short-term sensory memory, followed by the translation to long-term memory through memory consolidation. The hippocampus is responsible for analyzing these responses, and decisions are committed to long-term memory. The "encoding for STM storage in the brain relies primarily on acoustic encoding while encoding for long-term storage relies on semantic encoding" [53] (Table 5.4).

5.9 Memory Consolidation

The memory formation process can be decomposed into achievements, alliance, and reconsolidation phases in which evidence is gained, stored, and modified [54, 55]. Consolidation is the slow progressions of stabilizing a memory trace (s) after the early attainment. The consolidation process includes integrating a new memory into an existing memory network; however, memory may be updated through one of several episodes or reconsolidations [56, 57]. Memory contents are interconnected with both newly formed and previously stored memory items [58, 59].

Consolidation of memories requires hippocampal-neocortical communication, the sharp-wave ripples in transferring the hippocampal information to the neocortex. The process continues after observation, which enables temporary changes in activity and synaptic strength to become long-lasting. The later reactivation of neural activity allows for the induction of long-term synaptic changes [60]. The consolidation process operates a phenomenon known as LTP, which permits a synapse to upsurge in strength with enhancing numbers of signals conveyed between the two neurons [60].

LTP occurs when similar neurons fire together and become permanently sensitized to each other. The brain arranges and rearranges responses to experiences, producing new memories impelled by experience, education, or training. Memory reconsolidation is the means of past consolidated memories being made to recall and then being actively consolidated to strengthen, maintain, and modify memories that are already stored in the long-term memory [58, 59].

Table 5.4 Types of Encoding of Memory

- **Acoustic encoding:** Encoding of sound, words, and auditory input
- **Visual encoding:** Encoding of images and visual sensory information
- **Tactile encoding:** Encoding of touch, temperature
- **Semantic encoding:** Encoding of sensory input with specific meaning

5.10 Memory Storage

The storage of memory is creating relatively stable memory traces or records of knowledge in the brain. Such traces require neuronal networks that can engage in neuronal coding, which is the substrate of information storage and evoked when we remember specific information [61–63] (Figure 5.9). Memory storage is the process of recollecting the information in the brain, in the form of "sensory memory, short-term memory or the long-term memory." The more the information is repeated or used, the longer the contents are retained in long-term memory. After consolidation, long-term memories are stored throughout the brain. The memory components linked to visual stimuli are store in the visual cortex, and neurons in the amygdala store the associated emotion. The human brain can store unlimited amounts of information indefinitely [61–63].

After a brief cellular consolidation period, memory can be stored for a long time [64]. The memory storage has occurred at synapses, strengthened through LTP [65, 66]. The long-lasting changes at synapses empower memory storage in the brain [67]. Synaptic variations characterize maintenance of LTP, including increasing α-Amino-3-hydroxy-5-methyl-4-isoxazolepropionic acid (AMPA)-type glutamate receptors subunits' density in the postsynaptic and enlarged synapse morphology [68–70]. Perpetuating signaling that maintains LTP is envisaged as the molecular basis of memory storage [66, 71].

Although aging naturally results in memory loss and its impaired formation, it is unknown whether the synaptic underpinnings of memory storage differ with age [67]. Most past research studies regarding memory storage have targeted the hippocampus due to its critical role in memory [72]. The literature provides evidence that aging reduces the brain's ability, the hippocampal memory formation decreases with aging [73, 74]. Despite this decline due to age,

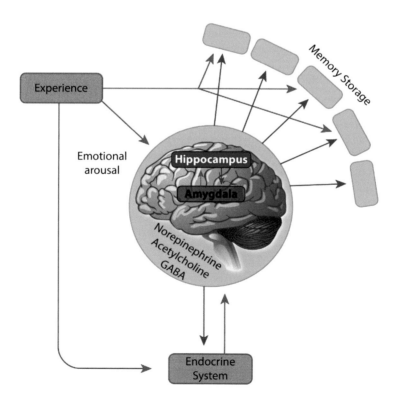

Figure 5.9 Memory storage.

hippocampal memory can still be formed and stored. However, this leaves the question of which exact synaptic mechanisms exist that enable memory storage in old age. Aging causes a deficit in LTP induction, which can be overcome with solid electrical stimulation [73, 75]. Therefore, it is likely that prolonged behavioral training can overcome the impairment due to age in LTP induction so that LTP may be the memory-storing mechanism, especially at younger ages. Alternatively, in old age, extended behavioral training might not be enough to induce LTP Instead, it may lead to an alternative synaptic change, such as forming multi-innervated dendritic spines. These multi-input synapses have been believed to store memory when LTP is blocked [76, 77].

The cortex and its allied structures, nuclei, and networks are a highly active component of the brain. The network is thoroughly familiar with synaptic connectivity and constantly changes across various events and time scales. Short-time scale synapses between the neurons change in an activity-dependent means leading to synaptic facilitation and depression. Assume that the network is optimized for storage at total capacity as suggested by Zhang et al. (2019) [78]; the network must forget some memories to store new memories and stay updated with the world's external environment. This function is related to how the synaptic connectivity changes as the network turn over its storage. It can be pursued in future work by adding synaptic mechanisms for forgetting and activity-independent fluctuations in synaptic strength. On a physiological basis, the synaptic connectivity is mainly genetically preconfigured and continuously tuned by a combination of activity-dependent plasticity rules and activity-independent cellular mechanisms [79].

5.11 Memory Recall and Retrieval

The recall or retrieval of memory refers re-accessing of events or information from the previous events, which have been earlier programmed and stored in the brain. It is a process of remembering. During the recall process, the brain "replays" procedural arrangements of neural activity produced in response to a particular event. The process in which memories are encoded, stored, and recalled is effectively an on-the-fly restoration of elements distributed throughout various brain areas [80]. Memory retrieval requires re-visiting the nerve pathways established when encoding the memory, and the pathways determine how swiftly memory can be recalled. Recall efficiently returns memory from long-term storage to short-term or working memory, where it can be accessed in a kind of encoding process [80].

The mechanisms allied to memory formation and retention remain a fundamental question in the field of neurosciences. The transience of some memories and the permanency of others demonstrate the brain's plasticity. The scientific literature on the hippocampus reveals that specific neurons form a briefly stable representation of places. Therefore, this brain region has become an essential focus for studying spatial memory and engram. It also assists as an empirically accessible representation for declarative (knowledge) and episodic (experience) memory in humans. However, how can a person remember the almost countless number of items, things with the limited storage capacity of the hippocampus? There is a piece of evidence that relevant representations are transferred to neocortical networks before forming long-lasting engrams. The hippocampus is then reset for the acquisition of new memories [81].

The literature shows that neuronal activity during sleep plays a significant role in these processes. Norimoto et al. (2020) [82] showed that sleep-associated activity patterns induce "negative" neuronal plasticity in the hippocampus, erasing remote memories. The findings further demonstrate that slow-wave states refine memory engrams by reducing recent memory-irrelevant neuronal activity [82]. A previous study conducted by Khodagholy et al. (2017) [83] reveals similar activity patterns in the neocortex [83]. The different stages of sleep and its rhythms, including multiple "NREMS oscillations and the REMS hippocampal theta rhythm, serve the multipurpose function of enabling memory consolidation and adaptive forgetting simultaneously. The same sleep rhythms that consolidate new memories in the cortex and hippocampus organize at the same time the adaptive forgetting of older memories in these brain regions" [84].

5.12 Curiosity-Driven Memory

Curiosity is highly essential for intellectual achievements; it primes the brain to learn and retain information. Memory processing depends on multiple elements, including encoding strength, emotion, and knowledge of future relevance. The primary function of curiosity in the learning shows that memory related to high-curiosity encoding states is retained better. This effect may be driven by activity within the dopaminergic circuit [85].

Gruber and Ranganath (2019) [86] reported that curiosity plays a vital role in learning and memory. Curiosity is related to modulations in activities in the dopaminergic circuit. These modulations influence memory consolidation and encoding for both targets of curiosity and incidental information encountered during curiosity states. The authors proposed the "Prediction, Appraisal, Curiosity, and Exploration (PACE) framework to explain curiosity and memory in cognitive processes, neural circuits, behavior, and subjective experience." Curiosity promotes memory encoding by increasing "attention, exploration, and information seeking and enhancing consolidation of information acquired while in a curious state through dopaminergic neuromodulation of the hippocampus" (Figure 5.10).

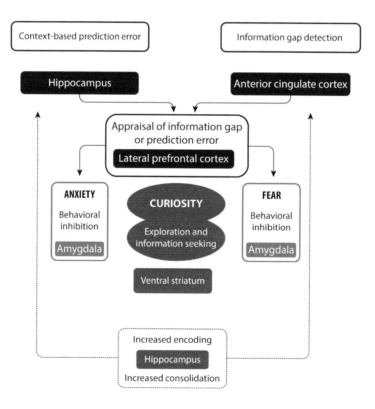

Figure 5.10 Role of curiosity in learning and memory.

Source: Modified with acknowledgement [86].

The hippocampus is concerned with converting recent memory to long-term memory. The hippocampus is involved in storing and categorizing afferent information related to recent memory [87]. A "lesion in the hippocampus results in the individual being unable to store long-term memory. The lesion to the amygdaloid nucleus, the hippocampus, produces more significant memory loss than injury to either one of these structures alone [88, 89]. Hippocampal abnormalities can produce profound deficits in real-world situations" [90].

The neurons and neuronal epigenome are susceptible to external events, and their "function is vital for producing stable behavioral outcomes, such as forming long-lasting memories. The growing evidence points to altered epigenome function in the aging brain as a contributing factor to age-related memory decline" [91]. Multiple physiological and pathological factors may impair memory; these factors include smoking [92], environment pollution [93], and RF-EMFR [94].

5.13 Effect of Electromagnetic Field Radiation on Memory

In the 21st century, there have been significant changes in the pattern of living standards and health and disease. In this most modern period of science and technology, the fastest-growing field is telecommunications, mainly mobile phones [95, 96]. In recent times, cell phones have become a necessity of daily life. Their role is not limited to making and receiving calls and has advanced functions for storing data, photography, scanning messages, listening to music, and watching videos. Moreover, modern cell phones provide internet access, banking services, sending and receiving documents, and geo-positioning system technology, allowing users to find locations and traffic conditions [96].

As per the World Bank report worldwide, about 7.68 billion inhabitants have a mobile phone, more than the world's population, as most people possess more than one cell phone [97]. The World Bank also reported that the number of mobile phone users is 104.45% of the world population [98]. The global telecommunication industry placed over 7 million base-station towers worldwide, and the number is increasing significantly with the introduction of fifth-generation technology [99]. The WHO now recognizes RF-EMFR emitted from mobile phones and base-station towers as a part of environmental pollution and possible cause of cancer [100].

The excessive use of the mobile phone has initiated a debate on their radiation impact on human health [101]. Since mobile phones are usually held close to the head during normal conversation, the brain is exposed to the RF-EMFR emitted from mobile phones. This has raised concerns about the potential adverse effect of electromagnetic radiation (EMR) on brain biology.

In recent years, the mobile phone industry has grown swiftly and made telecommunication more convenient, accessible, and faster [102]. The multifunctional smartphone technology appeals to both gender and all age groups, mainly the youth and adults. The excessive use of mobile phones has markedly increased and has become a primary tool in daily life [102].

The wide-ranging mobile phone usage has led to the emergent installation of MPBSTs in highly populated, crowded residential and commercial areas [102, 103] (Figure 5.11). Smartphones are a significant source of the eminence of electromagnetic waves. Mobile phones generated RF-EMFR ranges from 900–1,800 and 24–80 GHz in 5G mobile [103].

The WHO [104] reported that mobile phones are low-powered radiofrequency transmitters, operating at frequencies between 450 and 2,700 MHz with peak powers in the range of 0.1–2 W. People using mobile phones 30–40 cm away from the body while messaging or using a "hands-free" device will have lower exposure to radiofrequency fields than those holding the handset against their head [74].

Evidence from literature shows that the human nervous system is susceptible to RF-EMFR [102]. The impact of RF-EMFR changes brain activity through non-thermal and thermal effects

Figure 5.11 Mobile phone base-station tower located in a residential area.

due to temperature rise from energy absorption. Mobile phone industries have perceived a tremendous expansion. Simultaneously, the biological effects of RF-EMFR generated from mobile phones and MPBST have not been thoroughly investigated. There is comparatively little research available on cognitive characteristics, mainly on memory assessment, and evidence lacks recommendations on mobile phone hazards and memory impairment. Table 5.5 summarizes the various studies published in biomedical journals; these studies demonstrate the impact of RF-EMFR generated from mobile phones and MPBSTs on memory and cognitive functions.

The brain is a highly complex organ in the human body, which controls thoughts, actions, memory, experiences, and attitudes. Brain biology depends upon the neurons, receptors, neurotransmitters, and neuromodulator chemicals and their receptor pathways. The disruption of memory can affect cognitive capabilities and thus the quality of life. Early disorders of learning and memory hinder the development of children as the usual weakening of memory with aging. During the last three decades, the biological studies of the brain have focused on establishing a common conceptual framework that extends from cell and molecular biology to brain biology [1]. In the present scenario, several studies identified that "RF-EMFR" generated by mobile phones, base-station towers, and other electronic devices has a negative biological impact on memory. However, literature is also available that shows no effect on brain biology.

Meo et al. (2018) [107] conducted a study among school adolescents exposed to high RF-EMF and found reduced memory among the school adolescents exposed to high RF-EMF compared to students exposed to low RF-EMF. Similarly, Singh et al. (2016) [119] demonstrated that people living adjacent to the MPBSTs complained of headaches, sleep disturbances, and difficulty in concentration. It is a fact that once an individual has a headache, sleep disturbances, and difficulty in concentration, it will impair memory.

Calvente et al. (2016) [120] demonstrate that children exposed to high RF-EMF had lower cognitive scores compared to those with low exposure. Thomas et al. (2010) [121] found a significant drop-in response time in adolescents using mobile phones for long durations. However, Malek et al. (2015) [122], Haarala et al. (2003) [123], and Riddervold et al. (2008) [114] reported no impact of RF-EMF exposure generated from MPBSTs on human cognitive performance, including memory.

Table 5.5 Impact of RF-EMFR on Memory					
Author and Year of Publication	Sample Size	Age	Type of Study	Gender	Study Outcome
Abramson et al. (2009) [105]	317	11–14	Cross-sectional	M/F	Insufficient accuracy of working memory
Schoeni et al. (2015) [106]	439	12–17	Cross-sectional	M/F	Decreases in figural memory
Meo et al. (2019) [107]	217	13–16	Cross-sectional	M	Delayed spatial working memory
Santini et al. (2003) [108]	530	18–67	Questionnaire Study	M/F	Decrease memory 20.40% out of 420
Lalrinthara et al. (2014) [109]	38	–	Cross-sectional	M/F	Decrease memory 31% out of 38
Deniz et al. (2017) [110]	60	18–25	Cross-sectional	F	Lack of attention and memory
Foerster et al. (2018) [111]	843	12–17	Prospective cohort	M/F	Decreased figural memory
Kalafatakis et al. (2017) [112]	64	60–80	Cross-sectional	M/F	Negative impact on working memory
Sauter et al. (2011) [113]	30	25	Cross-sectional	M	No effect on cognition
Riddervold et al. (2008) [114]	80	15–40	Cross-sectional	M/F	No change in cognitive task
Haarala et al. (2004) [115]	64	–	Cross-sectional	M/F	No effect on cognitive function
Preece et al. (2005) [116]	18	10–12	Cross-sectional	M/F	Decreased reaction time
Eltiti et al. (2009) [117]	88	–	Cross-sectional	M	No effect on memory
Khan (2008) [118]	286	16–23	Questionnaire study	M/F	Memory disturbances 40.56%

Abdel-Rassoul et al. (2007) [124] conducted a study among subjects residing near MPBSTs. The exposed subjects experienced neuropsychiatric symptoms, including memory changes compared to their control group. The exposed subjects showed declined performance in STM. Moreover, Khan (2008) [118] reported that about 40.56% of the students complained of memory disturbances after long-term use of cell phones. Correspondingly, Kalafatakis et al. (2017) [112] reported that mobile phone use negatively impacts working memory performance.

Raika et al. (2017) [125] found a dose–response association between mobile phone usage and concentration difficulties, headache, fatigue, insomnia, and ear warming. The authors also found an association between vicinity to base stations and general health problems. Similarly, Deniz et al. (2017) [110] conducted a study on sixty healthy female medical school students aged 18–25 years. These students were divided into a low exposure group (30 students, less than 30 min day, and a high exposure group of 30 students with more than 90 min daily use of mobile phone close to the head. The authors concluded that a lack of attention and concentration was found among students who used mobile phones for a longer time than those who use phones relatively for less time. The neurocognitive tests of the high exposure group for evaluating attention were significantly more inadequate than the low exposure group. It is a well-established fact that attention is strongly associated with memory.

The literature demonstrates that children and adolescents who use excessive wireless devices, including cell phones, have various adverse neurobehavioral and neurodevelopmental problems. The impairment in memory, cognition, attention, learning, and behavioral disturbances has been reported in multiple studies. They are similarly manifested in autism and ADHD (attention deficit hyperactivity disorders) as a result of EMF and RFR exposures where both epigenetic drivers and genetic (DNA) damage are likely contributors [126].

In animal model studies, it has also been proven that exposure to radiation of 800–2,100 MHz leads to damage and decrease of neurons in the hippocampal region, which alters learning and memory [12, 88]. Nidhi et al. (2014) [127] indicated an extensive neurodegenerative change in radio wave exposure. These findings show that extensive neurodegeneration in selective areas of the cerebral cortex results in alterations in behavior related to memory and learning.

Saba Shahin et al. (2015) [128] performed a study to understand the mechanism of the dose–response effect of short-term (15 days) and long-term (30, 60 days) low-level radiation exposure on the hippocampus with particular reference to learning and memory in male mice. The authors found that short-term and long-term exposure to 2.45-GHz radiation increases oxidative stress, leading to enhanced apoptosis in hippocampal both neuronal and non-neuronal cells. They also found that learning and spatial memory deficit increases with the increased duration of radiation exposure. The authors further conclude that exposure to radiation leads to hippocampal neuronal and non-neuronal apoptosis associated with spatial memory loss.

In another study, Saba Shahin et al. (2018) [129] elucidates the radiation-induced stress factors on the hippocampal spatial memory formation pathway. The authors identified that 2.45-GHz-irradiated mice showed slow learning and increased working and reference memory errors. Similarly, Narges et al. (2018) [130] determined the impact of microwave radiation on memory, hippocampal synaptic plasticity, and hippocampal neuronal cell numbers. The authors found that radiation exposure decreased the number of neurons and learning and memory performance. Furthermore, Ragy (2015) [131] investigated the effect of EMR emitted from the mobile on male albino rats' brains. The study group animals were exposed to 900-MHz EMR for 1 hour per day for 60 days, and the withdrawal group was exposed for 1 hour per day for 60 days then left for 30 days without exposure. The authors identified that RF-EMFR produces impairments in bio-chemicals components and oxidative stress in albino rats' brains.

The potential mechanism that impairs the cognitive functions, memory, in subjects exposed to RF-EMF radiation generated from mobile phone and their base-station towers are sleep disturbances, behavioral problems [132], decreased cerebral blood flow [133], myelin sheath damage [134]. Neuronal function lessening [135, 136] leads to damage and decreased neurons in the hippocampal region [89]. All these mechanisms are associated with Rf-EMFR resulting in impairment of cognitive functions and memory.

Presently, significant literature worldwide has identified substantial support for the psychometric properties of mobile phones and base-station tower radiations associated with loss of cognitive functions resulting in memory impairment. However, literature also shows no effect of RF-EMFR on the brain.

Conclusion

In conclusion, learning and memory are highly crucial capabilities of the human brain. The brain has proficiency in receiving, analyzing, and filtering the information in a sequence, learned, remembered, and forgotten. Worldwide, people are facing health issues of environmental pollution. The addition of mobile phone technology has changed the face of the modern world. Presently, about 7.68 billion inhabitants possess a mobile phone, surpassing the total count of the world's population. The WHO recognizes RF-EMFR as a type of environmental pollution and a possible cause of various health hazards. The present chapter highlights that RF-EMFR generated by mobile phones, base-station towers, and numerous electronic devices negatively impacts cognitive functions and memory.

References

[1] Kandel E.R., Dudai Y., Mayford M.R. The molecular and systems biology of memory. *Cell.* 2014; 157(1): 163–86. doi: 10.1016/j.cell.2014.03.001.

[2] Evans K.K., Baddeley A. Intention, attention, and long-term memory for visual scenes: It all depends on the scenes. *Cognition.* 2018; 180: 24–37. doi: 10.1016/j.cognition.2018.06.022.

[3] Velez-Pardo C., Jimenez-Del-Rio M. From dark to bright to gray sides of memory: In search of its molecular basis & Alzheimer's disease. *Austin. J. Clin. Neurol.* 2015; 2(5): 1045.

[4] Tulving E. Memory systems and the brain. *Clin. Neuropharmacol.* 1992; 15(Pt A): 327A–8.

[5] Tulving E. Episodic memory and common sense: How far apart? Philos. *Trans. R. Soc. Lond. B. Biol. Sci.* 2001; 356: 1505–15.

[6] Brown R.E., Milner P.M. The legacy of Donald O. Hebb: More than the Hebb synapse. *Nat. Rev. Neurosci.* 2003; 4(12): 1013–9. doi: 10.1038/nrn1257.

[7] Cooper S.J., Donald O. Hebb's synapse and learning rule: A history and commentary. *Neurosci. Biobehav. Rev.* 2005; 28(8): 851–74.

[8] Rose Steven. Eric Kandel: Making memories. *The Lancet.* 2006; 367(9524): 1721–2.

[9] Rose Steven. Eric Kandel: Making memories. Eric R Kandel. In: *Search of Memory: The Emergence of a New Science of Mind.* New York: W. W. Norton. 2006, pp. 352–4.

[10] Miller G.A. The magical number seven, plus or minus two: Some limits on our capacity for processing information. *Psychological. Review.* 1956; 63: 81–97.

[11] Cowan N. George Miller's magical number of immediate memory in retrospect: Observations on the faltering progression of science. *Psychol. Rev.* 2015; 122(3): 536–41.

[12] Walter Schneider, Richard M. Shiffrin. Controlled and automatic human information processing: detection, search, and attention. *Psychological. Review.* 1977; 84(1): 1–66.

[13] Atkinson R.C., Shiffrin R.M. Human memory: A proposed system and its control processes. In: Spence K.W., Spence J.T., editors. *The Psychology of Learning and Motivation: Advances in Research and Theory.* New York: Academic Press. 1968, pp. 89–195.

[14] Craik F.I.M., Lockhart, R.S. Levels of processing: A framework for memory research. *J. Verbal Learn. Verbal Behav.* 1972; 1072(11): 671–84.

[15] Craik F.I.M., Tulving E. Depth of processing and the retention of words in episodic memory. *J. Exp. Psychol.* 1975; 104(3): 268–94.

[16] Rose N.S.L., Craik F.I., Buchsbaum B.R. Levels of processing in working memory: Differential involvement of frontotemporal networks. *J. Cogn. Neurosci.* 2015; 27(3): 522–32.

[17] Ferrari C., Cattaneo Z., Oldrati V., Casiraghi L., Castelli F., D'Angelo E., Vecchi T. TMS over the cerebellum interferes with short-term memory of visual sequences. *Sci. Rep.* 2018; 8(1): 6722. doi: 10.1038/s41598-018-25151-y.

[18] Pleger B., Timmann D. The role of the human cerebellum in linguistic prediction, word generation, and verbal working memory: Evidence from brain imaging, non-invasive cerebellar stimulation and lesion studies. *Neuropsychologia.* 2018 Jan; 115: 204–10.

[19] Shanahan L.K., Gjorgieva E., Paller K.A., Kahnt T., Gottfried J.A. Odor-evoked category reactivation in human ventromedial prefrontal cortex during sleep promotes memory consolidation. *Elife.* 2018; 7. pii: e39681. doi: 10.7554/eLife.39681.

[20] Chavez C.M., McGaugh J.L., Weinberger N.M. The basolateral amygdala modulates specific sensory memory representations in the cerebral cortex. *Neurobiol. Learn. Mem.* 2009; 91(4): 382–92.

[21] Tripathy S.P., Öğmen H. Sensory memory is allocated exclusively to the current event-segment. *Front. Psychol.* 2018; 9: 1435.

[22] Tripathy Srimant P., Öğmen Haluk. Sensory memory is allocated exclusively to the current event-segment. *Front. Psychol.* 2018; 9: 1435. doi: 10.3389/fpsyg.2018.01435. eCollection 2018.

[23] Rensink R.A. Limits to the usability of iconic memory. *Front. Psychol.* 2014; 5: 971. doi: 10.3389/fpsyg.2014.00971.

[24] Guyton and Hall. *Text Book of Medical Physiology*, 13th edition. Philadelphia: Elsevier. 2016, pp. 744–50.

[25] Crowder J., Carbone J. Occam learning through pattern discovery: Computational Mechanics in A.I. Systems. Proceedings of the 12th Annual International Conference on Artificial Intelligence, Las Vegas, Nevada, U.S.A. 2011.

[26] Alan Baddeley. Working memory: Theories, models, and controversies. *Annu. Rev. Psychol.* 2012; 63: 1–29.

[27] Wilson B.A., Baddeley A. Spontaneous recovery of impaired memory span: Does comprehension recover? *Cortex.* 1993; 29(1): 153–9.

[28] Hanten G.L., Martin R.C. A developmental phonological short-term memory deficit: A case study. *Brain. Cogn.* 2001; 45(2): 164–88.

[29] Majerus S., Leclercq A.L., Grossmann A., Billard C., Touzin M., Van der Linden M., Poncelet M Serial order short-term memory capacities and specific language impairment: No evidence for a causal association. *Cortex.* 2009; 45(6): 708–20.

[30] Majerus S., Heiligenstein L., Gautherot N., Poncelet M., Van der Linden M. Impact of auditory selective attention on verbal short-term memory and vocabulary development. *J. Exp. Child. Psychol.* 2009; 103(1): 66–86.

[31] Shallice T., Warrington E.K. Independent functioning of verbal memory stores: A neuropsychological study. *Q. J. Exp. Psychol.* 1970; 22: 261–73.

[32] Vallar G., Papagno C. Neuropsychological impairments of verbal short-term memory. In: Baddeley A.D., Kopelman M.D., Wilson B.A., editors. *The Handbook of Memory Disorders*, 2nd edition. Chichester, UK: Wiley. 2002, pp. 249–270.

[33] Scoville W.B., Milner B. Loss of recent memory after bilateral hippocampal lesions. *J. Neurol. Neurosurg. Psychiatry.* 1957; 20: 11–21.

[34] John J., Richard L.L., Derek E.N., Cindy A.L., Marc G.B., Katherine S.M. The mind and brain of short-term memory. *Annu. Rev. Psychol.* 2008; 59: 193–224. doi: 10.1146/annurev.psych.59.103006.093615.

[35] Squire L.R. Memory and brain systems: 1969–2009. *J Neurosci.* 2009; 29: 12711–6.

[36] Van Strien N.M., Cappaert N.L., Witter M.P. The anatomy of memory: An interactive overview of the parahippocampal-hippocampal network. *Nat. Rev. Neurosci.* 2009; 10: 272–82.

[37] Quam C., Wang A., Maddox W.T., Golisch K., Lotto A. Procedural-memory, working-memory, and declarative-memory skills are each associated with dimensional integration in sound-category learning. *Front. Psychol.* 2018; 9: 1828. doi: 10.3389/fpsyg.2018.01828.

[38] Schultz H., Sommer T., Peters J. The role of the human entorhinal cortex in a representational account of memory. *Front. Hum. Neurosci.* 2015; 9: 628. doi: 10.3389/fnhum.2015.00628.

[39] Van Abswoude F., van der Kamp J., Steenbergen B. The roles of declarative knowledge and working memory in explicit motor learning and practice among children with low motor abilities. *Motor. Control.* 2018: 1–18. doi: 10.1123/mc.2017-0060.

[40] Levine B.1., Turner G.R., Tisserand D., Hevenor S.J., Graham S.J., McIntosh A.R. The functional neuroanatomy of episodic and semantic autobiographical remembering: A prospective functional M.R.I. Study. *J. Cogn. Neurosci.* 2004; 16(9): 1633–46.

[41] Devitt A.L., Addis D.R., Schacter D.L. Episodic and semantic content of memory and imagination: A multilevel analysis. *Mem. Cognit.* 2017; 45(7): 1078–94.

[42] Horzyk A., Starzyk J.A., Graham J. Integration of semantic and episodic memories. *IEEE. Trans. Neural. Netw. Learn. Syst.* 2017; 28(12): 3084–95.

[43] Uttl B., White C.A., Cnudde K., Grant L.M. Prospective memory, retrospective memory, and individual differences in cognitive abilities, personality, and psychopathology. *PLoS. One.* 2018; 13(3): e0193806. doi: 10.1371/journal.pone.0193806.

[44] McDaniel M.A., Einstein G.O. Prospective memory. *Psychol. Int. Encycl. Social & Behav. Sci.* 2001; 12241–4.

[45] Graf P., Uttl B. Prospective memory: A new focus for research. *Conscious. Cogn.* 2001; 10(4): 437–50.

[46] Graf P., Uttl B. Prospective memory: A new focus for research. *Conscious. Cogn.* 2001; 10: 437–50. doi: 10.1006/ccog.2001.0504.

[47] Uttl B. Transparent meta-analysis of prospective memory and aging. *PloS One.* 2008; 3: e1568. doi: 10.1371/journal.pone.0001568.

[48] Lee S.D., Ong B., Pike K.E., Kinsella G.J. Prospective memory and subjective memory decline: A neuropsychological indicator of memory difficulties in community-dwelling older people. *J. Clin. Exp. Neuropsychol.* 2018; 40(2): 183–97.

[49] Squire L.R. Memory systems of the brain: A brief history and current perspective. *Neurobiol. Learn. Mem.* 2004; 82(3): 171–7.

[50] Carpenter K.L., Wills A.J., Benattayallah A., Milton F. A Comparison of the neural correlations underlie rule-based and information-integration category learning. *Hum. Brain. Mapp.* 2016; 37(10): 3557–74.

[51] Hayne H., Boniface J., Barr R. The development of declarative memory in human infants: Age-related changes in deferred imitation. *Behav. Neurosci.* 2000; 114: 77–83.

[52] Roediger H.L. Encoding. In: Roediger H.L., Dudai Y., Fitzpatrick S.M., editors. *Science of Memory: Concepts.* Oxford: Oxford University Press. 2007, pp. 121–4.

[53] Maril A., Davis P.E., Koo J.J., Reggev N., Zuckerman L., Ehrenfeld L., Mulkern R.V., Waber D.P., Rivkin M.J. Developmental fMRI study of episodic verbal memory encoding in children Neurology. 2010; 75(23): 2110–16.

[54] Buzsáki G. Neural syntax: Cell assemblies, synapsembles, and readers. *Neuron.* 2010; 68(3): 362–85.

[55] Nadel L. Memory formation, consolidation, and transformation. *Neurosci. Biobehav. Rev.* 2012; 36(7): 1640–5.

[56] Besnard A., Caboche J., Laroche S. Reconsolidation of memory: A decade of debate. *Prog. Neurobiol.* 2012; 99(1): 61–80.

[57] Lee J.L.C. Reconsolidation: Maintaining memory relevance. *Trends Neurosci.* 2009; 32(8): 413–20.

[58] Tronson N.C., Taylor J.R. Molecular mechanisms of memory reconsolidation. *Nat. Rev. Neurosci.* 2007; 8(4): 262–75.

[59] Nader K., Einarsson E.Ö. Memory reconsolidation: An update. *Ann. N. Y. Acad. Sci.* 2010; 1191(1): 27–41.

[60] Hasselmo M.E. Encoding: Models linking neural mechanism to behavior. In: Roediger H.L., Dudai Y., Fitzpatrick S.M., editor. *Science of Memory.* Oxford: Oxford University Press. 2007, pp. 123–31.

[61] Jensen O. Maintenance of multiple working memory items by temporal segmentation. *Neuroscience.* 2006; 139: 237–49.

[62] Jensen O., Lisman J.E. Hippocampal sequence-encoding driven by a cortical multi-item working memory buffer. *Trends. Neurosci.* 2005; 28: 67–72.

[63] Jensen O., Gelfand J., Kounios J., Lisman J.E. Oscillations in the alpha band (9–12 Hz) increase memory load during retention in a short-term memory task. *Cereb. Cortex.* 2002; 12: 877–82.

[64] Dudai Y. The neurobiology of consolidations, or how stable is the engram? *Annu. Rev. Psychol.* 2004; 55: 51–86.

[65] Morris R.G., Moser E.I., Riedel G., Martin S.J., Sandin J., Day M., O'Carroll C. Elements of a neurobiological theory of the hippocampus: The role of activity-dependent synaptic plasticity in memory. *Philos. Trans. R. Soc. Lond. B Biol. Sci.* 2003; 358: 773–86.

[66] Lisman J. Criteria for identifying the molecular basis of the engram (CaMKII, PKMzeta) Mol. *Brain.* 2017; 10: 55.

[67] Aziz W., Kraev I., Mizuno K., Kirby A., Fang T., Rupawala H., Kasbi K., Rothe S., Jozsa F., Rosenblum K., Stewart M.G., Giese K.P. Multi-input Synapses, but Not LTP-strengthened synapses, correlate with hippocampal memory storage in aged mice. *Curr. Biol.* 2019 Nov 4; 29(21): 3600–10.e4. doi: 10.1016/j.cub.2019.08.064.

[68] Matsuzaki M., Honkura N., Ellis-Davies G.C., Kasai H. Structural basis of long-term potentiation in single dendritic spines. *Nature.* 2004; 429: 761–6.

[69] Medvedev N.I., Popov V.I., Rodriguez Arellano J.J., Dallérac G., Davies H.A., Gabbott P.L., Laroche S., Kraev I.V., Doyère V., Stewart M.G. The N-methyl-D-aspartate receptor antagonist C.P.P. Alters synapse and spine structure and impairs long-term potentiation and long-term depression induced morphological plasticity in the dentate gyrus of the awake rat. *Neuroscience.* 2010; 165: 1170–81.

[70] Bourne J.N., Harris K.M. Coordination of size and number of excitatory and inhibitory synapses results in balanced structural plasticity along mature hippocampal CA1 dendrites during L.T.P. *Hippocampus.* 2011; 21: 354–73.

[71] Sacktor T.C., Hell J.W. The genetics of PKMζ and memory maintenance. *Sci. Signal.* 2017; 10: eaao2327.

[72] Eichenbaum H. On the integration of space, time, and memory. *Neuron.* 2017; 95: 1007–18.

[73] Foster T.C. Dissecting the age-related decline on spatial learning and memory tasks in rodent models: N-methyl-D-aspartate receptors and voltage-dependent Ca2+ channels in senescent synaptic plasticity. *Prog. Neurobiol.* 2012; 96: 283–303.

[74] Leal S.L., Yassa M.A. Neurocognitive aging and the hippocampus across species. *Trends Neurosci.* 2015; 38: 800–12.

[75] Murphy G.G., Fedorov N.B., Giese K.P., Ohno M., Friedman E., Chen R., Silva A.J. Increased neuronal excitability, synaptic plasticity, and learning in aged Kvbeta1.1 knockout mice. *Curr. Biol.* 2004; 14: 1907–15.

[76] Radwanska K., Medvedev N.I., Pereira G.S., Engmann O., Thiede N., Moraes M.F., Villers A., Irvine E.E., Maunganidze N.S., Pyza E.M. Mechanism for long-term memory formation when synaptic strengthening is impaired. *Proc. Natl. Acad. Sci. U.S.A.* 2011; 108: 18471–475.

[77] Giese K.P., Aziz W., Kraev I., Stewart M.G. Generation of multi-innervated dendritic spines as a novel mechanism of long-term memory formation. *Neurobiol. Learn. Mem.* 2015; 124: 48–51.

[78] Zhang D., Zhang C., Stepanyants A. Robust associative learning is sufficient to explain the structural and dynamical properties of local cortical circuits. *J. Neurosci.* 2019; 39(35): 6888–904.

[79] Shang J. Local cortical circuit features arise in networks optimized for associative memory storage. *J. Neurosci.* 2020; 40(13): 2590–2. doi: 10.1523/JNEUROSCI.2560-19.2020.

[80] Amin H., Malik A.S. Human memory retention and recall processes. A review of EEG and fMRI studies. *Neurosciences (Riyadh)* 2013; 18(4): 330–44.

[81] Andreas Draguhn. Making room for new memories. *Science.* 2018; 359(6383): 1461–2. doi: 10.1126/science.aat1493.

[82] Norimoto H., Makino K., Gao M., Shikano Y., Okamoto K., Ishikawa T., Sasaki T., Hioki H., Fujisawa S., Ikegaya Y. Hippocampal ripples down-regulate synapses. *Sci.* 2018 30; 359(6383): 1524–7. doi: 10.1126/science.aao0702.

[83] Khodagholy D., Gelinas J.N., Buzsáki G. Learning-enhanced coupling between ripple oscillations in association cortices and hippocampus. *Sci.* 2017; 358(6361): 369–72. doi: 10.1126/science.aan6203.

[84] Langille J.J. Remembering to forget: A dual role for sleep oscillations in memory consolidation and forgetting. *Front. Cell. Neurosci.* 2019; 12; 13: 71. doi: 10.3389/fncel.2019.00071.

[85] Stare C.J., Gruber M.J., Nadel L., Ranganath C., Gómez R.L. Curiosity-driven memory enhancement persists over time but does not benefit from post-learning sleep. *Cogn. Neurosci.* 2018; 9(3–4): 100–15. doi: 10.1080/17588928.2018.1513399.

[86] Gruber M.J., Ranganath C. How curiosity enhances hippocampus-dependent memory: The prediction, appraisal, curiosity, and exploration (PACE) framework. *Trends. Cogn. Sci.* 2019; 23(12): 1014–25. doi: 10.1016/j.tics.2019.10.003.

[87] Lisman J., Buzsáki G., Eichenbaum H., Nadel L., Ranganath C., Redish A.D. Viewpoints: How the hippocampus contributes to memory, navigation, and cognition. *Nat. Neurosci.* 2017; 20(11): 1434–47. doi: 10.1038/nn.4661.

[88] Ryan Splittgerber. In: *Snell's Clinical Neuroanatomy. Introduction and Organization of the Nervous System*, 8th edition. Philadelphia: Wolters and Kluwer. 2019, pp. 305–6.

[89] Krishnakishore G., Venkateshu K.V., Sridevi N.S. Effect of 1800–2100 MHz electromagnetic radiation on learning-memory and hippocampal morphology in Swiss albino mice. *J. Clin. Diagnostic Res.* 2019; 13(2): AC14–7.

[90] Rubin R.D., Watson P.D., Duff M.C., Cohen N.J. The role of the hippocampus inflexible cognition and social behavior. *Front. Hum. Neurosci.* 2014; 8: 742. doi: 10.3389/fnhum.2014.00742.

[91] Creighton S.D., Stefanelli G., Reda A., Zovkic I.B. Epigenetic mechanisms of learning and memory: implications for aging. *Int. J. Mol. Sci.* 2020 Sep 21; 21(18): 6918. doi: 10.3390/ijms21186918. PMID: 32967185; PMCID: PMC7554829.

[92] Hawkins K.A., Emadi N., Pearlson G.D., Taylor B., Khadka S., King D., Blank K. The effect of age and smoking on the hippocampus and memory in late middle age. *Hippocampus.* 2018 Nov; 28(11): 846–9. doi: 10.1002/hipo.23014. Epub 2018 Sep 5. PMID: 30070068.

[93] Ghasemi E., Afkhami Aghda F., Rezvani M.E., Shahrokhi Raeini A., Hafizibarjin Z., Zare Mehrjerdi F. Effect of endogenous sulfur dioxide on spatial learning and memory and hippocampal damages in the experimental model of chronic cerebral hypoperfusion. *J. Basic. Clin. Physiol. Pharmacol.* 2020 Jan 31; 31(3). http//j/jbcpp.2020.31.issue-3/jbcpp-2019-0227/jbcpp-2019-0227.xml. doi: 10.1515/jbcpp-2019-0227. PMID: 32004146.

[94] Hao D., Yang L., Chen S., Tong J., Tian Y., Su B., Wu S., Zeng Y. Effects of long-term electromagnetic field exposure on spatial learning and memory in rats. *Neurol. Sci.* 2013 Feb; 34(2): 157–64. doi: 10.1007/s10072-012-0970-8. Epub 2012 Feb 24. PMID: 22362331.

[95] D'Silva M.H., Swer R.T., Anbalagan J., Bhargavan R. Effect of ultrahigh-frequency radiation emitted from 2G cell phone on developing lens of chick embryo: A histological study. *Adv Anat.* 2014: 1–9.

[96] Al-Khlaiwi T., Meo S.A. Association of mobile phone radiation with fatigue, headache, dizziness, tension and sleep disturbance in Saudi population. Saudi. Med. J. 2004; 25: 732–6.

[97] The World Bank. Mobile cellular subscriptions. Available at: https://data.worldbank.org/indicator/IT.CEL.SETS?year_high_desc=true. Cited date Oct 20, 2018.

[98] The World Bank. Mobile cellular subscriptions (per 100 people). Available at: https://data.worldbank.org/indicator/IT.CEL.SETS.P2. Cited date Oct 22, 2018.

[99] Blog. How many global base stations are there anyway. Available at: www.mobileworldlive.com/blog/blog-global-base-station-count-7m-or-4-times-higher/. Cited date Nov 2, 2018.

[100] World Health Organization (WHO). Clarification of the mooted relationship between mobile telephone base stations and cancer. Available At: www.who.int/mediacentre/news/statements/statementemf/en/. Cited date Oct 1, 2018.

[101] Buckus R., Strukčinskienė B., Raistenskis J., Stukas R., Šidlauskienė A., Čerkauskienė R., Cretescu I. A technical approach to the evaluation of radiofrequency radiation emissions from mobile telephony base stations. *Int. J. Environ. Res. Pub. Health.* 2017; 14: E244.

[102] Meo S.A., Alsubaie Y., Almubarak Z., Almutawa H., AlQasem Y., Hasanato R.M. Association of exposure to radio-frequency electromagnetic field radiation (RF-EMFR) generated by mobile phone base stations with glycated hemoglobin (HbA1c) and risk of type 2 diabetes mellitus. *Int. J. Environ. Res. Pub. Health.* 2015; 12(11): 14519–28.

[103] Zhang J.P., Zhang K.Y., Guo L., Chen Q.L., Gao P., Wang T., Ding G.R. Effects of 1.8 GHz radiofrequency fields on the emotional behavior and spatial memory of adolescent mice. *Int. J. Environ. Res. Health.* 2017; 14: E1344.

[104] World Health Organization. Electromagnetic fields and public health: Mobile phones. Available at: www.who.int/news-room/fact-sheets/detail/electromagnetic-fields-and-public-health-mobile-phones. Cited date Mar 12, 2020.

[105] Abramson M.J., Benke G.P., Dimitriadis C., Inyang I.O., Sim M.R., Wolfe R.S., Croft R.J. Mobile telephone use is associated with changes in cognitive function in young adolescents. *Bioelectromagnetics.* 2009; 30: 678–86.

[106] Schoeni A., Roser K., Röösli M. Memory performance, wireless communication and exposure to radiofrequency electromagnetic fields: A prospective cohort study in adolescents. *Environ. Int.* 2015; 85: 343–51.

[107] Meo S.A., Almahmoud M., Alsultan Q., Alotaibi N., Alnajashi I., Hajjar W.M. Mobile phone base station tower settings adjacent to school buildings: Impact on students' cognitive health. *Am. J. Mens. Health.* 2018: 7: 1557988318816914.

[108] Santini R., Santini P., Le Ruz P., Danze J.M., Seigne M. Survey study of people living in the vicinity of cellular phone base stations. *Electromagn. Biol. Med.* 2003; 22: 41–9.

[109] Lalrinthara P., Lalrntluanga S., Zaithanzauva P., Lalngneia P.C. R.F. Radiation from mobile phone towers and their effects on the human body, Indian. *J. Radio. Space. Phys.* 2014; (43): 186–9.

[110] Deniz O.G., Kaplan S., Selçuk M.B., Terzi M., Altun G., Yurt K.K., Aslan K., Davis D. Effects of short and long term electromagnetic fields exposure on the human hippocampus. *J. Microsc. Ultrastruct.* 2017; 5: 191–7.

[111] Foerster M., Arno Thielens, Wout Joseph, Marloes Eeftens, Martin Röösli. A prospective cohort study of adolescents' memory performance and individual brain dose of microwave radiation from wireless communication. *Environ. Health Perspect.* 2018; 126: 1 13.

[112] Kalafatakis F., Bekiaridis-Moschou D., Gkioka E., Tsolaki M. Mobile phone use for 5 minutes can cause significant memory impairment in humans. *Hell. J. Nucl. Med.* 2017; 20 Suppl: 146–54.

[113] Sauter C., Dorn H., Bahr A., Hansen M.L., Peter A., Bajbouj M., Danker-Hopfe H. Effects of exposure to electromagnetic fields emitted by GSM 900 and WCDMA mobile phones on cognitive function in young male subjects. *Bioelectromagnetics.* 2011; 32: 179–90.

[114] Riddervold I.S., Pedersen G.F., Andersen N.T., Pedersen A.D., Andersen J.B., Zachariae R., Mølhave L., Sigsgaard T., Kjaergaard S.K. Cognitive function and symptoms in adults and adolescents to R.F. radiation from UMTS base stations. *Bioelectromagnetics.* 2008; 29: 257–67.

[115] Haarala C., Ek M., Björnberg L., Laine M., Revonsuo A., Koivisto M., Hämäläinen H. 902 MHz mobile phone does not affect short term memory in humans. *Bioelectromagnetics.* 2004; 25: 452–6.

[116] Preece A.W., Goodfellow S., Wright M.G., Butler S.R., Dunn E.J., Johnson Y., Manktelow T.C., Wesnes K. Effect of 902 MHz mobile phone transmission on cognitive function children. *Bioelectromagnetics.* 2005 (Suppl 7): S138–43.

[117] Eltiti S., Wallace D., Ridgewell A., Zougkou K., Russo R., Sepulveda F., Fox E. Short-term exposure to mobile phone base station signals does not affect cognitive functioning or physiological measures in individuals who report sensitivity to electromagnetic fields and controls. *Bioelectromagnetics.* 2009; 30: 556–63.

[118] Khan M.M. Adverse effects of excessive mobile phone use. *Int. J. Occup. Med. Environ. Health.* 2008; 21: 289–93.

[119] Singh K., Nagaraj A., Yousuf A., Ganta S., Pareek S., Vishnani P. Effect of electromagnetic radiations from mobile phone base stations on general health and salivary function. *J Int Soc Prev Community Dent.* 2016; 6: 54–59.

[120] Calvente I., Pérez-Lobato R., Núñez M.I., Ramos R., Guxens M., Villalba J., Fernández M.F. Does exposure to environmental radiofrequency electromagnetic fields cause cognitive and behavioral effects in 10-yearold boys? *Bioelectromagnetics.* 2016; 37: 25–36.

[121] Thomas S., Benke G., Dimitriadis C., Inyang I., Sim M.R., Wolfe R., Croft R.J., Abramson M.J. Use of mobile phones and changes in cognitive function in adolescents. *Occup. Environ. Med.* 2010; 67: 861–6.

[122] Malek F., Rani K.A., Rahim H.A., Omar M.H. Effect of short-term mobile phone base station exposure on cognitive performance, body temperature, heart rate and blood pressure of Malaysians. *Sci. Rep.* 2015; 5: 13206.

[123] Haarala C., Björnberg L., Ek M., Laine M., Revonsuo A., Koivisto M. Effect of a 902 MHz electromagnetic field emitted by mobile phones on human cognitive function: A replication study. *Bioelectromagnetic.* 2003; 24: 283–8.

[124] Abdel-Rassoul G., El-Fateh O.A., Salem M.A., Michael A., Farahat F., El-Batanouny M. Neurobehavioral effects among inhabitants around mobile phone base stations. *Neurotoxicology.* 2007; 28: 434–40.

[125] Durusoy Raika, Hassoy Hür, Özkurt Ahmet, Karababa Ali Osman. Mobile phone use, school electromagnetic field levels and related symptoms: A cross-sectional survey among 2150 high school students in Izmir. *Environ. Health.* 2017 Jun 2; 16(1): 51. doi: 10.1186/s12940-017-0257-x.

[126] Sage Cindy, Ernesto Burgio. Electromagnetic fields, pulsed radiofrequency radiation, and epigenetics: How wireless technologies may affect childhood development. *Child. Dev.* 2018; 89(1): 129–36. doi: 10.1111/cdev.12824.

[127] Saikhedkar Nidhi, Bhatnagar Maheep, Jain Ayushi, Sukhwal Pooja, Sharma Chhavi, Jaiswal Neha. Effects of mobile phone radiation (900 MHz radiofrequency) on structure and functions of rat brain. Neurol. Res. 2014; 36(12): 1072–9. doi: 10.1179/1743132814Y.0000000392.

[128] Shahin Saba, Banerjee Somanshu, Singh Surya Pal, Chaturvedi Chandra Mohini. 2.45 GHz microwave radiation impairs learning and spatial memory via oxidative/Nitrosative stress-induced p53-dependent/independent hippocampal apoptosis: Molecular basis and underlying mechanism. *Toxicol Sci.* 2015; 148(2): 380–99. doi: 10.1093/toxic/kfv205.

[129] Shahin Saba, Banerjee Somanshu, Singh Surya Pal, Chaturvedi Chandra Mohini. From the Cover: 2.45-GHz microwave radiation impairs hippocampal learning and spatial memory: Involvement of local stress mechanism-induced suppression of iGluR/ERK/CREB signaling. *Toxicol Sci.* 2018; 161(2): 349–74. doi: 10.1093/toxsci/kfx221.

[130] Karimi Narges, Bayat Mahnaz, Haghani Masoud, Saadi Hamed Fahandezh, Ghazipour Gholam Reza. 2.45 GHz microwave radiation impairs learning, memory, and hippocampal synaptic plasticity in the rat. *Toxicol. Ind. Health.* 2018; 34(12): 873–83. doi: 10.1177/0748233718798976.

[131] Ragy M.M. Effect of exposure and withdrawal of 900-MHz-electromagnetic waves on brain, kidney, and liver oxidative stress and biochemical parameters in male rats. *Electromagn. Biol. Med.* 2015; 34(4): 279–84.

[132] Hardell L., Carlberg M., Hedendahl L.K. Radiofrequency radiation from nearby base stations gives high levels in an apartment in Stockholm, Sweden: A case report. *Oncology. Letters.* 2018; 15: 7871–83.

[133] Huber R., Treyer V., Borbély A.A., Schuderer J., Gottselig J.M., Landolt H.P., Achermann P. Electromagnetic fields, such as mobile phones, alter regional cerebral blood flow and sleep and waking E.E.G. *J. Sleep. Res.* 2012; 11: 289–95.

[134] Kim J.H., Yu D.H., Huh Y.H., Lee E.H., Kim H.G., Kim H.R. Long-term exposure to 835MHz RF-EMF induces hyperactivity, autophagy, and demyelination in the cortical neurons of mice. *Sci. Rep.* 2017; 20: 41129.

[135] Kim J.H., Kim H.J., Yu D.H., Kweon H.S., Huh Y.H., Kim H.R. Changes in numbers and size of synaptic vesicles of cortical neurons induced by exposure to 835 MHz radiofrequency-electromagnetic field. *PLoS One.* 2017; 12: e0186416.

[136] García-Lázaro H.G.L., Ramirez-Carmona R., Lara-Romero R., Roldan-Valadez E. Neuroanatomy of episodic and semantic memory in humans: A brief review of neuroimaging studies. *Neurol. India.* 2012; 60(6): 613–7. doi: 10.4103/0028-3886.105196.

Electromagnetic Field Radiation and Autism Spectrum Disorder

6.1 Introduction

Autism spectrum disorder (ASD) is a complex neurodevelopmental condition characterized by "deficits in socialization, communication, and repetitive or unusual behavior" [1, 2] (Figure 6.1). ASD was first labeled as "autism" in 1943 by Leo Kanner, a child psychiatrist at Johns Hopkins University, School of Medicine, Baltimore, USA. He also reported the various major behavioral features associated with communication, social interaction, and tendency toward limited interest and unusual repetitive behaviors [3].

The types of ASD include autism, Asperger's syndrome or Asperger's disorder (AD), and pervasive developmental disorder not otherwise specified (Figure 6.2). In addition, to these disorders, two more pervasive developmental disorders exist, including Rett's syndrome and childhood disintegrative disorder, which are qualitatively distinct from the other subtypes [4]. ASD tends to originate in the first 5 years of childhood and continue into adulthood [5]. However, the wearying effects of ASD continue to exist in the later stages of life, resulting in individuals requiring high levels of lifelong psychological and physical support [6]. Autism has a profound impact on individuals and their families, alongside widespread social and economic effects.

6.2 Prevalence of ASD

ASD is a global health problem in both developing and developed countries. During the last two decades, the global prevalence of ASD increased from 2 to 6/10,000 in the 1990s

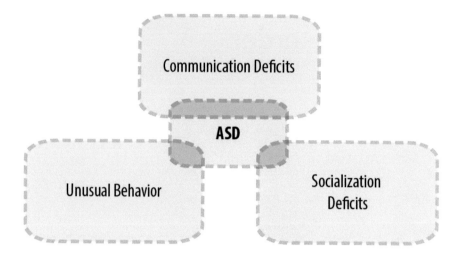

Figure 6.1 Characteristics of autism spectrum disorder (ASD).

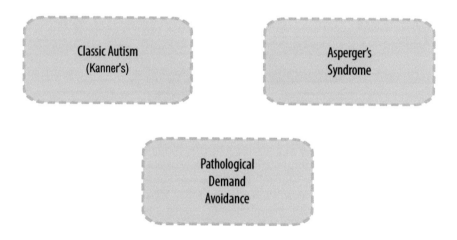

Figure 6.2 Types of autism spectrum disorder.

to current estimates of 260/10,000 or 2.6% [7]. As per the WHO report [5] worldwide, 1 in 160 children has an ASD. This estimation represents an average numeral, and prevalence varies considerably across societies and states. The recent literature identified the increasing prevalence of ASD [5]. The community-based prevalence of autism in the general public varies, although the systematic appraisals and large-scale epidemiological studies demonstrate the estimated rates of between 0.7 and 1.1% [8, 9]. However, current literature estimates that ASD prevalence varies from 2.4 and 9.9% [10]. The occurrence of ASD is four times higher in males than in females [11].

For the most updated information and understanding of the prevalence of ASD, the literature was recorded from various reliable sources. As per the Centers for Disease Control and Prevention Report, USA, 2020, the overall findings indicate considerable variability in ASD prevalence across the communities. The current prevalence of ASD is higher than previous estimates reported from the various networks. In Hispanic children, ASD prevalence was lower than in white or black children. The black and Hispanic children with ASD were evaluated at an older age than white children and were more likely to have cognitive function impairment [12].

Autism affects all the ethnic and socioeconomic groups of both genders residing in developed and developing nations. There is a significant difference in the prevalence proportion between the gender. In March 2020, Maenner et al. (2020) [12] reported that approximately 1 in 54 children is diagnosed with ASD (Figure 6.3). In male children, the prevalence of ASD is 1 in 34, and in girls, ASD is 1 in 144 (Figure 6.4). The girls are four times more likely to be identified with ASD than boys. About 31% of children with ASD have a cognitive function, intellectual function impairment [12]. All these facts show that the prevalence of ASD is swiftly increasing (Figure 6.5).

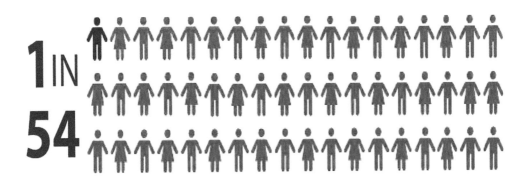

Figure 6.3 Prevalence of autism spectrum disorder among children.

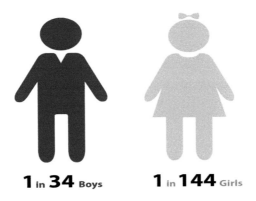

Figure 6.4 Autism spectrum disorder among male and female children.

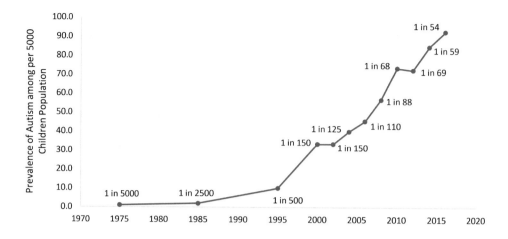

Figure 6.5 Rising pattern of autism spectrum disorder from 1970 to 2016.

6.3 Clinical Characteristics of ASDs

The core clinical characteristics of ASD are mainly linked to poor social interaction, communication with restricted behavior [1, 13–15]. The clinical signs and symptoms are frequently noticeable by the age of 3, but classic language development might delay identifying the condition [1]. Core domains of ASD characteristics present in most of the affected children are shown in Table 6.1 and Figure 6.6.

ASD is a multifactorial disorder; hence, understanding the etiopathology for ASD is highly complex. The numerous factors contribute to the development of ASD, including parents'

Table 6.1 Core Domains of Autism
1. Socialization
• Delayed using non-verbal behaviors
• Poor peer interactions
• No friendships
• Lack of enjoyment
• No social interaction
2. Communication
• Delay in verbal language
• Delay in expressive language and conversation
• Delayed creative and social activities
3. Restricted behavior
• Restricted interest
• Limited routines with perseverative behavior
• Specific repetitive motor gestures
• Self-stimulatory behavior
• Unusual visual exploration
• Disruptive, irritable, aggressive behavior
• Attentional problems
• Cognitive, intellectual disability
• Self-injurious behavior [1, 13–15]

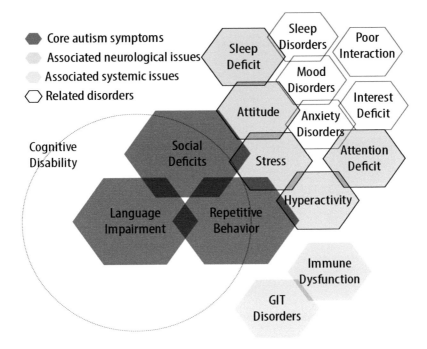

Figure 6.6 Clinical characteristics of ASD.

6.4 Environmental Pollution and ASD

Over the last two decades, the global concern on environmental pollution on human health has become an essential aspect of the public health debate. Researchers focused on discovering the link between environmental pollution and ASD. However, despite decades of extensive research, the etiopathology and increasing prevalence of ASD has not been well-identified [7]. The genetic, epigenetic, electromagnetic field radiation (RF-EMFR), immune abnormalities, and neuroinflammation are the possible factors in the etiopathology of ASD (Figure 6.7). The increasing prevalence of ASD indicates the possible involvement of various pollutants in the environment [16]. Environmental pollution exposure is increasingly being identified as a potential risk factor for ASD. The possibility that the prenatal environment, including maternal exposure to various risk factors such as smoking and alcohol during pregnancy, may affect fetal programming in offspring. Recent research has also linked certain factors associated with environmental pollution, heavy metals, toxic waste, and water pollutants, increasing the risk of ASD [16]. Evidence suggests that environmental factors may contribute to the etiology of the ASD disorder [17].

Alongside the environment, some maternal and paternal factors can cause changes in brain development and result in ASD [18] (Figure 6.8). It has also been demonstrated that non-genetic factors, mainly environmental pollution, particularly heavy metals and particulate matter (PM2.5), have been considered a leading risk factor during prenatal and infant ages. These factors may negatively influence neurodevelopment, and higher center programming significantly increases the risk for ASD [19, 20].

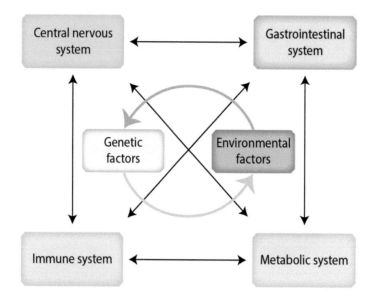

Figure 6.7 Etiopathology of autism spectrum disorder.

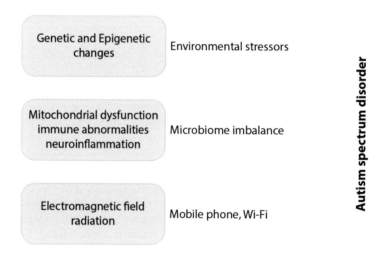

Figure 6.8 Factors associated with autism spectrum disorder.

The increasing prevalence of autism has raised importance in the possible contribution of pollutants in the environment [16]. The environment has a great relationship with the pattern of health and diseases. Environmental exposure has been identified as a risk factor in the rising incidence of ASD. The probability that prenatal environment, including drug, infection, inflammations, tobacco, and alcohol intake during the gestational period may affect fetal programming during the offspring period [21] (Figure 6.9). The recent research has also established an association with "air pollution, pesticide, polychlorinated biphenyls (PCBs), poly-brominated diphenyl ethers (PBDEs), heavy metals, toxic waste, water pollutants and in-house flooring material to increase the risk of ASD [22]. Furthermore, exposure to various chemicals, medical trials, nutritional causes, or strain involving the parents before conception affects the newly

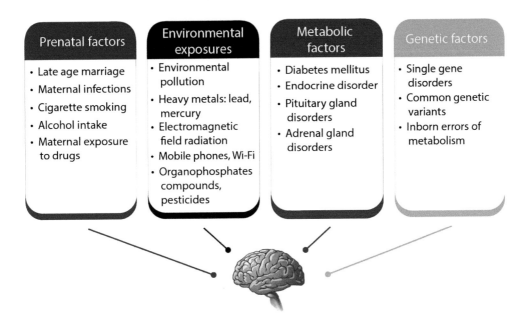

Figure 6.9 Etiopathology of autism spectrum disorder.

born infants. These facts show that multiple environmental factors possibly play a role in the development of ASD disorders" [19].

Flores-Pajot et al., 2016] [16] analyzed seven cohorts and five cross-sectional studies on outdoor air pollution and its association with autism in children. Their meta-analysis findings support the evidence that exposure to ambient air pollution was related to an increased risk of autism. The authors also reported that genetic, environmental factors, air pollution increases the risk of ASD. "It has been believed that neuroinflammatory processes and immune dysfunction linked with autism can result following early-life exposure to various environmental pollutants" [23, 24].

The WHO [25] identifies the RF-EMFR emitted from mobile phones and base-station towers as a part of environmental pollution. Moreover, the current literature established a link between RF-EMFR and ASD. The RF-EMFR is a critical topic; hence, emphasis has been placed on finding an association between RF-EMFR and ASD.

6.5 Electromagnetic Field Radiation and ASD

Recently, some studies have established a link between environmental pollution and oxidative stress with autism, but the primary etiopathology remains unknown. Despite significant developments in the medical sciences, many aspects of the etiopathogenesis of ASD remain obscure, and its prevalence seems to increase intensely without a plausible explanation. There is a debate in the science community to formulate alternative hypotheses for understanding what is behind this phenomenon. In the last two decades, there is a significant change in the environment. In this perspective, one possible hypothesis may be the marked increase in RF-EMFR pollution caused by the massive deployment of modern wireless technologies, including mobile phones and base-station towers (Figure 6.10). The fact that the diffusion of wireless technologies coincides chronologically with ASD prevalence does not necessarily imply a correlation. However,

Figure 6.10 Installation of mobile phone base station towers in a residential area.

the hypothesis that EMFR pollution constitutes at least a predisposing pre-and postnatal factor to ASD deserves to be scrutinized [26].

In the 21st century, there have been significant changes in the pattern of health and disease. The pattern of life has been shifting toward technology-based living standards. The mobile phone industry is the fastest growing industry globally; it provides mobile communications and allied services. In recent times, cell phones have become a necessity of daily life. As per the World Bank report, about 7.68 billion people own a mobile phone, which is more than the world's population combined, portraying how most individuals possess more than one cell phone. The World Bank also reported that the number of mobile phone subscribers is 104.45% of the world population [27]. The telecom industry placed over 7 million base-station towers worldwide, and the figure is markedly increased with the introduction of third-generation technology [28]. The WHO [25] recognizes RF-EMFR emitted from mobile phones and base-station towers as a part of environmental pollution and a possible cause of various debilitating diseases, including cancer [25]. Mobile phones generate 900–1,800- and 2,100-MHz radiofrequency radiation, a growing concern about the effects of radiofrequency radiation [29].

The RF-EMFR has become a leading cause of persistent electromagnetic pollution. The primary source for RF-EMFR that contaminates the atmosphere is regular devices such as mobile phones, tablets, laptops, computers, television, and wireless devices [30]. RF-EMFR is a known pollutant that affects various elements of the environment [30].

More recently, many studies identified that RF-EMFR alters brain biology and affects various organs, including the nervous system. The research links these exposures to increased oxidative stress [29], cellular damage, mitochondria dysfunction [31], genotoxicity, and BBB disruption [32].

The RF-EMFR exposure increases stress at the cellular, biochemical, immune, and electrical levels. Infants are more vulnerable as their skull bones and brain tissues are in the growing and developing age, so the chances of neurological impairments are much more significant. With the globally increasing prevalence of ASD, it is vital to understand the mechanisms that have contributed to ASD. Depending on genetic or environmental risk factors, ASD either develops before or after birth, affecting brain development [26].

The RF-EMFR generated from mobile phones, base-station towers, and other wireless sources cause potentially harmful health impacts. The recent literature demonstrates that

RF-EMFR causes impaired neurodevelopmental and neurobehavioral changes. The most frequent symptoms include memory impairment, reduced learning, cognition capabilities [33], and behavioral problems. The clinical presentation was similar to ASD and attention deficit hyperactivity disorders due to RF-EMF exposures, where epigenetic drivers and genetic (DNA) damage are the most probable contributors [34].

Thornton (2006) [35] identified that RF-EMFR generated from mobile phones is the leading source of temporal noise in the environment and reported that temporal interruption from the atmosphere plays a negative role in neuronal dysfunction, which may cause a pattern of deficits associated with ASD. The author also suggested that the developing nervous system is highly susceptible and prone to temporal noise in the infant age that can impede the initial calibration of neuronal networks. The mirror neuron system and temporal disruption might delay or disrupt vital calibration processes [35]. It is also not an over-statement to call mobile phones "a major source of distraction for parents," especially during the infant's initial days of life when most care is needed. By diminishing their focus, eye contact, and communication, parents miss emerging cues in an attempt to establish reciprocal relationships. It impairs the development of combined attention between infants and their caregivers in a critical developmental stage and causes developmental delay with ASD features.

It is well acknowledged in the literature that RF-EMFR enhances cell apoptosis and activates oxidative stress. It has also been reported that during the gestational period, exposure to chemicals triggers oxidative stress, which then strengthens autistic-like traits, including delayed motor function and increased vocalization rate. Moreover, apart from oxidative stress, dysregulated immune responses during the fetal period can participate in the pathogenesis of ASD [36].

6.6 Mechanism: RF-EMFR and ASD

Although the epidemiological evidence is limited and findings are debatable, various biological mechanisms are proposed to reveal that RF-EMFR causes brain malformation and aberrant neurodevelopment. There are multiple mechanisms directly and indirectly involved through which RF-EMFR which can cause ASD (Figure 6.11). The most probable mechanism behind this is RF-EMFR impairing cell membrane biology and making the cell membranes leaky,

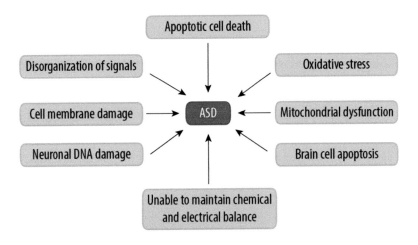

Figure 6.11 Mechanism through which RF-EMFR can cause ASD.

Table 6.2 RF-EMFR Affect Brain Function and Cause Several Neurological Disorders

- RF-EMFR
- Cell membrane damage
- Unable to maintain chemical and electrical balance
- Oxidative stress
- DNA strand breaks in the brain
- Brain cell apoptosis
- Mitochondrial dysfunction
- Entropy, disorganization of signals in the brain
- Radiation-induced apoptotic cell death

[32–37, 39–41]

preventing the cell from maintaining a chemical and electrical balance between intracellular and extracellular body fluids and their ionic concentration [37, 38]. RF-EMFR affects brain function and causes neurological disorders (Table 6.2). The mechanism behind this is oxidative stress, cellular stress, mitochondrial dysfunction, production of heat shock proteins, DNA strand breaks in the brain, and brain cell apoptotic death. These basic mechanisms are the leading cause of neurological disorders and ASD in children [38].

When mitochondria are inefficient, electrical communication and synapses in the brain cannot perform normal functions and cause neuronal function impairment. RF-EMFR can damage the DNA, proteins-rich complex integration of brain waves, increase the entropy, and deregulate the signals in the brain that become less synchronized or coordinated. These mechanisms reduce the coordination in brain biology and have been identified in ASDs.

6.7 RF-EMFR, Male Sperm Cell Biology, and ASD

The effects of RF-EMFR emitted by mobile phone devices on male reproductive biology are currently subject to controversy. RF-EMFR, non-ionizing radiation, impairs the male liver [42] and sperm cell biology [43]. The RF-EMFR exposure is considered environmental pollution, affecting the sperm cell biology, which could be responsible for causing autism to increase to an epidemic level today [39].

The ROS are the most fascinating potentially toxic by-products, responsible for interrupting various biological and biochemical functions. The low ROS levels play an essential role as second messengers, activating and signaling the cascades. This process causes cell proliferation and other cell dysfunctions. However, uncontrolled ROS production or impaired ROS scavenging causes oxidative damage to membrane phospholipids, nucleic acids, and proteins. These mechanisms are involved in the disruption of normal cellular processes and also triggers severe cell dysfunctions. The literature demonstrated that exposure of the cell to a 50-Hz RF-EMFR induced an increase in intracellular ROS levels [44–47], triggering a temporary mitochondrial permeability transition [46].

The literature identified that RF-EMR induces mitochondrial dysfunction leading to elevated ROS, decreased sperm cell motility, and increased DNA damage [48, 49]. It has been reported that the long-term exposure to RF-EMFR generated during mobile communication can cause "hypospermatogenesis and maturation arrest in the spermatozoa" [43].

The literature demonstrates an incontrovertible link between RF-EMFR and impairment of cellular mechanism. It enhances mitochondrial ROS, minimizing the "motility and vitality of sperm cells, DNA fragmentation in spermatozoa and defects in embryonic development" [43]. These findings have substantial implications for the safety of extensive mobile phone use, potentially affecting fertility and the offspring's health and well-being [50].

The magnitude of paternal damaged DNA has been linked with ASD. A child might inherit slightly fragmented DNA, which may impair brain functions; however, moderate fragmentation from their father has lifelong social, cognitive, and mental health problems and could lead to a child with social unpredictability in society. The current literature demonstrates that exposure to RF-EMFR is now contributing to behavioral problems [51–54] and the development of autism disorder in children.

6.8 RF-EMFR: Pregnancy and Autism

Although the literature established a causal link with the effects of RF-EMFR generated from cell phones on ASD, the literature highlights that early embryos are highly susceptible to exposure to environmental pollution, including RF-EMFR. The adverse effects during pregnancy on early fetal development are discussed in the literature. The RF-EMFR could conceivably result in fetal biological impairment. The RF-EMFR can impair the biological and molecular mechanism during the neurodevelopmental stages. The pregnant female is easily exposed to RF-EMFR. The distance from outside the body to inside the uterus is very close, and fetal tissues are easily exposed to low-frequency EMR [34]. These radiations penetrate the pregnant female and affect fetal development.

Moreover, during the pregnancy, the mother's exposure to environmental pollution and RF-EMFR increases the incidences of fetal damage [55]. The fetus may not be well protected from RF-EMFR during the intrauterine period, including pelvic tissue arrangements and amniotic fluid. The pelvic structure permits deep penetration of the RF-EMFR to be absorbed within the developing fetus. It has been reported that the incidence of ASD has been increased since the invention of the mobile phone. It shows a relationship between the neurotoxicity of cell phone radiation and ASD [56, 57]. The potential mechanisms involved in the pathogenesis of ASD are oxidative stress, neuroinflammation, glutamate excitotoxicity. These mechanisms increase the prevalence of autism. It is a well-established fact that exposure to RF-EMFR can cause oxidative stress in human and animal models and cause autism [58]. The low-intensity RF-EMFR generated from a cellular phone is a common environmental risk factor for pregnant females, fetuses, infants, children, and adults. The cell phone radiations enhance the formation of free radicals through the Fenton reaction as the catalytic process. Through this process, iron converts hydrogen peroxides, a product of oxidative respiration in the mitochondria, into hydroxyl free radical, which is very potent and can induce damage of macromolecules, such as membrane phospholipids, DNA, and protein. The RF-EMFR impairs mitochondrial metabolism and modulates glutathione, glutamate, and gamma-aminobutyric acid (GABA), which are substances related to the pathophysiology of autism [59–61].

Fragopoulou et al. (2012) [62] conducted a study on proteomic analysis of brain regulatory proteins and their association with exposure to EF-EMF. The authors found that these RF-EMFR led to either downregulation or overexpression of 143 proteins [62]. The altered proteins include neural function-related proteins, alpha-synuclein, glia maturation factor-beta, cytoskeletal proteins, heat shock proteins, apolipoprotein E, brain metabolism proteins, aspartate 64 neurotoxins aminotransferase, and glutamate dehydrogenase. The authors pointed out that oxidative stress was consistent with some changes in proteomic markers. Alteration in blood and brain glutathione status and deficiencies of reduced glutathione is increasingly associated with autism. The literature demonstrates that antioxidants, including vitamin C and vitamin E, decrease oxidative impacts on the endometrium. This is why vitamin C and E decrease the adverse effects of exposure to EMF of 900 MHz on the endometrium [63, 64].

The near-field RF-EMFR has been the most extensive study relative to mobile phones, base stations, and other sources of RF-EMFR. In this new era of technology, infants, children, and adults all use mobile phones and other sources of wireless communications, hence are exposed to increasing near-field and far-field RF-EMFR. Many people are continuously under-exposure to RF-EMR while living in or near the vicinity of base-station towers. These people are more prone to develop autism in their children.

The first mechanism occurs when mitochondria dysfunction triggers cellular dysfunction, and the BBB is ruptured following the EMR exposure. The second mechanism is interference with DNA following EMR exposure. The third step involves alterations in mRNA folding and consequent "transcription of under stress" condition causing the structure of mitotic daughter cells to be altered. These mechanisms demonstrate an environmentally induced genetic change that could elucidate the possible pathology present in autistic children. From biological and clinical disease perspectives, these pathways affect all critical levels of neuro-behavioral functioning. The compound effect of cellular dysfunction due to EMR, disruption of intercellular mechanism, and cell organelles dysfunction compromise the biology and physiology of the cells. Also, RF-EMFR present in the environment shows that these fields can pass through the cell membrane and cause intracellular heating. These findings suggest that RF-EMFR contributes to the etiopathology of ASD in conjunction with genetic and environmental factors and offers a mechanical explanation for the association between concurrent increases in the incidence of autism and exposure to RF-EMFR [55].

ASDs involve multi-level factors of underlying biology that match the physiological impacts of RF-EMFR exposures. The children with ASD have identified oxidative stress, free radical cellular injury, cellular stress proteins, and increased intracellular calcium due to genetics or environmental exposures. The various vital but vulnerable mechanisms are disrupted by RF-EMFR [40–41]. The science literature established a pathophysiological involvement in cellular processes allied with ASD and biological effects of RF-EMFR exposure contributing to persistently disrupted homeostasis. The literature established to support dramatic increases in the prevalence of ASDs parallel in time and biological and physiological mechanisms with the excessive use of RF-EMFR generated from mobile phone and other wireless technologies [40–41].

RF-EMFR is generated from mobile phones, their base-station towers, wireless communication, and other technologies. Evidence in the literature suggests that prolonged exposure to RF-EMFR has serious health effects, and RF-EMFR is associated with a risk of ASD [65]. Because environmental health scientists try to control various global severe issues, including climatic change and chemical toxicants in public health, a crucial need has arisen to address electrosmog. A genuine approach based on evidence is needed for the risk assessment and regulation of RF-EMFR, which will help the global population. A few governmental health authorities have recently taken directives to minimize public exposure to RF-EMFR by regulating children's mobile phone usage and wireless allied devices. However, this should be a coordinated and sincere international effort so that it can benefit the population.

Conclusion

ASDs are lifelong, devastating neurodevelopmental disorders presenting in early childhood that severely impair the affected individuals' social abilities and personal autonomy. During the last two decades, there has been an increase in the global prevalence of ASD. It is a multifactorial disorder, and multiple conditions contribute to the development of ASD, including parents' age, health status, mother, fetal and childhood exposures, underlying genetic, diet, and environmental causes that range from risk to develop ASD. The increasing prevalence of ASD indicates the possible involvement of various pollutants in the environment and RF-EMFR. None can deny that mobile phones and their allied services have become essential for daily activities, but while connecting individuals to the rest of the world, this technology can create gaps between family members living under one roof due to a lack of interaction amongst themselves. The excessive use of cell phones and social media minimizes parents' attention toward their young children, preventing them from learning or practicing social skills, leading to children developing autistic features amongst vulnerable groups. These findings propose that ASD is multifactorial, involving complex interactions between genetics, environmental, and RF-EMFR. All these factors act synergistically during crucial neurodevelopment periods and increase the prevalence of ASD.

References

[1]   Levy S.E., Mandell D.S., Schultz R.T. Autism. *The. Lancet.* 2009; 374: 1627–38. doi: 10.1016/S0140-6736(09)61376-3.

[2]   Rapin I. The autistic-spectrum disorders. *N Engl J Med.* 2002; 347: 302–3.

[3]   Kanner L. Autistic disturbances of affective contact. *Nerv. Child.* 1943: 217–50.

[4]   Witwer A.N., Lecavalier L. Examining the validity of autism spectrum disorder subtypes. *J. Autism. Dev. Disord.* 2008; 38(9): 1611–24. doi: 10.1007/s10803-008-0541-2.

[5]   World Health Organization. WHO. Autism spectrum disorders. Available at: www.who.int/news-room/fact-sheets/detail/autism-spectrum-disorders. Cited date Dec 12, 2108.

[6]   Howlin P., Goode S., Hutton J., Rutter M. Adult outcome for children with autism. *J. Child Psychol. Psychiatry.* 2004; 45(2): 212–29. doi: 10.1111/j.1469-7610.2004.00215.x.

[7]   May T., Sciberras E., Brignell A., Williams K. Autism spectrum disorder: Updated prevalence and comparison of two birth cohorts in a nationally representative Australian sample. *BMJ. Open.* 2017; 7(5): e015549.

[8]   Baxter A.J., Brugha T.S., Erskine H.E., Scheurer R.W., Vos T., Scott J.G. The epidemiology and global burden of autism spectrum disorders. *Psychol. Med.* 2015; 45(3): 601–13.

[9]   Brugha T.S., Spiers N., Bankart J., Cooper S.A., McManus S., Scott F.J., Smith J., Tyrer F. Epidemiology of autism in adults across age groups and ability levels. *Br. J. Psychiatry.* 2016; 2009(6): 498–503.

[10] Tromans S., Chester V., Kiani R., Alexander R., Brugha T. The prevalence of autism spectrum disorders in adult psychiatric inpatients: A systematic review. *Clin. Pract. Epidemiol. Ment. Health.* 2018; 14: 177–87.

[11] Mosca-Boidron A.L., Gueneau L., Huguet G., Goldenberg A., Henry C., Gigot N. A de novo microdeletion of SEMA5A in a boy with autism spectrum disorder and intellectual disability. *Eur. J. Hum. Genet.* 2016; 24(6): 838–43. doi: 10.1038/ejhg.2015.211.

[12] Maenner M.J., Shaw K.A., Baio J., Washington A., Patrick M., DiRienzo M., Christensen D.L., et al. Prevalence of autism spectrum disorder among children aged 8 years-autism and developmental disabilities monitoring network, 11 sites, the United States., 2016. US department of health and human services/centers for disease control and prevention. *MMWR.* 2020; 69(4): 1–12.

[13] Newschaffer C.J., Croen L.A., Daniels J. The epidemiology of autism spectrum disorders. *Annu. Rev. Public. Health.* 2007; 28: 235–58.

[14] Hartley S.L., Sikora D.M., McCoy R. Prevalence and risk factors of maladaptive behavior in young children with autistic disorder. *J. Intellect. Disabil. Res.* 2008; 52: 819–29.

[15] Limoges E., Mottron L., Bolduc C., Berthiaume C., Godbout R. Atypical sleep architecture and the autism phenotype. *Brain.* 2005; 128: 1049–61.

[16] Flores-Pajot M.C., Ofner M., Do M.T., Lavigne E., Villeneuve P.J Childhood autism spectrum disorders and exposure to nitrogen dioxide, and particulate matter air pollution: A review and meta-analysis. *Environ. Res.* 2016; 151: 763–76.

[17] Rossignol D.A., Frye R.E. A review of research trends in physiological abnormalities in autism spectrum disorders: Immune dysregulation, inflammation, oxidative, stress, mitochondrial dysfunction, and environmental toxicant exposures. *Mol. Psychiatry.* 2012; 17: 389–401.

[18] Chen G., Jin Z., Li S., Jin X., Tong S., Liu S., Yang Y., Huang H., Guo Y. Early life exposure to particulate matter air pollution (PM1, PM2.5, and PM10) and autism in Shanghai, China: A case-control study. *Environ. Int.* 2018; 121(Pt 2): 1121–27. doi: 10.1016/j.envint.2018.10.026.

[19] Gadad B.S., Hewitson L., Young K.A., German D.C. Neuropathology and animal models of autism: Genetic and environmental factors. *Autism. Res. Treat.* 2013: 731935. doi: 10.1155/2013/731935.

[20] Lyall K., Schmidt R.J., Hertz-Picciotto I. Maternal lifestyle and environmental risk factors for autism spectrum disorders. *Int. J. Epidemiol.* 2014; 43(2): 443–64. doi: 10.1093/ije/dyt282.

[21] Reynolds L.C., Inder T.E., Neil J.J., Pineda R.G., Rogers C.E. Maternal obesity and increased risk for autism and developmental delay among very preterm infants. *J. Perinatol.* 2014; 34: 688–92. doi: 10.1038/jp.2014.80.

[22] Rossignol D.A., Genuis S.J., Frye R.E. Environmental toxicants, and autism spectrum disorders: A systematic review. *Transl. Psychiatry.* 2014; 4: e360. doi: 10.1038/tp.2014.4.

[23] Hertz-Picciotto I., Park H.Y., Dostal M., Kocan A., Trnovec T., Sram R. Prenatal exposures to persistent and non-persistent organic compounds and effects on immune system development. *Basic. Clin. Pharmacol. Toxicol.* 2008; 102: 146–54. doi: 10.1111/j.1742-7843.2007.00190.x.

[24] Abrahams B.S., Geschwind D.H. Advances in autism genetics: On the threshold of a new neurobiology. *Nat. Rev. Genet.* 2008; 9(5): 341–55.

[25] WHO. Clarification of the mooted relationship between mobile telephone base stations and cancer. Available At: www.who.int/mediacentre/news/statements/statementemf/en/. Cited date Oct 1, 2019.

[26] Posar A., Visconti P. To what extent do environmental factors contribute to the occurrence of autism spectrum disorders? *J Pediatr Neurosci.* 2014; 9(3): 297–8. doi: 10.4103/1817-1745.147610.

[27] The World Bank. Mobile cellular subscriptions (per 100 people). Available at: https://data.worldbank.org/indicator/IT.CEL.SETS.P2. Cited date Oct 22, 2020.

[28] Blog. How many global base stations are there anyway. Available at: www.mobileworldlive.com/blog/blog-global-base-station-count-7m-or-4-times-higher/. Cited date Nov 2, 2018.

[29] Alkis M.E., Bilgin H.M., Akpolat V., Dasdag S., Yegin K., Yavas M.C., Akdag M.Z. Effect of 900-, 1800-, and 2100-MHz radiofrequency radiation on DNA and oxidative stress in the brain. *Electromagn. Biol. Med.* 2019; 38(1): 32–47. doi: 10.1080/15368378.2019.1567526.

[30] Redlarski G., Lewczuk B., Żak A., Koncicki A., Krawczuk M., Piechocki J., Jakubiuk K., Tojza P., Jaworski J., Ambroziak D.3., Skarbek Ł.3., Gradolewski D.3. The influence of electromagnetic pollution on living organisms: Historical trends and forecasting changes. *Biomed. Res. Int.* 2015; 2015: 234098. doi: 10.1155/2015/234098.

[31] Santini S.J., Cordone V., Falone S., Mijit M., Tatone C., Amicarelli F., Di Emidio G. Role of Mitochondria in the oxidative stress induced by electromagnetic fields: Focus on reproductive systems. *Oxid. Med. Cell. Longev.* 2018; 2018: 5076271. doi: 10.1155/2018/5076271.

[32] Tang J., Zhang Y., Yang L., Chen Q., Tan L., Zuo S., Feng H., Chen Z., Zhu G. Exposure to 900 MHz electromagnetic fields activates the map-1/ERK pathway and causes blood–brain barrier damage and cognitive impairment in rats. *Brain. Res.* 2015; 1601: 92–101.

[33] Meo S.A., Almahmoud M., Alsultan Q., Alotaibi N., Alnajashi I., Hajjar W.M. Mobile phone base station tower settings adjacent to school buildings: Impact on students' cognitive health. *Am. J. Mens. Health.* 2019; 13(1): 1557988318816914. doi: 10.1177/1557988318816914.

[34] Sage C., Burgio E. Electromagnetic fields, pulsed radiofrequency radiation, and epigenetics: How wireless technologies may affect childhood development. *Child. Dev.* 2018; 89(1): 129–36. doi: 10.1111/cdev.12824.

[35] Thornton I.M. Out of time: A possible link between mirror neurons, autism, and electromagnetic radiation. *Med. Hypotheses.* 2006; 67(2): 378–82.

[36] Desai N.R., Kesari K.K., Agarwal A. Pathophysiology of cell phone radiation: Oxidative stress and carcinogenesis focus on the male reproductive system. *Reprod. Biol. Endocrinol.* 2009; 7: 114.

[37] Beneduci A., Filippelli L., Cosentino K., Calabrese M.L., Massa R., Chidichimo G. Microwave induced shift of the main phase transition in phosphatidylcholine membranes. *Bioelectrochemistry.* 2012; 84: 18–24.

[38] Kesari K.K.L., Meena R., Nirala J., Kumar J., Verma H.N. Effect of 3G cell phone exposure with the computer-controlled 2-D stepper motor on non-thermal activation of the hsp27/p38MAPK stress pathway in rat brain. *Cell. Biochem. Biophys.* 2014; 68(2): 347–58.

[39] Aitken R.J., De Iuliis G.N., McLachlan R.I. Biological and clinical significance of DNA damage in the male germline. *Int. J. Androl.* 2009; 32(1): 46–56. doi: 10.1111/j.1365-2605.2008.00943.x.

[40] Herbert M.R., Sage C. Autism and EMF? Plausibility of a pathophysiological link—Part I. *Pathophysiology.* 2013; 20(3): 191–209. doi: 10.1016/j.pathophys.2013.08.001.

[41] Herbert M.R., Sage C. Autism and EMF? Plausibility of a pathophysiological link part II. *Pathophysiology.* 2013; 20(3): 211–34. doi: 10.1016/j.pathophys.2013.08.002.

[42] Meo S.A., Arif M., Rashied S., Husain S., Khan M.M., Al-Masri A., Usmani A.M., Husain A., Al-Drees A. Morphological changes induced by mobile phone radiation in liver and pancreas in Wistar albino rats. *Eur. J. Anat.* 2010; 14(3): 105–9.

[43] Meo S.A., Arif M., Rashied S., Khan M.M., Vohra M.S., Usmani A.M., Imran M.B., Al-Drees A.M. Hypospermatogenesis and spermatozoa maturation arrest in rats induced by mobile phone radiation. *J. Coll. Physicians. Surg. Pak.* 2011; 21(5): 262–5. doi: 05.2011/JCPSP.262265.

[44] Ay$e I.G.İ., Zafer A., $ule O., I$il I.$.T., Kalkan T. Differentiation of K562 cells under ELF-EMF applied at different time courses. *Electromagn. Biol. Med.* 2010; 29(3): 122–30.

[45] Mannerling A.C., Simkó M., Mild K.H., Mattsson M.O. Effects of 50-Hz magnetic field exposure on superoxide radical anion formation and HSP70 induction in human K562 cells. *Radiat. Environ. Bioph.* 2010; 49(4): 731–41.

[46] Patruno A., Tabrez S., Pesce M., Shakil S., Kamal M.A., Reale M. Effects of extremely low-frequency electromagnetic field (ELF-EMF) on catalase, cytochrome P450 and nitric oxide synthase in erythro-leukemic cells. *Life. Sci.* 2015; 121: 117–23.

[47] Calcabrini C., Mancini U., De Bellis R. Effect of extremely low-frequency electromagnetic fields on antioxidant activity in the human keratinocyte cell line NCTC 2544. *Biotechnol. Appl. Biochem.* 2017; 64(3): 415–22. doi: 10.1002/bab.1495.

[48] Feng B., Qiu L., Ye C., Chen L., Fu Y., Sun W. Exposure to a 50-Hz magnetic field induced mitochondrial permeability transition through the ROS/GSK-3β signaling pathway. *Int. J. Radiat. Biol.* 2016; 92(3): 148–55.

[49] Houston B.J., Nixon B., King B.V., De Iuliis G.N., Aitken R.J. The effects of radiofrequency electromagnetic radiation on sperm function. *Reproduction.* 2016; 152(6): R263–R276.

[50] De Iuliis G.N.L., Newey R.J., King B.V., Aitken R.J. Mobile phone radiation induces reactive oxygen species production and DNA damage in human spermatozoa in vitro. *PLoS One.* 2009; 4(7): e6446. doi: 10.1371/journal.pone.0006446.

[51] Hozefa A., Divan I., Leeka K., Carsten O., Jørn O. Prenatal and postnatal exposure to cell phone use and behavioral problems in children. *Epidemiology.* 2008; 19(4): 523–32. doi: 10.1097/EDE.0b013e318175dd47.

[52] Tamir S.A., Geliang G., Xiao-Bing G., Hugh S.T. Fetal Radiofrequency radiation exposure from 800–1900 MHZ-rated cellular telephones affects neurodevelopment and behavior in mice. *Sci. Rep.* 2012; 2: 312. doi: 10.1038/srep00312.

[53] Jiang Zhaocai, Mingyan Shi. Prevalence and co-occurrence of compulsive buying, problematic internet and mobile phone use in college students in Yantai, China: Relevance. Of. BMC. *Public. Health.* 2016; 16(1): 1211. doi: 10.1186/s12889-016-3884-1.

[54] Sahu M., Gandhi S., Sharma M.K. Mobile phone addiction among children and adolescents: A systematic review. *J. Addict. Nurs.* 2019; 30(4): 261–8.

[55] Tamara J.M., George L.C. Wireless radiation in the etiology and treatment of autism: Clinical observations and mechanisms. *J Aust Coll Nutr & Env Med.* 2007; 26(2): 3–7.

[56] Byrd R.H., Hribar M.E., Nocedal J. An interior-point algorithm for large-scale nonlinear programming. *SIAM J. Optim.* 1999; 9(4): 877–900.

[57] Bertrand J., Mars A., Boyle C., Bove F., Yeargin-Allsopp M., Decoufle P. Prevalence of autism in a United States population: The brick township, New Jersey, investigation. *Pediatrics.* 2001; 108(5): 1155–61. doi: 10.1542/peds.108.5.1155.

[58] Phillips J.L., Singh N.P., Lai H. Electromagnetic fields, and DNA damage. *Pathophysiology.* 2009; 16(2): 79–88.

[59] Brown H.M., Oram-Cardy J., Johnson A. A meta-analysis of the reading comprehension skills of individuals on the autism spectrum. *J. Autism Dev. Disord.* 2013; 43(4): 932–55.

[60] Choudhury P.R., Lahiri S., Rajamma U. Glutamate mediated signaling in the pathophysiology of autism spectrum disorders. Pharmacology. *Biochemistry and. Behavior.* 2012; 100(4): 841–9.

[61] Essa M.M., Braidy N., Waly M.I., Al-Farsi M., et al. Impaired antioxidant status and reduced energy metabolism in autistic children. *Res. Autism Spectr. Disord.* 2013; 7(5): 557–65.

[62] Fragopoulou A.F., Samara A., Antonelou M.H. Brain proteome response following whole-body exposure of mice to mobile phone or wireless DECT base radiation. *Electromagn. Biol. Med.* 2012; 31(4): 250–74.

[63] Guney M., Ozguner F., Oral B., Karahan N., Mungan T. 900 MHz radiofrequency-induced histopathologic changes and oxidative stress in rat endometrium: Protection by vitamins E and C. *Toxicol. Ind. Health.* 2007; 23(7): 411–20.

[64] Glinton K.E., Elsea S.H. Untargeted metabolomics for autism spectrum disorders: Current status and future directions. *Front. Psychiatry.* 2019; 10: 647. doi: 10.3389/fpsyt.2019.00647.

[65] Baste V., Oftedal G., Møllerløkken O.J., Mild K.H., Moen B.E. Prospective study of pregnancy outcomes after parental cell phone exposure: The norwegian mother and child cohort study. *Epidemiology.* 2015; 26(4): 613–21.

Electromagnetic Field Radiation: A Risk to Develop Brain Cancer

7.1 Electromagnetic Field Radiation (EMFR)

Electrical appliances have achieved rapid developments during the last three decades, especially with wireless equipment and smartphones. The smartphone technology has left behind many other essential and frequently used electronic devices, such as radio, telephone, telegraph, fax, documents scanning machine, etc. With the rapid proliferation of wireless technologies, smartphones, and installing MPBSTs in residential and thickly populated areas, the public has felt concerned over RF-EMFR generated from these novel technology devices. For the thoughtful understanding of RF-EMFR and its impact on brain biology, it is vital to comprehend the electromagnetic field and its natural and human-made sources.

7.1.1 Electromagnetic Field (EMF)

Electromagnetic field (EMF) is the unification of invisible electric and magnetic fields of forces. The EMF is generated through natural phenomena and daily-life activities of human beings with electricity. EMF is a fragment of the physical electromagnetic spectrum that extends from the extreme high of powerful gamma rays to the low of static magnetic fields. When current flows, electric and magnetic fields both rise. The power or intensity of electric fields is closely associated with the force of flow; however, the strength of magnetic fields is proportional to the rate of flow [1].

Electric fields are generated by differences in voltage, "higher the voltage, stronger is the resultant field." While magnetic fields' production occurs when an electric current flows, "greater the current, stronger is the magnetic field." An electric field exists

DOI: 10.1201/9781003212461-7   129

even when no current is flowing. While current flows, the strength of the electric field remains constant while the magnetic field's strength fluctuates with power consumption [2]. A moving charge always has both a magnetic and an electric field. They are two different fields with nearly the same characteristics. Therefore, they are interrelated in a field called the electromagnetic field [3].

RF-EMFR is the transfer of energy by radiofrequency waves. The RF-EMFR is the frequency or the consistent wavelength with different frequencies that behave with the human body in various ways. Electromagnetic waves are a series of consistent waves that travel at incredible speeds, approximately at the speed of light. The "frequency" defines "the number of oscillations per second," while the "wavelength" demonstrates "the distance between one wave point to the next wave point." Therefore, the wavelength and frequency are unable to separate. The higher the frequency, the shorter the wavelength. RF-EMFR is produced from both natural and artificial sources [2].

7.1.2 Natural Sources of Electromagnetic Field Radiation

Although RF-EMFR exists everywhere in the surroundings, it is mainly invisible to human senses, including vision, hearing, smell, taste, and touch. The natural sources which emit low levels of RF-EMFR include the sun, earth, rocks, lodestone, and mountains (Table 7.1; Figure 7.1). Electric fields are generated when electric charges build up in the atmosphere associated with thunderstorms. The earth's magnetic field causes a compass needle to orient itself in a north–south direction and is also used for navigation by fish and birds (WHO) [2, 3].

7.1.3 Human-Made Sources of Electromagnetic Field Radiation

In this modern era of science and technology, the exposure of human beings to human-made EMFs has dramatically increased. The RF-EMFR sources are accompanied by public

Figure 7.1 Natural sources of electromagnetic field radiation.

Table 7.1 Natural Sources of Electromagnetic Field Radiation [4]	
Cosmic radiation	Sun Earth Atmosphere "Celestial events in the universe."
Terrestrial radiation	Earth's crust "Natural deposits of uranium, potassium, and thorium." minerals, building materials, etc.
Inhalation	Radon: The decay of uranium-238 produces radioactive gas, which moves from the ground to the surrounding atmosphere. Thoron: Gas produced by thorium. Radon and thoron largely depend on location and the composition of soil and rock.
Ingestion	Some traces of radioactive minerals may be present in some food and drinking water. Moreover, vegetables grow in soil and water, which contain radioactive minerals. Potassium-40 and carbon-14 may contain non-radioactive isotopes.

Figure 7.2 Magnetic Mountain, Saudi Arabia: natural source of electromagnetic field radiation.

debate about possible adverse effects on human health. Besides the natural sources, the EMFR is also generated by human-made sources such as power sockets, radio, television, computers, laptops, antennas, broadcasting devices, mobile phones, wireless networks, cordless phones, and satellite communication devices (Figure 7.2). The other sources of RF-EMFR include microwave ovens, radar, radio stations, or MPBSTs [2]. These all can generate non-ionizing and ionizing EMFR.

7.2 Ionizing and Non-Ionizing Electromagnetic Field Radiation

The EMF properties are determined by the wavelength and frequency of the waves. The electromagnetic waves are carried by particles which are called quanta. The quanta of higher frequency waves possess more significant levels of energy than lower frequency fields. Certain types of electromagnetic waves carry enough energy per quantum to break bonds between molecules and ionize atoms. In the electromagnetic spectrum, cosmic rays, gamma rays, and X-rays have this characteristic and are considered ionizing radiation. The fields in which the quanta do not possess sufficient energy to ionize atoms are known as "non-ionizing radiation." The RF-EMFR is included in the non-ionizing radiation category as it does not contain the required amount of energy to break chemical bonds or remove electrons [5–7]. (Figure 7.3).

In daily-life activities, human beings are exposed to different types of EMFR, which is characterized by numerous physical properties. It is essential to understand the direct and indirect impact of EMFR on the pathological and physiological processes occurring at the cellular, organ, and body system levels. Biological science studies suggest that low-frequency magnetic fields generated from daily use electrical appliances potentially negatively impact the various biological and physiological processes [3].

Home electrical appliances can generate low-frequency EMFR 50–60 Hz daily, negatively impacting various body organisms. However, electronic devices such as mobile phones, televisions, or radios can emit radiofrequency electromagnetic radiation (RF-EMFR) with much higher frequencies, ranging from 300 MHz to 300 GHz [3]. In this new millennium, the most frequently used device in all age groups, genders, and regions worldwide is smartphones.

7.3 Mobile Phones: Modern Technology of Modern Era

On April 3, 1973, an engineer Martin Cooper, a Motorola industry member, made the world's first mobile phone call. After ten years, the first commercial cell phone system was launched in the United States in 1983. In early 1990, the "digital system, Global System for Mobile Communication (GSM), operated using dual-band 900 and 1800 MHz. In early 2003, 3G, the third generation of mobile phones, or UMTS (Universal Mobile Telecommunication System) using

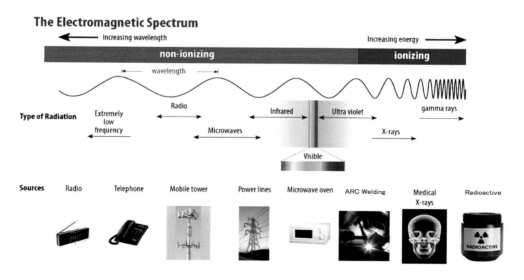

Figure 7.3 Non-ionizing and ionizing electromagnetic field radiation.

1900/2100 MHz RF fields" technology, was launched worldwide. The fourth-generation (4G), which operated at 800/2,700 MHz, was introduced in 2010. More recently, in 2019, the comprehensive development of fifth-generation (5G) wireless technology for the digital cellular network was established. Its low band uses a similar frequency range as its predecessor, 4G. However, 5G also contains much higher frequencies of about 25–39 GHz, with approximate speeds of 1–2 Gbit/s [8]. The second-, third-, and fourth-generation mobile network technology emits RF-EMFR in a range of about 800–2,700 MHz [9]. However, the 5G high-frequency spectrum includes 24.25–52.6 GHz [10, 11] and 80 GHz [9].

7.3.1 Worldwide Mobile Phone Users

In these last two decades of science and technology, the usage of mobile phones has markedly increased amongst all age groups, individuals, urban and rural areas, and both developing and developed nations. Mobile phones have become an essential and common means of everyday communication [12] (Figure 7.4). Their role in daily life is limited to making and receiving calls and has advanced functions for storing data, photography, listening to music, and watching videos. Moreover, modern cell phones provide internet access, online banking, scanning/transfer of documents, and geo-positioning system technology, allowing users to find locations and traffic conditions [12].

As per the World Bank Report 2019 [13], the world population is 7.594 billion; however, 7.858 billion mobile phone subscribers globally [14]. The mobile phone subscription is more than the world population as most people have two or more subscriptions [14]. This figure is increasing since the introduction of smartphone technology.

In addition to the excessive usage of mobile phones, people are also markedly using other information technology-associated devices, such as laptops. YouTube gets about 30 million visitors daily, with five billion videos watched per day on the platform. Twitter has 321 million monthly active users. Moreover, 5.8 billion people are daily actively searching Google, and there are 71,780 searches per second. The global email users are 3.9 billion, and 124.5 billion emails are sent and received per day. Furthermore, regarding WhatsApp, one million people are registered on the platform daily, with 300 million daily active users worldwide and 2 billion

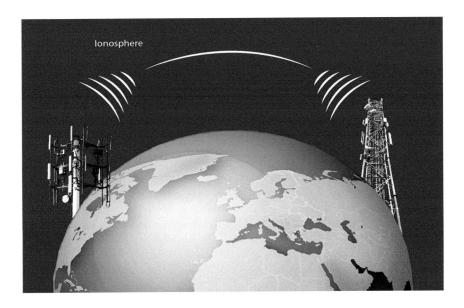

Figure 7.4 Mobile phone tower.

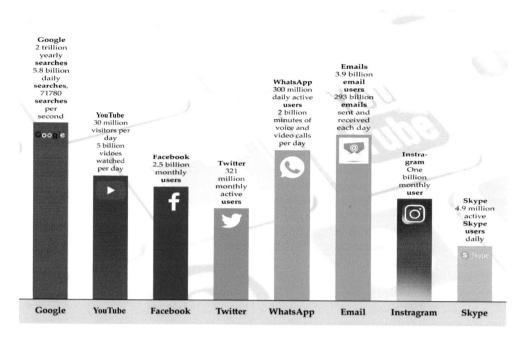

Figure 7.5 Worldwide users of information technology allied devices.

minutes of voice and video calls per day on WhatsApp (Figure 7.5). All these activities are available on mobile phones, and therefore this, directly and indirectly, exposes people to RF-EMFR.

7.3.2 Radiation Generated from Mobile Phones

Science literature acknowledged many natural and synthetic sources of RF-EMFR. Mobile phones are widely used and are the most substantial sources of human exposure to RF-EMFR [12]. The "Global System for Mobile Communication" (GSM) was operated between 900 and 1,800 MHz. Marie-Christine et al. (2013) [15]. The third-generation 3G or "Universal Mobile Telecommunication System" (UMTS) used 1,900/2,100 MHz. The fourth-generation (4G) operates at 800/2,700 MHz. More recently, in 2019, the vast development of fifth-generation (5G) wireless technology uses about 25–39 GHz frequencies. Moreover, the literature also demonstrates that the second-, third-, and fourth-generation mobile phones emit RF-EMFR in a range of about 800–2,700 MHz [9], and 5G, high-frequency technology emits 24.25–52.6 GHz [10, 11] and 24–80 GHz [9]. Similar radiation levels are produced by various other wireless digital systems, such as data communication [16].

The RF-EMFR affects human biology through thermal and non-thermal effects that lead to an increase in the temperature of tissues and organs, thereby damaging body cells. Low RF-EMFR generated by electrical devices was classified as possibly carcinogenic to humans by the "International Agency for Research on Cancer" (IARC) in 2002. Furthermore, in 2011, WHO and IARC classified these radiations as possibly increasing the risk of malignant brain tumors [17, 18].

7.3.3 Mobile Phone and RF-EMFR

There has been an exponential evolution of mobile communication systems and associated devices in the recent decade, and the public is frequently being unintentionally exposed to radiofrequency radiation. Many scientific studies have attempted to determine whether EMFR has any harmful effects on human health [19, 20]. Public concerns continue to exist regarding RF-EMFR emitted by mobile phones and base-station towers. The RF-EMFR can be absorbed in all living organisms; the absorption of RF-EMF depends on the distance of the device, frequency, and duration of exposure. The frequency is essential since the 5G network operates at higher frequencies [21, 22].

The RF-EMFR has been one of the leading health problems associated with multiple human-made sources such as mobile phones, mobile phone towers, tablets, laptops, computers, television sets, radars, microwave ovens, high power lines, transformers, and various other wireless gadgets. The use of these devices that generate RF-EMFR radiation has increased tremendously during the last three decades, and human exposure is widespread. Mobile phones generate various radiofrequency, including 900–1,800 and 2,600-MHz radiofrequency radiation. The RF-EMFR generated from Wi-Fi, mobile phones, and electronic devices is growing public concerns over the harmful effects of radiofrequency radiation on human health [23].

The RF-EMFR has become a leading cause of persistent electromagnetic pollution that affects various environmental elements [24]. Mobile phones have become the most common source of RF-EMFR exposure to the brain. Moreover, nowadays, children are also frequently using mobile phones. Whether the exposure of RF-EMFR can cause cancer risk has been addressed in numerous case-control studies, and findings are heterogeneous [9].

7.4 RF-EMFR and Neurological Manifestations

The nervous system is susceptible and the most complex system of the human body. The brain is sensitive for controlling and coordinating the various physiological functions, but this system is also highly complex for thoughts, emotions, fear, lesion, and traumatic conditions. In the past three decades, the widespread use of mobile phones raises various concerns about possible hazardous health effects of the RF-EMFR emitted from mobile phones and their associated devices on the nervous system (Figure 7.6). People keep the phone close to ear, head, and indeed the

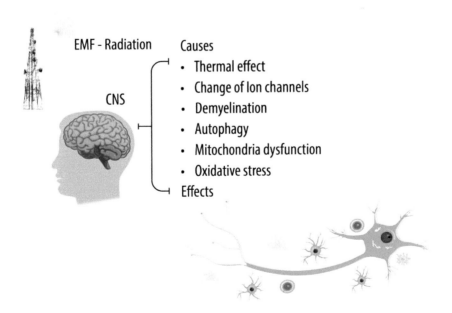

Figure 7.6 Effect of RF-EMFR on brain.

brain while using cell phones. The RF-EMFR affects the nervous system by two main mechanisms. These are thermal (heating) effects caused mainly by holding mobile phones close to the body and possible non-thermal effects [25].

Al-Khlaiwi and Meo (2004) [12] conducted a study on the use of mobile phones and health hazards. The authors reported that mobile phones could cause headaches, sleep disturbance, tension, fatigue, and dizziness. In another study, Meo and Al-Dress (2005) [26] reported that mobile phone users have complaints of impairment in hearing, earache, and warmth on the ear (34.59%), and impairment in vision or blurred vision (5.04%).

In contradiction to the study findings of Meo and Al-Dress (2005) [26], Harry and Mark (2007) [27] showed no hearing, tinnitus, or balance complaints among high or long-term users compared to low or short-term users. The most probable reason for this contradiction was the selection criteria. Harry and Mark (2007) [27] compared the findings between high and low mobile users. The best way was to compare the findings between mobile phone users and non-users. Considering this study was conducted in 2007, authors could have easily chosen individuals who were not avid mobile phone users.

Mobile phones can cause numerous neurological adverse health problems such as headache, discomfort, sleep disturbance, lack of concentration, worm on-ear, burning skin, peripheral neurophysiological changes, dizziness, impairment of STM, seizures in epileptic children, high blood pressure, and brain tumors [28–30].

Haarala et al. (2007) [31] performed behavioral tests in a case-control study and did not find adverse effects of cell phones on cognitive function. However, Huber et al. (2002) [32] identified that pulsed electromagnetic field exposure by cell phones affects the regional Cerebral Blood Flow (CBF). Hossman and Herman 2003 [33] concluded that pulsed RF radiation emitted from mobile phones damages brain cells, leading to headaches, mood swings, lethargy, and depression. Similarly, Salford et al. 2003 [34] have also identified that RF-EMFR from mobile phones can accelerate dementia and early Alzheimer's.

7.5 Prevalence of Brain Cancer

The CNS cancer is responsible for substantial morbidity and mortality worldwide. The prevalence significantly increased between 1990 and 2020. It is topographical and area disparity in the prevalence of CNS cancer, which might reflect "variances in diagnoses and reporting practices or social, environmental, and genetic risk factors." A recent study published in 2019 reported 330,000 CNS cancer cases and 227,000 (205,000 to 241,000) deaths globally. The age-standardized incidence rates of CNS cancer increased globally by 17.3% between 1990 and 2016, 4.63 per 100,000 person-years. Most incident cases of CNS cancer in both genders were reported from East Asia (108,000), followed by Western Europe (49,000), and then South Asia (31,000). Furthermore, the top three countries with the highest number of incidences of CNS cancer are China, the USA, and India [35].

The epidemiological and experimental studies are elucidating the numerous explanations of the primary etiopathological facts of brain cancers. Brain tumors comprise 2% of brain cancers [36]. The highest percentage of all brain tumors that originate in the brain and CNS are benign growths occurring in the meninges [37]. Most malignant primary brain tumors are found in the cerebral cortex [37], with the highest percentage developing in the frontal lobe [36] [Figure 7.7].

There is a gender variation in the incidences of brain cancer both in males and females regardless of the people's age. The epidemiological studies are uncertain why this difference of occurrence is variable between the age groups. Approximately 55% of malignant brain tumors

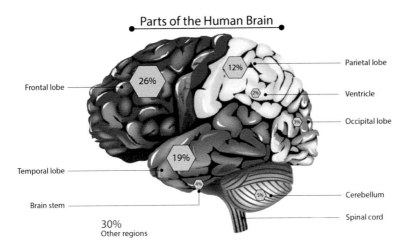

Figure 7.7 Percentage of malignant primary brain tumors in various parts of the brain.

Source: Modified and adapted from Ostrom et al. (2015) [37].

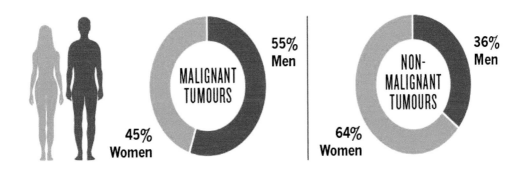

Figure 7.8 Malignant brain tumors among males and females.

Source: Modified and adapted from the study Ostrom et al. (2015) [37].

occurred in males as compared to 45% in females. However, 36% of non-malignant brain tumors occurred in males compared to 64% in females (Figure 7.8) [36, 37].

The researchers identified that social, cultural, and genetic factors might significantly affect this incidence. Moreover, external causes such as environment, food, ionizing and non-ionizing radiation are the significant causes of cancer in the various regions of the brain. According to the WHO Report 2018 [38], more than 300,000 children from infancy to 19 years are diagnosed with cancer each year. Approximately eight children out of ten are from low- and middle-income countries where the survival rate is about 20% [39]. The International Agency for Research on Cancer (IARC) [6, 7] reported that worldwide there were 257,000 new cases of brain or nervous system cancer with an incidence of over 3 cases per 100,000 people [6, 7]. Brain cancer is most common in white individuals, and its incidence is highest in northern Europe, approximately 10 cases per 100,000, followed by the United States, Canada, and Australia [40, 41].

7.6 RF-EMFR and Brain Cancer

Since the last two decades, the widespread use of cell phones and other wireless devices has grown in all segments of society, all age groups, both genders, and the developing and developed world. The entire global population is at risk of exposure to non-ionizing and ionizing radiation resulting from workplace exposure such as mobile phones, base-station towers, television, and various electronic and wireless devices. Pregnant females, infants, and children are vulnerable since these RF radiations continuously react with the developing embryo, negatively affecting cells, and thermal heating causes various health hazards, including the brain [42].

The worldwide dissemination of mobile phones and wireless devices has been under the heated debate of the science community, and public concern about exposure to RF-EMFR emitted from the devices and possible effects on health, especially cancer. Long-term exposure to RF-EMFR poses a significant health hazard that has increased the risk of impairment of various body functions. There are reliable pieces of evidence that exposures can result in neurobehavioral decrements, and individuals may develop "electro-hypersensitivity" syndrome commonly characterized as "idiopathic environmental intolerance." This is mainly found in children with excessive use of wireless technologies. The children are highly vulnerable as all the body tissues and organs, including the nervous system, are developing phases. In children, hyper conductivity of brain tissue and greater penetration of RF-EMFR relative to head size and their exposure increase risk for a brain tumor. Long-term exposure to RF-EMFR at low intensities poses a "risk of brain cancer both in humans and animals. The mechanism includes induction of reactive oxygen species, gene expression alteration, and DNA damage through epigenetic, genetic, and environmental processes" [43].

The science community had established the first few links with cancer and exposure to EMFs in 1979 when Wertheimer and Leeper (1979) [44] reported that children dying from cancer resided more often in homes exposed to higher RF-EMFR. The children were more vulnerable who had spent more time at the same address, and it appeared to be a dose-related response. In May 2011, the WHO's International Agency for Research on Cancer (IARC) classified RF-EMFR as a possible human carcinogen to humans. The working group consisted of 31 scientists from 14 states worldwide, and the meeting was held in Lyon, France. These members investigated the potential carcinogenic hazards of RF-EMFR emitted from mobile phones and other allied devices [17].

The WHO, "International Agency for Research on Cancer (IARC), classifies the EMF produced by mobile phones as possibly carcinogenic to humans [17]. In this light, numerous research studies are conducted worldwide to evaluate the potential long-term risks associated with a mobile phone with brain cancer. However, considering the evidence of free-radical damage confirmed in studies on humans and animals, the scientific community believes that EMF-induced oxidative stress can cause damage to cellular targets such as DNA and trigger processes that lead to brain cancer" [45, 46].

WHO identified the possible adverse effects of mobile phones on the developing tissues as a major research priority. The WHO now recognizes that RF-EMFR generated from mobile phones, base-station towers, telecommunication, and other various power devices are a part of environmental pollution and a possible cause of cancer [2, 17].

The epidemiological studies have also reported that long-term and excessive use of a mobile phone is associated with an increased risk for brain tumors. In contrast, some studies contradict and have reported complete denial. The literature also demonstrates that RF-EMFR may induce gene/protein expression alterations in various cells/tissues associated with potentially harmful health outcomes and brain cancer [47].

WHO International Agency for Research on Cancer (IARC) (2011) [17] categorized the radiofrequency radiation (RFR) generated from mobile phones and other wireless devices as possible human carcinogenic. The science community published many epidemiological, case-control, and cohort studies on brain cancer and RF-EMFR [41, 48]. This section was endeavored to discuss both animal and human model studies that show the effect of RF-EMFR on brain cancer.

7.6.1 Animal Model Studies on RF-EMFR and Neuronal Cell Deaths and Brain Cancer

To understand the scientific facts and findings and reach appropriate conclusions, gathering and articulating the animal model basic sciences data in clinical reasoning is vital. The clinical concepts and allied knowledge cannot be compiled or encapsulated without understanding the basic sciences literature. This section highlights the basic science findings with animal model studies on RF-EMFR and Brain Cancer (Table 7.2).

Table 7.2 Effect of RF-EMFR on Brain Biology and Cancer

Author and Year of Study	Type of Study	Sample Size	Study Outcomes
Wyde et al. (2018) [50]	Cross-sectional	90 Sprague-Dawley rats per group	Malignant glioma 8 (8.88%); Glial cell hyperplasia 6 (6.66%), malignant glioma 1 (1.11%); glial cell hyperplasia 1 (1.11%) compared to control (0.0%).
Falcioni et al. (2018) [55]	Cross-sectional	Rats exposed to 1.8 GHz GSM for 19 hours/day	Increase incidence of malignant glial tumors in female rats at the dose of 50 V/m.
Nidhi et al. (2014) [56]	Cross-sectional	Six rats in each group exposed to 900 MHz for 4 hours/day for 15 days	Histological findings showed neurodegenerative and neuronal cells damage in the hippocampal and cerebral cortex, change in behavior, and memory
Saba Shahin et al. (2015) [51]	Cross-sectional	Mice exposed to 2.45-GHz radiation 2 hours/day for 15, 30, and 60 days	2.45-GHz radiation exposure leads to hippocampal neuronal and non-neuronal apoptosis and memory loss.
Saba Shahin et al. (2018) [52]	Cross-sectional	Mice exposed to 2.45 GHz 2 hours/day for 15, 30, and 60 days	2.45-GHz-irradiated mice showed poor learning with many working and reference memory errors.
Narges et al. (2018) [53]	Cross-sectional	Mice exposed to 2.45-GHz radiation 2 hours/day for 40 days	The morphology showed that neuronal density in the hippocampal area was decreased by radiation.
Ragy (2015) [54]	Cross-sectional	40 albino rats exposed to 900-MHz EMR for 1 hour/day for 60 days	EMFR from mobile phone cause impairments in bio-chemicals and cause oxidative stress.
Adamantia et al. (2012) [57]	Cross-sectional	18 mice, 6 in each group	Brain plasticity alterations indicative of oxidative stress in the nervous system with apoptosis.
Adamantia et al. (2018) [58]	Cross-sectional	14 mice, 8 in control, 6 exposed, for 2 hours to GSM 1,800 MHz mobile phone radiation	EF-EMFR causes hippocampal lipidome & transcriptome changes, which indicates brain proteome changes and memory deficits.
Hao et al. (2013) [59]	Cross-sectional	32 Wistar rats in the exposed group and the control group	The hippocampal neurons showed irregular firing, which indicates that EMFR influences learning and memory in rats.

Source: Basic sciences data based on the animal model studies.

Melnick et al. (2019) [49] conducted a study on RF-EMFR generated from cell phones in rats and mice. The study findings showed a significant increase in incidences and trends of gliomas and glial cell hyperplasia is in the brain of exposed male rats. They also identified a marked increase in DNA damage in the brains of exposed rats and mice. Their study results were in agreement with the "International Agency for Research on Cancer (IARC)" as they classified RFR as a "possible human carcinogen" with increased risks of gliomas and acoustic neuromas among long-term users of cell phones.

Wyde et al. (2018) [50] conducted a study on the prevalence of brain lesions in male and female rats exposed to experimentally generated 900-MHz RF fields. The low incidence of malignant gliomas and glial cell hyperplasia was found in male rats exposed to GSM modulated RFR compared to control groups. Saba Shahin et al. (2015) [51] performed a study to understand the mechanism of the dose–response effect of short-term (15 days) and long-term (30, 60 days) low-level radiation exposure on hippocampus regarding learning and memory in male mice. The authors found that short-term and long-term exposure to 2.45-GHz radiation increases oxidative stress, leading to mitochondrial dysfunction, inflammatory events, and neurodegenerative pathology and apoptosis in hippocampal both neuronal and non-neuronal cells. They also found that learning and spatial memory deficit increases with the increased duration of radiation exposure. In another study, Saba Shahin et al. (2018) [52] elucidates the radiation-induced stress factors on the hippocampal spatial memory formation pathway. The authors identified that 2.45-GHz-irradiated mice showed slow learning and increased working and reference memory errors.

Similarly, Narges et al. (2018) [53] determined the impact of RF-EMFR radiation on memory, hippocampal synaptic plasticity, and hippocampal neuronal cell numbers. The authors found that radiation exposure decreased the number of neurons and learning and memory performance. Furthermore, Ragy (2015) [54] investigated the effect of EMR emitted from a mobile on male albino rats' brains. The animals were exposed to 900-MHz EMR for 1 hour a day for 60 days. The authors identified that RF-EMFR emitting from mobile phones produces impairments in bio-chemicals components and oxidative stress in albino rats' brains.

Nidhi et al. (2014) [56] investigated the effects of long-term usage of mobile phones on the "cytological makeup of the hippocampus in rat brains and antioxidant status and cognitive behavior, particularly on learning and memory." The authors identified anxiety, behavioral changes, and poor learning in the exposed group compared to controls. Furthermore, the "histological examination showed neurodegenerative cells in hippocampal sub-regions and the cerebral cortex." Their study findings further indicated that extensive neurodegeneration and neuronal damage result in alterations in memory and learning behavior.

Adamantia et al. (2012) [57] investigated the long-term effects of EMFs on the proteome of the cerebellum, hippocampus, and frontal lobe in mice. The findings demonstrated that long-term irradiation significantly altered the expression of 143 proteins in total downregulation. The authors noticed that protein expression changes were related to brain plasticity alterations, which was indicative of oxidative stress in the nervous system involved apoptosis. These findings potentially explain the reasons for human health hazards such as headaches, sleep disturbance, fatigue, memory deficits, and long-term brain tumor induction under similar exposure conditions. [57]

Adamantia et al. (2018) [58] conducted a study on 14 adult male mice, 6 were exposed for 2 hours to GSM 1,800 MHz mobile phone radiation, and 8 were in the sham-exposed group. The study provides evidence that mobile phone radiation induces hippocampal lipidome and transcriptome changes that may explain the memory deficits. Similarly, Hao et al. (2013) [59] examined the "effect of RF-EMFR on spatial learning and memory in rats. The hippocampal neurons showed irregular firing patterns and more spikes with a shorter inter-spike interval the experiment period". It indicates that the EMFR influences learning and memory in rats. Most of the animal model study supported the hypothesis that RF-EMFR can cause various hazards on the brain,

7.6.2 Human Model Studies on RF-EMFR and Brain Cancer

A balanced approach was employed to discuss both the animal and human model studies to reach appropriate conclusions (Tables 7.2 and 7.3). A large number of human model studies are available on the effects of RF-EMFR and Brain. Lagorio and Röösli (2014) [60] conducted a meta-analysis of studies on intracranial tumors and mobile phone use. A meta-regression of the studies showed that mobile phone use could cause the occurrence of intracranial tumors. Similarly, Moulder et al. (2005) [61] analyzed the scientific literature on radiofrequency emissions from mobile phones and wireless communication. The authors found a linkage between radiofrequency emissions and malignancy.

Moreover, Cardis et al. (2010) [62] conducted a case-control study and found an increased risk of glioma with exposure to RF-EMFR. In another study, Cardis et al. (2011) [63] recruited 553 glioma and 676 meningioma cases and 1762 and 1911 controls for analysis. They found an increased risk of glioma and meningioma in long-term mobile phone users with high RF exposure and increased meningioma risk.

Wang et al. (2018) [64] conducted a meta-analysis and found a significant association with the risk of glioma in long-term users of cell phones. Similarly, Sato et al. (2016) [65] assessed the incidence of malignant neoplasms of the CNS. The findings showed a relative risk of 1.4 among the mobile phones more. Lerchl et al. (2015) [66] investigated the cancerogenic effects of exposure to electromagnetic fields (RF-EMF) emitted by mobile phones and base stations, and their study result suggested the tumor-promoting effects of RF-EMF [67].

Hardell et al. (2013) [68] reported an increased association between long-term use of mobile, cordless phones, glioma, and acoustic neuroma. The authors found that the ipsilateral use of phones resulted in a higher risk than contralateral mobile and cordless phone use. These findings support the hypothesis that RF-EMFs play a role in the initiation and promotion of carcinogenesis. In another study, Hardell and Carlberg (2015) [69] analyzed malignant brain tumors with patients diagnosed with brain cancer compared to age and gender-matched controls. It was identified that mobile phone use increased the risk of glioma; the highest findings were found for ipsilateral mobile or cordless phone users.

Coureau et al. (2014) [70] conducted a multicenter case-control study and analyzed the relationship between mobile phone exposure and primary central nervous system tumors. No association with brain tumors was observed when comparing regular mobile phone users with non-users. However, a positive association was statistically significant in the heaviest users when considering lifelong cumulative duration. This additional data support previous findings concerning a possible association between heavy mobile phone use and brain tumors. [70]

Prasad et al. (2017) [76] demonstrated that mobile phones emit RF-EMFR and are classified as possibly carcinogenic. The authors analyzed 14 case-control studies and found evidence of mobile phone use with risk to brain tumors, especially in long-term mobile phone users. Similarly, Gong et al. (2014) [77] suggested an increased risk of glioma among long-term ipsilateral users of mobile phones. In another study, French et al. (2001) [78] established a mechanism in which mobile phone radiofrequency radiation-induced cancer. The mechanism was the chronic activation of the heat shock response, upregulation of heat shock proteins (Hsps). The authors suggested that "repeated exposure to mobile phone radiation acts as repetitive stress leading to continuous expression of Hsps in exposed cells and tissues, which affects their

Table 7.3 Effect of RF-EMFR on Brain Biology and Cancer

Author and Year of Study	Type of Study	Sample Size	Study Outcomes
Cardis et al. (2010) [62]	Case control study	2708 glioma 2409 meningioma	There was no increased incidence of glioma or meningioma. However, a rising risk of glioma and meningioma was found with high exposure levels of 10 years.
Cardis et al. (2011) [63]	Case control study	553 glioma 676 meningioma & 3673 controls	Increased risk of glioma and meningioma in long-term mobile phone users with high RF exposure
Hardell et al. (2013) [68]	Population-based case-control study	men and women aged 18–75 years	593 (87%) cases of malignant brain tumor, 1,368 (85%) of controls. "Association between use of mobile, cordless phones, and malignant brain tumors. The risk was higher for ipsilateral use tumors in the temporal lobe".
Hardell et al. (2015) [69]	Population-based case-control	20–80	1,498 (89%) cases; 3,530 (87%) controls. Mobile phone use increased the risk of glioma with more than 25 years of latency.
Frank deVocht (2016) [71]	Population-based case-control		Malignant neoplasms of the temporal lobe were increased faster by 35% cases during 2005–2014; additional 188 cases annually.
Grell et al. (2016) [72]	Population-based case-control	30–59 years	Total glioma: 2,700; 1,530 tumors with localization, 933 regular phone users. Mobile phone allied with glioma closer to the ear on the side of the head.
He Gao et al. (2019) [73]	Population-based case-control	48518, 5.9 year follow-up period	716 (1.47%) incident cancer cases were identified, overall no association between personal radio use and risk of cancers.
Aydin et al. (2011) [74]	352 case-patients: 646 control	7–19 years	Regular users of mobile phones were not associated with brain tumors compared with non-users.
Coureau et al. (2014) [70]	Multicenter case-control study	253 gliomas, 194 meningiomas	A positive link was in excessive users "considering life-long cumulative duration (≥896 hours for gliomas; meningiomas. No association of brain tumors was found when comparing regular mobile phone users with non-users for gliomas and meningiomas".
Patrizia Frei et al. (2011) [75]	Nationwide cohort study	10,729 casestumor	A little evidence for a causal association was seen, but no increased risks of tumors in the brain.

Source: Basic and clinical sciences data based on the human studies.

normal regulation and results" in conversion into cancer cells. This hypothesis provides evidence of an association between mobile phone use and cancer [78].

Momoli et al. (2017) [79] conducted a case-control study in Canada as part of the Interphone Study. The study population was limited to Canadian citizens, with age ranges between 30–59 years. The total participants were 1,058; among them, control was 653, and cases were 405. The subjects were matched for age, gender, and participants' region. The authors analyzed the data from the 13 countries as a part of the interphone case-control study. The authors concluded that there was little evidence of increased risk of meningioma, acoustic neuroma, and parotid gland tumors concerning mobile phone use. Mortazavi et al. (2019) [80] also established a relationship between exposure to RF-EMFs generated by mobile phones and brain cancer. They demonstrated that the magnitude of exposure to RF-EMFs plays an essential role in RF-induced carcinogenesis. Some evidence indicates a similar pattern with ionizing radiation, and the carcinogenesis of non-ionizing RF-EMF may have a nonlinear dose–response relationship (Table 7.4).

7.6.3 Meta-Analysis and Systematic Studies on RF-EMFR and Brain Cancer

We also analyzed the meta-analysis and systematic review literature on mobile phones and brain cancer in addition to animal and human model studies (Table 7.5). Gao et al. (2019) [73] conducted a study on RF-EMFR generated from mobile phones and investigated an association of monthly personal radio use and risk of cancer among 48,518 police officers and staff members. During the median follow-up period of 5.9 years, 716 incident cancer cases were identified among the users. However, overall, there was no association between personal radio use and the risk of cancers. There was 1.06 per doubling of minutes of personal radio use linked to head and neck cancers among personal radio users.

Yang et al. (2017) [90] performed a meta-analysis with 11 studies comprising 6,028 cases and 11,488 controls. Their results suggest that long-term mobile phone use may be associated with an increased risk of glioma in regular and long-term users. Hardell et al. (2008) [91] evaluated the long-term use of mobile phones and the risk for brain tumors in case-control studies, identified ten studies on glioma, six studies for ipsilateral use, and four studies for contralateral use. The authors reported a consistent association between mobile phone use and ipsilateral glioma and acoustic neuroma among mobile phones with more than 10-years. Michael Kundi (2010) [92] conducted a meta-analysis of 33 epidemiologic studies, mostly 25 studies were on brain tumors. Although there were some methodologic limitations, the overall concluding evidence speaks in favor of increased risk. In another study, Michael and Hardell (2017) [93] analyzed the multiple studies and found an increased risk for glioma in the temporal lobe. The highest risk was identified with 20 years latency group.

Bortkiewicz et al. (2017) [94] performed a meta-analysis of 24 studies with a sample size of 26846 cases and 50013 controls. Their analytic findings supported the hypothesis that long-term use of mobile phones was related to an increased risk of intracranial tumors. The authors also demonstrated that mobile phone use over ten years was significantly associated with a higher risk of various intracranial tumors. Another meta-analysis was performed by Wang and Guo (2016) [95] with 11 studies, 2603 control, and 5460 cases. The authors showed a significant association between mobile phones for more than five years and the risk of glioma [95].

Yakymenko et al. (2011) [100] reported that exposure to radiofrequency radiation for more than ten years was linked to carcinogenic. In addition, Alexiou and Sioka (2105) [101] demonstrated that although long-term mobile phone use can be associated with increased risk of intracranial tumors. Leng et al. (2016) [102] conducted a study and found an increased risk for pituitary tumors related to mobile phone use. However, Schoemaker et al. (2009) [103] and Shrestha et al. (2015) [104] did not find risk associated with self-reported short or medium-term

Table 7.4 Increased Odd Ratio Values in the Interphone Studies Relationships between Mobile Phone Use and Tumors

Author and Year Tumor type	The Period Since MP Use	Tumor/Control Cases	Odds Ratio 95%CI	Ipsilateral Tumor		Contralateral Tumors	
				Cases/Control	OR 95%CI	Cases/Control	OR 95%CI
Lonn et al. (2004) [81]	since ≥ 10	14/29	1.9 (0.9–4.1)	12/15	3.9 (1.6–9.5)	4/17	0.8 (0.2–2.9)
Acoustic neuromas	for ≥ 10	11/26	1.6 (0.7–3.6)	9/12	3.1 (1.2–8.4)	4/16	0.8 (0.2–3.1)
Schoemaker et al. (2005) [82]	since ≥ 10	47/212	1.0 (0.7–1.5)	31/124	1.3 (0.8–2.0)	20/105	1.0 (0.6–1.7)
Acoustic neuromas	for ≥ 10	31/131	1.1 (0.7–1.8)	23/72	1.8 (1.1–3.1)	12/73	0.9 (0.5–1.8)
Lonn et al. (2005) [83]	since ≥ 10	25/38	0.9 (0.5–1.5)	15/18	1.6 (0.8–3.4)	11/25	0.7 (0.3–1.5)
Gliomas	for ≥ 10	22/33	0.9 (0.5–1.6)	14/15	1.8 (0.8–3.9)	9/23	0.6 (0.3–1.4)
Meningiomas	since ≥ 10	12/36	0.9 (0.4–1.9)	5/18	1.3 (0.5–3.9)	3/22	0.5 (0.1–1.7)
	for ≥ 10	8/32	0.7 (0.3–1.6)	4/15	1.4 (0.4–4.4)	3/23	0.5 (0.1–1.8)
Lonn et al. (2006) [84] Gliomas	since ≥ 10	7/15	1.4 (0.5–3.9)	6/9	2.6 (0.9–7.9)	1/9	0.3 (0.0–2.3)
Klaeboe et al. (2007) [85]	since ≥ 6	70/73	0.8 (0.5–1.2)	39/37	1.3 (0.8–2.1)	32/42	0.8 (0.5–1.4)
Gliomas	for ≥ 6	55/61	0.7 (0.4–1.2)	30/30	1.2 (0.7–2.1)	27/34	0.9 (0.5–1.5)
Lahkola et al. (2007) [86]	since ≥ 10	143/220	0.95 (0.74–1.23)	77/117	1.39 (1.01–1.92)	67/121	0.98 (0.71–1.37)
Gliomas	for ≥ 10	88/134	0.94 (0.69–1.78)	43/74	1.14 (0.76–1.72)	41/71	1.01 (0.67–1.53)
Lahkola et al. (2008) [87]	since ≥ 10	73/212	0.91 (0.67–1.25)	33/113	1.05 (0.67–1.65)	24/117	0.62 (0.38–1.03)
Meningiomas	for ≥ 10	42/130	0.85 (0.57–1.26)	21/73	0.99 (0.57–1.73)	13/68	0.64 (0.33–1.23)
Interphone (2110) [88] (Gliomas)	≥ 1640 calls	160/113	1.82 (1.15–2.89)	100/62	1.96 (1.22–3.16)	39/31	1.25 (0.64–2.42)
Sadetzki et al. (2008) [89]	> 5479 calls	86/157	1.13 (0.79–1.61)	121/159	1.58 (1.11–2.24)	46/135	0.78 (0.51–1.19)

OR: odd ratio.

144

Table 7.5 Association of RF-EMFR with Brain Tumors (Meta-Analysis Studies)

Name of Author and Year	Type of Study	Sample Size	Outcomes
Röösli et al. (2019) [9]	Meta-analysis	. . . . . .	There is no increased brain or salivary gland tumor; uncertainty remains regarding long-term (>15 years) mobile phone use and brain tumor.
Lagorio and Röösli (2014) [60]	Meta-analysis	47 studies	Mobile phone use can cause the occurrence of intracranial tumors
Wang et al. (2018) [64]	Meta-analysis	10 studies	The increased association was found with the risk of glioma among long-term mobile phone users (≥10 years).
Prasad et al. (2017) [76]	Meta-Analysis	22 case-control studies	Fourteen case-control studies showed no increase in brain tumor risk for more than ten years of mobile phone use. However, the overall result showed a significant 1.33 times increase in risk.
Yang et al. (2017) [90]	Meta-Analysis	11 studies with 6,028 cases and 11,488 controls	Long-term mobile phone use may be linked with an increased risk of glioma.
Hardell et al. (2008) [91]	Meta-Analysis	10 studies	Mobile phone use for more than 10-years causes an ipsilateral side glioma and acoustic neuroma of the brain.
Michael Kundi (2010) [92]	Meta-Analysis	25 studies	With some methodologic limitations, overall concluding evidence speaks in favor of an increased risk.
Michael andHardell (2017) [93]	Review	Nine reports, viewpoints	Increased risk for glioma in the temporal lobe, the highest risk with 20 years' latency group.
Bortkiewicz et al. (2017) [94]	Meta-Analysis	24 studies, 26,846 cases, and 50,013 controls.	long-term use of mobile phones is related to increased risk of intracranial tumors.
Wang and Guo (2016) [95]	Meta-Analysis	11 studies, 12,603 control, and 5,460 cases	A significant relationship between long-term mobile phone use >5years and glioma risk.
Angelo et al. (2011) [96]	Meta-Analysis	----	They found almost doubling of the risk of head tumors induced by long-term mobile phone use or latency.
Seung-Kwon et al. (2009) [97]	Meta-Analysis	23 case-control studies total 37,916; 12,344 cases and 25,572 controls	Mobile phone use of 10 years or longer was linked with a risk of tumors on the same side of the head when compared with never or rare use of mobile phones.

(Continued)

145

Table 7.5 (Continued)			
Name of Author and Year	Type of Study	Sample Size	Outcomes
Vini Khurana et al. (2009) [98]	Meta-Analysis	11 studies	Cell phone use for ten years doubles the risk of a brain tumor on the same "ipsilateral" side of the head as that preferred for cell phone use.
Leng et al. (2016) [99]	Meta-Analysis	24 studies	In Asian people, cell phone use and glioma had some link, with a bit of relationship with meningioma incidence. In children and teenagers, mobile phone use was allied with the incidence of brain tumors.

use of mobile phones. Michael and Hardell (2017) [93] performed a study based on nine epidemiology and laboratory studies. They found an increased risk for glioma in the temporal lobe, with the highest risk in mobile phone users over twenty years.

Angelo et al. (2011) [96] established "a cause-effect relationship between long-term mobile phone use or latency and a statistically significant increase of ipsilateral head tumor risk. The authors found almost doubling of the risk of head tumor induced by long-term mobile phone use or latency". In another study, Seung-Kwon et al. (2009) [97] analyzed the findings of twenty-three "case-control studies, which involved 37,916 participants, 12,344 patients, and 25,572 controls with never or rarely having used a mobile phone. Mobile phone use of 10 years or longer was associated with a risk of tumors in 13 studies. The authors also found possible evidence linking mobile phone use to an increased risk of tumors on the same side of the head".

Vini Khurana et al. (2009) [98] performed a meta-analysis based on 11 long-term epidemiologic studies. Their results indicate that using a cell phone for about ten years doubles the risk of a brain tumor on the same "ipsilateral" side of the head. Similarly, Hardell et al. (2008) [91] also evaluated the long-term use of mobile phones and the risk for brain tumors. They conclude a consistent association between mobile phone use and ipsilateral glioma and acoustic neuroma among subjects who used mobile phones for more than 10-years. Leng et al. (2016) [99] performed a meta-analysis based on 24 long-term epidemiologic studies. Their results indicate that using a cell in Asian people, cell phone use, and glioma had some link, with a bit of relationship with meningioma incidence. In children and teenagers, mobile phone use was allied with the incidence of brain tumors.

Even though the many available studies support the hypothesis on the adverse effects of RF-EMFR on brain and brain cancer, the literature also endorses RF-EMFR generated from mobile phones, MPBSTs, and other electronic devices do not correlate with brain cancer. Muscat et al. (2001) [105] conducted a case-control study on 469 men and women with brain cancer and 422 matched controls without brain cancer. The findings suggest that the use of cellular telephones was not associated with a risk of brain cancer. Similarly, Inskip et al. (2001) [106] conducted a study and examined the effects of cellular telephones on intracranial tumors. There was no evidence that the risks were higher among persons who used cellular telephones for 60 or more minutes per day or regularly for five or more years. These data do not support the hypothesis that the use of hand-held cellular telephones causes brain tumors. Correspondingly, Croft et al. (2008) [107], Ahlbom et al. (2009) [108], de Vocht (2019) [109] reported no evidence with mobile phone use on acoustic neuroma and meningioma.

Takebayashi et al. (2008) [110] performed a study on brain tumors related to mobile phones among 322 cases and 683 matched controls. There was no significant trend toward an increasing odd ratio concerning specific absorption rate (SAR) derived exposure. Sorahan et al. (2014) [111] also observed no evidence on exposure to magnetic fields as a risk factor for gliomas.

Turner et al. (2014) [112] assess the association between exposure to extremely low-frequency magnetic fields (ELF) and brain tumors in the large-scale INTEROCC case-control study. The authors recruited 3,761 patients with brain tumors (1,939 gliomas and 1,822 meningiomas) and 5,404 controls. There was no association between lifetime ELF exposure and glioma or meningioma risk. In another study, Turner et al. (2017) [113] steered a population-based, more extensive INTERPHONE case-control study among people exposed to extremely low-frequency magnetic fields (ELF). The study analysis was based on 1,939 glioma cases, 1,822 meningioma cases, and 5,404 controls. There was no evidence for interactions between occupational ELF and glioma or meningioma risk observed.

In a sizeable sample-sized study, Karipidis et al. (2018) [114] recruited 16,825 brain cancer cases from all of Australia, 10,083 males, and 6,742 females. The authors reported that there had been no increase in any brain tumor histological type or glioma that can be attributed to mobile phones. In an animal model study, Bua et al. (2018) [115] evaluates the possible carcinogenic effects of extremely low-frequency electromagnetic fields (ELFEMF) exposure to Sprague-Dawley rats. The authors did not find an increased incidence of neoplasia in any organ, including the brain. These studies' findings demonstrate no relationship between RF-EMFR and brain cancer. The animal, human model studies, and conclusions based on the meta-analysis literature support the hypothesis that excessive exposure to RF-EMFR correlates with brain tumors (Table 7.6).

Table 7.6 Association of RF-EMFR on Brain Tumors

Author and Year of Study	Type of Study	Sample Size	Study Outcomes
Muscat et al. (2001) [105]	Case-control study	469 patients with brain cancer and 422 controls	No evidence with risk of brain cancer and use of hand-held cellular telephones.
Inskip et al. (2001) [106]	Case-control study	782 patients	No evidence for risks among persons who used cellular telephones for 60 or more min per day or five years use of mobile phone.
Cardis et al. (2010)[62]	Case-control study	2,708 glioma 2,409 meningioma	No rise in the risk of glioma or meningioma.
Croft et al. (2008) [107]	Review	----	No evidence was found between mobile phones and related technologies and head and neck tumors.
Patrizia Frei et al. (2011) [75]	Nationwide cohort study	10,729 brain tumor	No increased risks of tumors in the central nervous system.

(Continued)

Table 7.6 (Continued)

Author and Year of Study	Type of Study	Sample Size	Study Outcomes
Ahlbom et al. (2009) [108]	Review	----	No overall increased risk within ten years of use of the mobile phone for any tumor of the brain.
de Vocht (2019) [109]	---	---	No evidence for mobile phone use on acoustic neuroma and meningioma.
Takebayashi et al. (2008) [110]	case-control study	322 cases and 683 matched controls.	No significant trend toward an increasing odd ratio concerning specific absorption rate exposure to mobile phones.
Sorahan et al. (2014) [111]	Longitudinal study	73,051 employees	No evidence on exposure to magnetic fields as a risk factor for gliomas.
Turner et al. (2014) [112]	large-scale INTEROCC case-control study.	3,761 brain tumor, cases and 5,404 controls	No association was found between lifetime cumulative ELF exposure and glioma or meningioma.
Turner et al. (2017) [113]	A population-based, more extensive INTERPHONE case-control study	1,939 glioma cases, 1,822 meningioma cases, and 5,404 controls.	There is no clear evidence of glioma or meningioma risk in people who had occupational exposure to extremely low-frequency magnetic fields (ELF).
Karipidis et al. (2018) [114]	The population-based ecological study,	16,825 brain cancer cases, 10,083 males and 6,742 females	No increase in any brain tumor histological type or glioma attributed to mobile phones.
Bua et al. (2018) [115]	Male and female Sprague-Dawley rats	Cross-sectional study	No increase incidence of brain cancer during exposure to extremely low-frequency electromagnetic fields.

Note: All these studies did not find any association between RF-EMFR and brain tumor.

7.7 Mechanism of RF-EMFR and Brain Cancer

Human beings are exposed to various natural and human-made sources of RF-EMFR. Anthropogenic RF-EMFR exposure levels for different periods in the evolution of wireless communication technologies were variable with different RF-EMFR (Figure 7.9). Over the last two decades, many studies have been conducted worldwide to evaluate the potential health risks of mobile phones. Some of them have suggested that people can have a higher risk of brain cancer, specifically on the side of the head, while using the mobile [116]. So far, limited biological mechanisms are established to link mobile radiation and the risk of head cancer. Overall, the results detracted from the hypothesis that mobile phone use can cause the occurrence of brain tumors [60, 117].

There is a piece of evidence about the effect of long-term use of mobile phones and brain cancer. The most probable mechanism in which RF-EMFR causes brain cancer is the thermal effect of RF-EMFR. The RF-EMFR causes inflammation, breakdown of the brain defense, and impairment of the BBB. RF-EMFR increases the formation of free radicles, ROS, including peroxides, superoxide, and superoxide hydroxyl radical. These factors can cause oxidative stress. The changes in antioxidant enzymatic function damage macro-molecules, DNA strand breakdown, and cause apoptosis and tumor formation (Figure 7.10).

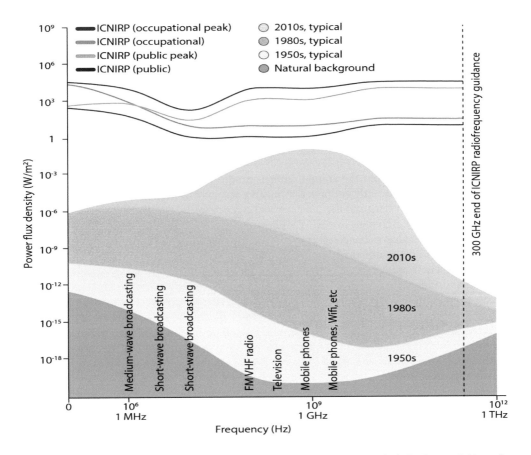

Figure 7.9 Anthropogenic RF-EMFR exposure levels for different periods in the evolution of wireless communication technologies.

Source: Adapted and modified with permission from the author and publisher [118].

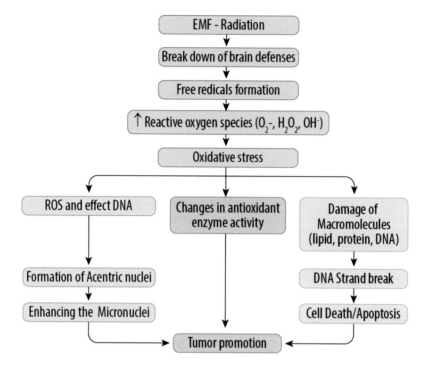

Figure 7.10 Mechanism of RF-EMFR and tumor promotion.

Conclusions

The evidence collected from the animal, human model studies, meta-analysis, and epidemiological literature significantly endorse the positive association between RF-EMFR and brain cancer exposure. The WHO and the International Agency for Research on Cancer (IARC) had already classified the extremely low RF-EMFR generated by mobile phones, base-station towers, and electrical devices as conceivably carcinogenic [6, 7]. Moreover World Health Organization also classified these radiations as increasing the risk of malignant brain tumors.

Presently, environmental health scientists highlight the global environmental, occupational, and industrial allied health hazards and climate change issues. An acute need to address the environmental pollution associated with Radio Frequency Electromagnetic Field Radiation (RF-EMFR), also known as "Electrosmog," affects human health and brain cancer. The governmental and health authorities must take directives to minimize public exposure to RF-EMFR by regulating mobile phone usage and avoiding installing the mobile phone base station towers in or near the thickly residential and high populated areas to minimize exposure to RF-EMFR.

References

[1] Heath Jr C.W. Electromagnetic field exposure and cancer: A review of epidemiologic evidence. *CA Cancer. J. Clin.* 1996; 46(1): 29–44. doi: 10.3322/canjclin.46.1.29.

[2] World Health Organization. Electromagnetic fields. Available at: www.who.int/peh-emf/about/WhatisEMF/en/. Cited date Dec 12, 2020.

[3] Lewczuk B., Redlarski G., Zak A., Ziółkowska N., Przybylska-Gornowicz B., Krawczuk M. Influence of electric, magnetic, and electromagnetic fields on the circadian system: Current stage of knowledge. *Biomed. Res. Int.* 2014; 2014: 169459.

[4] Canadian Nuclear Safety Commission: Types and sources of radiation. Available at: http://nuclearsafety.gc.ca/eng/resources/radiation/introduction-to-radiation/types-and-sources-of-radiation.cfm. Cited date Jun 2019.

[5] International Commission on Non-Ionizing Radiation Protection [ICNIRP]. Guidelines for limiting exposure to time-varying electric, magnetic, and electromagnetic fields (up to 300 GHz). *Heal. Phys.* 1998; 74: 494–522.

[6] International Agency for Research on Cancer. Non-ionizing radiation, part 1: Static and extremely low-frequency (ELF) electric and magnetic fields. *IARC Monographs on the Evaluation of Carcinogenic Risks to Humans.* 2002; 80: 1–395.

[7] International Agency for Research on Cancer. IARRC classifies radiofrequency electromagnetic fields as possibly carcinogenic to humans. Press Release No 2008, 2011.

[8] History of mobile phones. Available at: http//docshare02.docshare.tips/files/3715/37154758.pdf. Cited date Nov 16, 2019.

[9] Röösli M., Lagorio S., Schoemaker M.J., Schüz J., Feychting M. Brain, and salivary gland tumors and mobile phone use: Evaluating the evidence from various epidemiological study designs. *Annu. Rev. Public. Health.* 2019; 40: 221–38.

[10] Rappaport T.S., Sun S., Mayzus R., Zhao H., Azar Y., Wang K., et al. Millimeter wave mobile communications for 5G cellular: It will work. *IEEE Access.* 2013; 1: 335–49. doi: 10.1109/ACCESS.2013.2260813.

[11] Miller A.B., Sears M.E., Morgan L.L., Davis D.L., Hardell L., Oremus M., Soskolne C.L. Risks to health and well-being from radio-frequency radiation emitted by cell phones and other wireless devices. *front. Public. Health.* 2019; 7: 223. doi: 10.3389/fpubh.2019.00223.

[12] Al-Khlaiwi T., Meo S.A. Association of mobile phone radiation with fatigue, headache, dizziness, tension and sleep disturbance in Saudi population. *Saudi Med. J.* 2004; 25: 732–36.

[13] The World Bank. Population total. Available at: https://data.worldbank.org/indicator/SP.POP.TOTL. Cited date Feb 12, 2019.

[14] The World Bank. Mobile phone subscription. Available at: https://data.worldbank.org/indicator/IT.CEL.SETS.P2. Cited date Jan 12, 2020.

[15] Marie-Christine G., Sven K., Niels K. Experimental and numerical assessment of low-frequency current distributions from UMTS and GSM mobile phones. *Phys. Med. Biol.* 2013; 58. doi: 10.1088/0031-9155/58/23/8339.

[16] Bhatt, C.R.R. Assessment of personal exposure from radiofrequency electromagnetic fields in Australia and Belgium using on-body calibrated expositors. *Environ. Res.* 2016; 151: 547–63.

[17] World Health Organization (WHO): International Agency for Research on Cancer (IARC). Classifies radiofrequency electromagnetic fields as possibly carcinogenic to humans. 2011. Available at: www.iarc.fr/wp-content/uploads/2018/07/pr208_E.pdf. Cited date Jan 2, 2018.

[18] World Health Organization. WHO-clarification of the mooted relationship between mobile telephone base stations and cancer. Available At: www.who.int/ mediacentre/news/statements/statementemf/en/. Cited date Oct 1, 2018.

[19] Feychting M., Ahlbom A., Kheifets L. EMF and health. *Annu. Rev. Public. Health.* 2005; 26(1): 165–89.

[20] Roosli M. Radiofrequency electromagnetic field exposure and non-specific symptoms of ill health: A systematic review. *Environ. Res.* 2008; 107(2): 277–87.

[21] Colombi D., Thors B., Tornevik C. Implications of emf exposure limits on output power levels for 5g devices above six GHz. *IEEE. Antennas. Wirel. Propag. Lett.* 2015; 14: 1247–9.

[22] Pi A., Khan F. An introduction to millimeter-wave mobile broadband systems. *IEEE. Commun. Mag.* 2011; 49: 101–7.

[23] Alkis M.E., Bilgin H.M., Akpolat V., Dasdag S., Yegin K., Yavas M.C., Akdag M.Z. Effect of 900–1800 and 2100-MHz radiofrequency radiation on DNA and oxidative stress in the brain. Electromagn. *Biol. Med.* 2019; 38(1): 32–47.

[24] Redlarski G., Lewczuk B., Żak A., Koncicki A., Krawczuk M. Piechocki J., Jakubiuk K., Tojza P., Jaworski J., Ambroziak D.3., Skarbek Ł.3., Gradolewski D.3. The influence of electromagnetic pollution on living organisms: Historical trends and forecasting changes. *Biomed. Res. Int.* 2015; 2015: 234098.

[25] Westerman R., Hocking B. Diseases of modern living: Neurological changes associated with mobile phones and radiofrequency radiation in humans. *Neurosci. Lett.* 2004; 361: 13–6.

[26] Meo S.A., Al-Dress A.M. Mobile phone related-hazards and subjective hearing and vision symptoms in the Saudi population. *Int. J. Occup. Med. Environ. Health.* 2005; 18(1): 53–7.

[27] Davidson Harry C., Lutman Mark E. Survey of mobile phone use and their chronic effects on the hearing of a student population. *Int. J. Audiol.* 2007; 46(3): 113–8.

[28] Balikci K., Cem Ozcan I., Turgut-Balik D., Balik H.H. A survey study on some neurological symptoms and sensations experienced by long-term mobile phone users. *Pathol. Biol.* 2005; 53: 30–4.

[29] Saini R., Saini S., Sharma S. Neurological dysfunction and mobile phones. *J. Neurosci. Rural. Pract.* 2010; 1(1): 57–8. doi: 10.4103/0976-3147.63110.

[30] Tettamanti G., Auvinen A., Åkerstedt T., Kojo K., Ahlbom A., Heinävaara S., Elliott P. Long-term effect of mobile phone use on sleep quality: Results from the cohort study of mobile phone use and health (COSMOS). *Environ. Int.* 2020 Jul; 140: 105687. doi: 10.1016/j.envint.2020.105687.

[31] Haarala C., Takio F., Rintee T., Laine M., Koivisto M., Revonsuo A., Hämäläinen H. Pulsed and continuous mobile phone exposure over left versus right hemisphere: Effect on human cognitive function. *Bioelectromagnetics.* 2007; 28(4): 289–95. doi: 10.1002/bem.20287.

[32] Huber R., Treyer V., Borbély A.A., Schuderer J., Gottselig J.M., Landolt H.P., Werth E., Berthold T. Electromagnetic fields, such as those from mobile phones, alter regional cerebral blood flow and sleep and waking EEG. *J. Sleep. Res.* 2002; 11(94): 289–95.

[33] Hossman K.A., Hermann D.M. Effects of electromagnetic radiation of mobile phones on the central nervous system. *Bioelectromagnetics.* 2003; 24: 49.

[34] Salford L.G., Brun Arne E., Eberhardt Jacob L., Lars Malmgren, Persson Bertil R.R. Nerve cell damage in mammalian brain after exposure to microwaves from GSM mobile phones. *Environ. Health. Perspect.* 2003; 111(7): 881–4.

[35] GBD: Brain and Other CNS Cancer Collaborators. Global, regional, and national burden of brain and other CNS cancer, 1990–2016: A systematic analysis for the global burden of disease study 2016. *Lancet. Neurol.* 2019; 18(4): 376–93.

[36] Gould J. Breaking down the epidemiology of brain cancer. *Nature.* 2018; 561(7724): S40–1.

[37] Ostrom Q.T., Gittleman H., Fulop J., Liu M., Blanda R., Kromer C., Wolinsky Y., Kruchko C., Barnholtz-Sloan J.S. CBTRUS statistical report: primary brain and central nervous system tumors diagnosed in the United States in 2008–2012. *Neuro. Oncol.* 2015 (Suppl 4): 62. doi: 10.1093/neuonc/nov189.

[38] World Health Organization. Available at: https: Cancer. //www.who.int/health-topics/cancer#tab=overview. Cited date Feb 22, 2020.

[39] International Childhood Cancer Day. [ICCD] Available at: www.internationalchildhoodcancerday.org/. Cited date Feb 28, 2020.

[40] Rebecca Leece, Jordan X.U., Ostrom Quinn T., Chen Yanwen, Kruchko Carol, Barnholtz-Sloan Jill S. The global incidence of malignant brain and other central nervous system tumors by histology, 2003–2007. *Neuro. Oncol.* 2017; 19(11): 1553–64.

[41] Savage N. Searching for the roots of brain cancer. *Nature.* 2018; 561(7724): S50–1.

[42] Zarei S., Vahab M., Oryadi-Zanjani M.M., Alighanbari N., Mortazavi S.M. Mother's exposure to electromagnetic fields before and during pregnancy is associated with risk of speech problems in offspring. *J. Biomed. Phys. Eng.* 2019; 9(1): 61–8.

[43] Belpomme D., Hardell L., Belyaev I., Burgio E., Carpenter D.O. Thermal and non-thermal health effects of low-intensity non-ionizing radiation: An international perspective. *Environ. Pollut.* 2018; 242(Pt A): 643–58. doi: 10.1016/j.envpol.2018.07.019.

[44] Wertheimer N., Leeper E. Electrical wiring configurations and childhood cancer. *Am. J. Epidemiol.* 1979; 109(3): 273–84.

[45] Kıvrak E.G., Yurt K.K., Kaplan A.A., Alkan I., Altun G. Effects of electromagnetic fields exposure on the antioxidant defense system. *J. Microsc. Ultrastruct.* 2017; 5(4): 167–76.

[46] Meral I., Mert H., Mert N., Deger Y., Yoruk I., Yetkin A. Effects of 900-MHz electromagnetic field emitted from cellular phone on brain oxidative stress and some vitamin levels of guinea pigs. *Brain. Res.* 2007; 1169: 120–4.

[47] Leszczynski D., Joenväärä S., Reivinen J., Kuokka R. Non-thermal Activation of the hsp27/p38MAPK stress pathway by mobile phone radiation in human endothelial cells: Molecular mechanisms for cancer- and blood–brain barrier-related effects differentiation. 2002; 70(2–3): 120–9. doi: 10.1046/j.1432-0436.2002.700207.x.

[48] Miller A.B., Morgan L.L., Udasin I., Davis D.L. Cancer epidemiology update, following the 2011 IARC evaluation of radiofrequency electromagnetic fields (Monograph 102). *Environ. Res.* 2018; 167: 673–83.

[49] Melnick R.L. Commentary on the utility of the national toxicology program study on cell phone radiofrequency radiation data for assessing human health risks despite unfounded criticisms aimed at minimizing the findings of adverse health effects. *Environ. Res.* 2019; 168: 1–6.

[50] Wyde M., Cesta M., Blystone C., Elmore S., Foster P., Hooth M. Bio. Rxiv. *Report of Partial Findings from the National Toxicology Program Carcinogenesis Studies of Cell Phone Radiofrequency Radiation in HSD: Sprague Dawley® SD Rats.* Whole Body Exposure. North Carolina, USA: United States National Toxicology Program. 2018, p. 055699.

[51] Saba Shahin, Somanshu Banerjee, Surya Pal Singh, Chandra Mohini Chaturvedi. 2.45 GHz microwave radiation impairs learning and spatial memory via oxidative/nitrosative stress-induced p53-dependent/independent hippocampal apoptosis: Molecular basis and underlying mechanism. *Toxicol. Sci.* 2015; 148(2): 380–99. doi: 10.1093/toxsci/kfv205.

[52] Saba Shahin, Somanshu Banerjee, Vivek Swarup, Surya Pal Singh, Chandra Mohini Chaturvedi. From the cover: 2.45-GHz microwave radiation impairs hippocampal learning and spatial memory: involvement of local stress mechanism-induced suppression of iGluR/ERK/CREB signaling. *Toxicol. Sci.* 2018; 161(2): 349–74. doi: 10.1093/toxsci/kfx221.

[53] Karimi Narges, Bayat Mahnaz, Haghani Masoud, Fahandezh Saadi Hamed, Ghazipour Gholam Reza. 2.45 GHz microwave radiation impairs learning, memory, and hippocampal synaptic plasticity in the rat. *Toxicol. Ind. Health.* 2018; 34(12): 873–83. doi: 10.1177/0748233718798976.

[54] Ragy M.M. Effect of exposure and withdrawal of 900-MHz-electromagnetic waves on brain, kidney, and liver oxidative stress, and biochemical parameters in male rats. Electromagnetic. *Biology. And. Medicine.* 2015; 34(4): 279–84.

[55] Falcioni L., Bua L., Tibaldi E., Lauriola M., De Angelis L., et al. Report of final results regarding brain and heart tumors in Sprague-Dawley rats exposed to prenatal life natural death to mobile phone radiofrequency field representative of a 1.8 GHz GSM base station environmental emission. *Environ. Res.* 2018; 165: 496–503.

[56] Saikhedkar Nidhi, Bhatnagar Maheep, Jain Ayushi, Sukhwal Pooja, Sharma Chhavi, Jaiswal Neha. Effects of mobile phone radiation (900 MHz radiofrequency) on structure and functions of rat brain. *Neurological. Research.* 2014; 36(12): 1072–9. doi: 10.1179/1743132814Y.0000000392.

[57] Adamantia F.F., Samara A., Antonelou M.H., Xanthopoulou A., Papadopoulou A., Vougas K. Brain proteome response following whole body exposure of mice to mobile phone or wireless DECT base radiation. *Electromagn. Biol. Med.* 2012; 31(4): 250–74. doi: 10.3109/15368378.2011.631068.

[58] Adamantia F.F, Polyzos A., Papadopoulou M.-D., Sansone A., Manta A.K., Balafas E., et al. Hippocampal lipidome and transcriptome profile alterations triggered by acute exposure of mice to GSM 1800 MHz mobile phone radiation: An exploratory study. *Brain. Behav.* 2018; 8(6): e01001. doi: 10.1002/brb3.1001.

[59] Hao D., Yang L., Chen S., Tong J., Tian Y., Su B., Wu S., Zeng Y. Effects of long-term electromagnetic field exposure on spatial learning and memory in rats. *Neurol. Sci.* 2013 Feb; 34(2): 157–64.

[60] Lagorio S.1., Röösli M. Mobile phone use and risk of intracranial tumors: A consistency analysis. *Bioelectromagnetics.* 2014; 35(2): 79–90.

[61] Moulder J.E., Foster K.R., Erdreich L.S., McNamee J.P. Mobile phones, mobile phone base stations, and cancer: A review. *Int. J. Radiat. Biol.* 2005; 81(3): 189–203.

[62] Cardis E., Deltour I., Vrijheid M., Combalot E., Moissonnier M., Tardy H., et al., INTERPHONE study group. brain tumor risk in relation to mobile telephone use: Results of the INTERPHONE international case-control study. *Int. J. Epidemiol.* 2010; 39(3): 675–94.

[63] Cardis E., Armstrong B.K., Bowman J.D., Giles G.G., Hours M., Krewski D., McBride M., Parent M.E., Sadetzki S., Woodward A., et al. Risk of brain tumours to estimated RF dose from mobile phones: Results from five interphone countries. *Occup. Environ. Med.* 2011; 68(9): 631–40. doi: 10.1136/oemed-2011-100155.

[64] Wang P., Hou C., Li Y., Zhou D. Wireless phone use and risk of adult glioma: Evidence from a meta-analysis. *World. Neurosurg.* 2018; 115: e629–36.

[65] Sato Y., Kiyohara K., Kojimahara N., Yamaguchi N. Time trend in the incidence of malignant neoplasms of the central nervous system in relation to mobile phone use among young people in Japan. *Bioelectromagnetics.* 2016; 37(5): 282–9.

[66] Lerchl A., Klose M., Grote K., Wilhelm A.F., Spathmann O., Fiedler T., Streckert J., Hansen V., Clemens M. Tumor promotion by exposure to radiofrequency electromagnetic fields below exposure limits for humans. *Biochem. Biophys. Res. Commun.* 2015; 459(4): 585–90.

[67] Tillmann T., Ernst H., Strecker J., Zhou Y., Taugner F., Hansen V., Dasenbrock C. Indication of the cocarcinogenic potential of chronic UMTS-modulated radiofrequency exposure in an ethyl-nitrosourea mouse model. *Int. J. Radiat. Biol.* 2010; 86(7): 529–41.

[68] Hardell L., Carlberg M., Söderqvist F., Mild K.H. Case-control study of the association between malignant brain tumors diagnosed between 2007 and 2009 and mobile and cordless phone use. *Int. J. Oncol.* 2013; 43(6): 1833–45.

[69] Hardell L., Carlberg M. Mobile phone and cordless phone use and the risk for glioma—Analysis of pooled case-control studies in Sweden, 1997–2003 and 2007–2009. *Pathophysiology.* 2015; 22(1): 1–13.

[70] Coureau G., Bouvier G., Lebailly P., Fabbro-Peray P., Gruber A., Leffondre K., Guillamo J.S., Loiseau H., Mathoulin-Pélissier S., Salamon R., Baldi I. Mobile phone use and brain tumours in the CERENAT case-control study. *Occup. Environ. Med.* 2014; 71(7): 514–22.

[71] deVocht F. Inferring the 1985–2014 impact of mobile phone use on selected brain cancer subtypes using Bayesian structural time series and synthetic controls. *Environ. Int.* 2016; 97: 100–7.

[72] Grell K., Frederiksen K., Schüz J., Cardis E., Armstrong B., Siemiatycki J., et al. The intracranial distribution of gliomas to exposure from mobile phones: Analyses from the INTERPHONE Study. *Am. J. Epidemiol.* 2016; 184(11): 818–28.

[73] Gao He, Aresu Maria, Vergnaud Anne-Claire, McRobie Dennis, Spear Jeanette, Heard Andy, Kongsgård Håvard Wahl, Singh Deepa, Muller David C., Elliott Paul. Personal radio use and cancer risks among 48,518 British police officers and staff from the airwave health monitoring study. *Br. J. Cancer.* 2019; 120(3): 375–78.

[74] Aydin D., Feychting M., Schüz J., Andersen T.V., Poulsen A.H. Impact of random and systematic recall errors and selection bias in case-control studies on mobile phone use and brain tumors in adolescents (CEFALO study). *Bioelectromagnetics.* 2011; 32: 396–407.

[75] Frei P., Poulsen A.H., Johansen C., Jørgen H.O., Steding-Jessen M., Schüz J. Use of mobile phones and risk of brain tumours: Update of Danish cohort study. *BMJ.* 2011 Oct 19; 343: d6387. doi: 10.1136/bmj.d6387.

[76] Prasad M., Kathuria P., Nair P., Kumar A., Prasad K. Mobile phone use and risk of brain tumors: A systematic review of the association between study quality, source of funding, and research outcomes. *Neurol. Sci.* 2017; 38(5): 797–810.

[77] Gong X., Wu J., Mao Y., Zhou L. Long-term use of mobile phone and its association with glioma: A systematic review and meta-analysis. *Zhonghua Yi Xue Za Zhi.* 2014; 94(39): 3102–6.

[78] French P.W., Penny R., Laurence J.A., McKenzie D.R. Mobile phones, heat shock proteins, and cancer. *Differentiation.* 2001; 67(4–5): 93–7. doi: 10.1046/j.1432-0436.2001.670401.x.

[79] Momoli F., Siemiatycki J., McBride M.L., Parent M.É., Richardson L., Bedard D., Platt R., Vrijheid M., Cardis E., Krewski D. Probabilistic multiple-bias modeling applied to the Canadian data from the interphone study of mobile phone use and risk of glioma, meningioma, acoustic neuroma, and parotid gland tumors. *Am. J. Epidemiol.* 2017 Oct 1; 186(7): 885–93.

[80] Mortazavi S.M.J., Mortazavi S.A.R., Haghani M. Evaluation of the validity of a nonlinear J-shaped dose-response relationship in cancers induced by exposure to radiofrequency electromagnetic fields. *J. Biomed. Phys. Eng.* 2019; 9(4): 487–94.

[81] Lonn S., Ahlbom A., Hall P., Feychting M. Mobile phone use and the risk of acoustic neuroma. *Epidemiology.* 2004; 15: 653–9.

[82] Schoemaker M.J., Swerdlow A.J., Ahlbom A., Auvinen A., Blaazas K.G., Cardis E. Mobile phone use and risk of acoustic neuroma: Results of the interphone case-control study in five North European countries. *Br. J. Cancer.* 2005; 93: 842–8.

[83] Lonn S., Ahlbom A., Hall P., Feychting M., Swedish interphone study group: Long-term mobile phone use and brain tumor risk. *Am. J. Epidemiol.* 2005; 161: 526–35.

[84] Lonn S., Ahlbom A., Christensen H.C., Johansen C., Schuz J., Edstrom S. Mobile phone use and risk of parotid gland tumor. *Am. J. Epidemiol.* 2006; 164(7): 637–43.

[85] Klaeboe L., Blaasaas K.G., Tynes T. Use of mobile phones in Norway and risk of intracranial tumors. *Eur. J. Cancer. Prev.* 2007; 16(2): 158–64.

[86] Lahkola A., Auvinen A., Raitanen J., Schoemaker M.J., Christensen H.C., Feychting M., Johansen C., Klaeboe L., Lonn S., Swerdlow A.J., Tynes T., Salminen T. Mobile phone use and risk of glioma in 5 North European countries. *Int. J. Cancer.* 2007; 120: 1769–75.

[87] Lahkola A., Salminen T., Raitanen J., Heinavaara S., Schoemaker M.J., Christensen H.C. Meningioma, and mobile phone use—a collaborative case-control study in five North European countries. *Int. J. Epidemiol.* 2008; 37: 1304–13.

[88] Cardis E., Deltour I., Vrijheid M., Combalot E., Moissonnier M., Tardy H., et al. The interphone study group: Brain tumor risk in relation to mobile telephone use: Interphone international case-control study results. *Int. J. Epidemiol.* 2010; 39: 675–94.

[89] Sadetzki S., Chetrit A., Jarus-Hakak A., Cardis E., Deutch Y., Duvdevani S. Cellular phone use and risk of benign and malignant parotid gland tumors—a nationwide case-control study. *Am. J. Epidemiol.* 2008; 167: 457–67.

[90] Yang M., Guo W., Yang C., JianQin T., Qian H., ShouXin F., AiJun J., XiFeng X.U., Guan J. Mobile phone use and glioma risk: A systematic review and meta-analysis. *PLoS One.* 2017 May 4; 12(5): e0175136. doi: 10.1371/journal.pone.0175136. eCollection 2017.

[91] Hardell L., Carlberg M., Söderqvist F., Mild K.H. Meta-analysis of long-term mobile phone use and the association with brain tumours. *Int. J. Oncol.* 2008; 32(5): 1097–103.

[92] Michael Kundi. The controversy about a possible relationship between mobile phone use and cancer. *Cien. Saude. Colet.* 2010; 15(5): 2415–30. doi: 10.1590/s1413-81232010000500016.

[93] Carlberg M., Hardell L. Evaluation of mobile phone and cordless phone use and glioma risk using the Bradford hill viewpoints from 1965 on association or causation. biomed. *Res. Int.* 2017; 2017: 9218486. doi: 10.1155/2017/9218486.

[94] Bortkiewicz A., Gadzicka E., Szymczak W. Mobile phone use and risk for intracranial tumors and salivary gland tumors—A meta-analysis. *Int. J. Occup. Med. Environ. Health.* 2017; 30: 27–43.

[95] Wang Y., Guo X. Meta-analysis of association between mobile phone use and glioma risk. *J. Cancer. Res. Ther.* 2016; 12: C298–C300. doi: 10.4103/0973-1482.200759.

[96] Levis A.G., Minicuci N., Ricci P., Gennaro V., Garbisa S. Mobile phones and head tumours. The discrepancies in cause-effect relationships in the epidemiological studies—how do they arise? *Environ. Health.* 2011; 10: 59. doi: 10.1186/1476-069X-10-59.

[97] Seung-Kwon M., Woong J.U., McDonnell D.D., Yeon J.L., Kazinets G., Chih-Tao C., Joel M.M. Mobile phone use and risk of tumors: A meta-analysis. *J. Clin. Oncol.* 2009; 27(33): 5565–72. doi: 10.1200/JCO.2008.21.6366.

[98] Khurana Vini G., Teo Charles, Kundi Michael, Hardell Lennart, Carlberg Michael. Cell phones and brain tumors: A review including the long-term epidemiologic data. *Surg. Neurol.* 2009; 72(3): 205–14. doi: 10.1016/j.surneu.2009.01.019.

[99] Leng L. The relationship between mobile phone use and risk of brain tumor: A systematic review and meta-analysis of trails in the last decade. *Chin. Neurosurg. Jl. 2.* 2106; 38. https://doi.org/10.1186/s41016-016-0059-y.

[100] Yakymenko I., Sidorik E., Kyrylenko S., Chekhun V. Long-term exposure to microwave radiation provokes cancer growth: Evidence from radars and mobile communication systems. *Exp. Oncol.* 2011; 33(2): 62–70.

[101] Alexiou G.A., Sioka C. Mobile phone use and risk for intracranial tumors. *J Negat Results Biomed.* 2015; 14: 23.

[102] Leng L., Zhang Y. Etiology of pituitary tumors: A case-control study. *Turk. Neurosurg.* 2016; 26(2): 195–9. doi: 10.5137/1019-5149.JTN.5985-12.1.

[103] Schoemaker M.J., Swerdlow A.J. Risk of pituitary tumors in cellular phone users: A case-control study. *Epidemiology.* 2009; 20(3): 348–54.

[104] Shrestha M., Raitanen J., Salminen T., Lahkola A., Auvinen A. Pituitary tumor risk in relation to mobile phone use: A case-control study. *Acta. Oncol.* 2015; 54(8): 1159–65. doi: 10.3109/0284186X.2015.1045624.

[105] Muscat J.E., Malkin M.G., Thompson S., Shore R.E., Stellman S.D., McRee D., Neugut A.I., Wynder E.L. Handheld cellular telephone use and risk of brain cancer. *JAMA.* 2000 Dec 20; 284(23): 3001–7.

[106] Inskip P.D., Tarone R.E., Hatch E.E., Wilcosky T.C., Shapiro W.R., Selker R.G., Fine H.A., Black P.M., Loeffler J.S., Linet M.S. Cellular-telephone use, and brain tumors. *N Engl J Med.* 2001 Jan 11; 344(2): 79–86. doi: 10.1056/NEJM200101113440201. PMID: 11150357.

[107] Croft R.J., McKenzie R.J., Inyang I., Benke G.P., Anderson V., Abramson M.J. Mobile phones and brain tumors: A review of epidemiological research. *Australas Phys Eng Sci Med*. 2008 Dec; 31(4): 255–67. doi: 10.1007/BF03178595. PMID: 19239052.

[108] Ahlbom A., Feychting M., Green A., Kheifets L., Savitz D.A., Swerdlow A.J. ICNIRP (International Commission for Non-Ionizing Radiation Protection) standing committee on epidemiology. Epidemiologic evidence on mobile phones and tumor risk: A review. *Epidemiology*. 2009 Sep; 20(5): 639–52. doi: 10.1097/EDE.0b013e3181b0927d.

[109] de Vocht F. Analyses of temporal and spatial patterns of glioblastoma multiforme and other brain cancer subtypes to mobile phones using synthetic counterfactuals. *Environ Res*. 2019; 168: 329–35. doi: 10.1016/j.envres.2018.10.011.

[110] Takebayashi T., Varsier N., Kikuchi Y., Wake K., Taki M., Watanabe S., Akiba S., Yamaguchi N. Mobile phone use, exposure to the radiofrequency electromagnetic field, and brain tumor: A case-control study. *Br J Cancer*. 2008 Feb 12; 98(3): 652–9. doi: 10.1038/sj.bjc.6604214.

[111] Sorahan T. Magnetic fields and brain tumor risks in UK electricity supply workers. *Occup Med (Lond)*. 2014 Apr; 64(3): 157–65. doi: 10.1093/occupied/kqu003.

[112] Turner M.C., Benke G., Bowman J.D., Figuerola J., Fleming S., Hours M., Kincl L., Krewski D., McLean D., Parent M.E., Richardson L., Sadetzki S., Schlaefer K., Schlehofer B., Schüz J., Siemiatycki J., van Tongeren M., Cardis E. Occupational exposure to extremely low-frequency magnetic fields and brain tumor risks in the INTEROCC study. *Cancer Epidemiol Biomarkers Prev*. 2014 Sep; 23(9): 1863–72. doi: 10.1158/1055-9965.EPI-14-0102.

[113] Turner M.C., Benke G., Bowman J.D., Figuerola J., Fleming S., Hours M., Kincl L., Krewski D., McLean D., Parent M.E., Richardson L., Sadetzki S., Schlaefer K., Schlehofer B., Schüz J., Siemiatycki J., Tongeren M.V., Cardis E. Interactions between occupational exposure to extremely low-frequency magnetic fields and chemicals for brain tumor risk in the INTEROCC study. *Occup Environ Med*. 2017; 74(11): 802–9. doi: 10.1136/oemed-2016-104080.

[114] Karipidis K., Elwood M., Benke G., Sanagou M., Tjong L., Croft R.J. Mobile phone use and incidence of brain tumor histological types, grading or anatomical location: A population-based ecological study. *BMJ Open*. 2018 Dec 9; 8(12): e024489. doi: 10.1136/bmjopen-2018-024489.

[115] Bua L., Tibaldi E., Falcioni L., Lauriola M., De Angelis L., Gnudi F., Manservigi M., Manservisi F., Manzoli I., Menghetti I., Montella R., Panzacchi S., Sgargi D., Strollo V., Vornoli A., Mandrioli D., Belpoggi F. Results of lifespan exposure to continuous and intermittent extremely low-frequency electromagnetic fields (ELFEMF) administered alone to Sprague Dawley rats. *Environ Res*. 2018; 164: 271–9. doi: 10.1016/j.envres.2018.02.036.

[116] Volkow N.D., Tomasi D., Wang G.J., Vaska P., Fowler J.S., Telang F. Effects of cell phone radiofrequency signal exposure on brain glucose metabolism. *JAMA*. 2011; 305: 808–13.

[117] Schüz J., Jacobsen R., Olsen J.H., Boice J.D., Jr., McLaughlin J.K., Johansen C. Cellular telephone use and cancer risk: Update of a nationwide Danish cohort. *J. Natl. Cancer. Inst*. 2006; 98: 1707–13.

[118] Bandara P., Carpenter D.O. Planetary electromagnetic pollution: It is time to assess its impact. *Lancet. Planet. Health*. 2018; 2(12): e512–4. doi: 10.1016/S2542-5196(18) 30221-3.

Environmental Pollution: Mobile Phone Radiation and Hearing

8.1 Introduction

Environmental pollution contaminates the natural atmosphere and adversely affects the health of living organisms. Various pollutants and persistent substances are produced by human activity, including biological agents, physical substances, chemicals, heavy metals, and particulate matter. These pollutants represent significant risk factors for human disease [1].

The WHO [2] has reported that air pollution has become a growing global concern over the past few decades. Air pollution, predominantly from developed countries, poses a significant threat to global health, climate, and economies. The WHO estimates that 9 out of 10 people breathe air that surpasses their guideline limits on levels of pollutants. Ambient and household air pollution cause approximately seven million deaths each year [2]. In 2020, the number of deaths due to air pollution was over two million in South-East Asia, two million in the Western Pacific region, one million in the African region, 0.5 million in the Eastern Mediterranean, 0.5 million in European regions, and 0.3 million in America [2] (Figure 8.1).

The seven million deaths per year caused by air pollution are attributed to a variety of diseases affecting multiple organs, including ischemic heart disease (34%), pneumonia (21%), stroke (20%), COPD (19%) and lung cancer (6–7%) [2] (Figure 8.2).

The majority of diseases associated with air pollution in the Western Pacific and South-East Asia are attributable to poorly planned heavy industries, although this also represents a global issue. In developing nations, air pollution has become a leading environmental risk factor for premature deaths in the European region [3, 4].

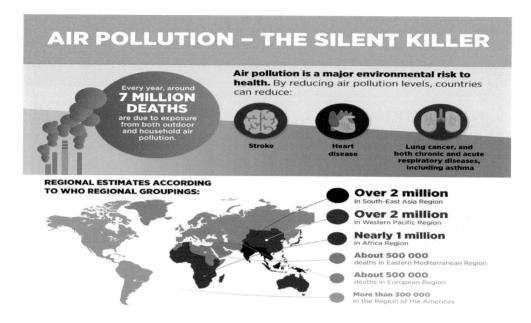

Figure 8.1 Air pollution: a significant environmental risk to health.

Source: Adapted with permission from WHO [2].

Environmental pollution has persisted as a significant concern over recent years with acute and chronic deleterious effects on human health [5]. In this highly advanced, technology-based, competitive era, people are impacted by significant changes related to urbanization, environmental pollution, and climate change. In addition, the heavy use of wireless communication technology, including mobile phones and Wi-Fi, has become a major source of environmental pollution through the generation of RF-EMFR, also known as "electromagnetic smog" [6]. Environmental pollution and RF-EMFR generated by mobile phones and electronic devices are reported to be associated with a variety of health hazards [7, 8].

8.2 Environmental Pollution and Pathophysiology of Hearing Impairment

Several scenarios whereby environmental pollution, including ambient particulate matter (PM) with aerodynamic diameters of PM 2.5 μm and PM 0.1 μm, could play a role in the pathophysiology and progression of diseases affecting various organs, including lungs, liver, heart, brain, and sensory organs. In humans, sensory organs allow people to respond to various stimuli, including sound, light, and information regarding the environment. The five well-described senses are vision, hearing, smell, taste, and touch. Moreover, humans also have a sense of equilibrium and possess a mixed sensory system known as somatosensation, allowing perception and responses to pain, pressure, temperature, and vibration [9]. The acuity and physiology of these senses are reduced with age and the environment, which can affect routine daily activities and reduce the ability of individuals to communicate and cooperate with the world around them [10].

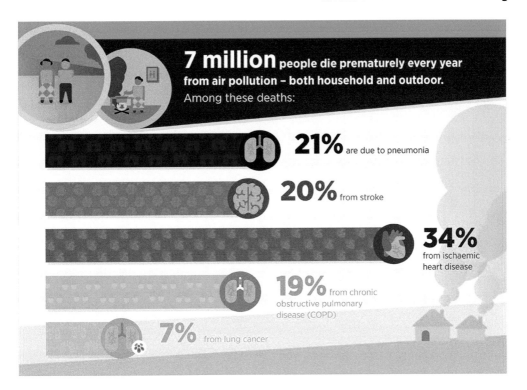

7 million people die prematurely every year from air pollution – both household and outdoor. Among these deaths:

21% are due to pneumonia

20% from stroke

34% from ischaemic heart disease

19% from chronic obstructive pulmonary disease (COPD)

7% from lung cancer

Figure 8.2 Mortality associated with air pollution.

Source: Adapted with permission from WHO [2].

All human organs are crucial for everyday healthy living. In addition to behavior and social well-being, vision and hearing are considered essential for decision-making and planning interventions. Hearing impairment is one of the most common disabilities in the human population and is associated with a range of adverse outcomes. Hearing impairment represents a significant barrier to everyday life by limiting communication, speech recognition, and language acquisition. Moreover, hearing impairment reduces physical functioning and limits the ability to socialize [11, 12].

Previous studies have reported a number of risk factors for cochlear damage and hearing impairment including loud noise [13, 14], viral infection [15, 16], genetic mutations [17], ototoxicity [18], autoimmune diseases [19], hypertension [20], diabetes mellitus [21], stroke [22], alcoholism [23], smoking [24], COPD [25], and occupational and environmental exposure to pollution [26].

Exposure to environmental pollutants such as "benzene, toluene, ethylbenzene, and xylene (a combination known as BTEX), which are soluble and volatile organic compounds found in crude oil and petroleum products" from oil refineries, are major risk factors for ototoxicity [27–30]. Furthermore, studies have demonstrated that long-term exposure to air pollution is associated with neurodegenerative diseases [31, 32], osteoporosis [33], and macular degeneration [34].

Environmental pollutants, predominantly consisting of airborne fine and ultrafine particulate matter (with aerodynamic diameters of less than PM 2.5 μm and less than PM 0.1 μm, respectively), can penetrate deep into the lungs and gastrointestinal tract (GIT). These particles have been shown to enter into the circulation via the lungs or GIT and deposit in many organs, including lungs, liver, heart, and brain, where they have deleterious effects on organ function [35–38].

Ultrafine PM may enter the brain through the BBB or nasal mucosa via the olfactory nerve and affect various brain structures and neuronal pathways [39–41].

PM particulates are reported to have prothrombotic effects by increasing levels of inflammatory mediators in the lungs and circulation. Exposure to pollutants such as petroleum and diesel exhaust particles results in aberrant platelet function and thrombus formation in both arteries and veins [42]. PM may exert prothrombotic effects by increasing circulating concentrations of fibrinogen, factor VIII, and tissue factor, thereby increasing the risk of both venous and arterial thrombosis [43]. The particulate matter with aerodynamic diameter PM 2.5 μm or less induces prothrombotic interactions. It affects multiple organs, impairs microcirculation, and results in ischemia and organ dysfunction [44, 45].

The hypothesis that environmental pollution has deleterious effects on various organs is supported by results demonstrating airborne PM has systemic effects, including increased oxidative stress, inflammation, and genotoxicity. These factors may play significant roles in the pathogenesis of various diseases [46], including inner ear dysfunction and hearing impairment.

Environmental and occupational exposure to pollutants, such as dust, fumes, organic and volatile substances, and PM, can impair the auditory system [47] by injuring the inner ear cochlea [48]. These pollutants induce oxidative stress in the inner ear, resulting in the formation of superoxides, such as 8-isoprostane, in the organ of Corti, spiral ganglion neurons, and stria vascularis [49]. Environmental pollution can impair microcirculation in the cochlea of the inner ear resulting in hypoxia and organ dysfunction [50, 51]. Hypoxia is an established cause of damage to cochlear hair cells and neurons in the inner ear and impairment of hearing [50, 51]. These mechanisms, including increased stress, inflammation, hypoxia, ischemia, and prothrombotic effects, contribute to disease progression in different inner ear and brain structures, resulting in hearing impairment.

8.3 Biology of the Ear and Hearing

Hearing is one of the primary senses and is vital for communication. The significant effect of hearing impairment is impaired communication, which can adversely affect relationships with family and friends and cause difficulties in the workplace. Hearing is a conscious appreciation of vibration perceived as a sound wave. The human ear consists of: the external ear (containing the pinna or concha and ear canal); the middle ear (containing the tympanic membrane and the three ossicles, which are the malleus, incus, and stapes); and the inner ear (containing the cochlea, connected to three semicircular canals by the vestibule which provides the sense of balance (Figure 8.3). The cochlea transfers information to the brain through the eighth cranial nerve, known as the vestibular cochlear nerve (VIII). The inner ear contains a specialized organ, known as the organ of Corti, which contains the sensory receptor cells of the ear and lies on the basilar membrane adjacent to the limbus. These cells are responsible for initiating impulses that are perceived as hearing and are supplied by fibers of the cochlear nerve [52].

The important excitatory event in auditory transduction is the deflection of cilia protruding from the surface of hair cells. When cilia are deflected, a transmitter substance is released from hair cells [53]. The hair cells are arranged in a pattern that resembles the shape of the letter,

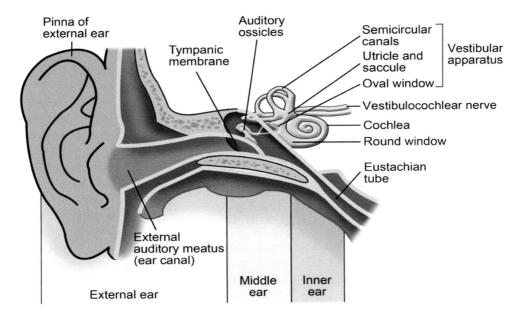

Figure 8.3 The ear structures.

'W', with the base of the letter directed laterally. Hair cells are divided into two types: outer and inner hair cells. Outer hair cells are arranged in parallel in three rows and are approximately 20,000 in number. Each outer hair cell contains 75–100 sensory hairs. Inner hair cells are arranged in a single row and are approximately 3,500 in number. Each inner hair cell contains 40–60 sensory hairs. Another essential structure is the tectorial membrane. This flap of tissue rests on the tips of the cilia of hair cells and is attached to one end of the limbus. Any movement of the basilar membrane causes the free end of the tectorial membrane to drag or push the cilia of hair cells. The spiral ganglion contains the cell bodies of the auditory nerve (cranial nerve VIII), which synapses on hair cells. The cell bodies of the afferent neurons arborize around the bases [54, 55].

8.4 Mechanism of Hearing

Hearing is the sense by which sounds are perceived. Alternatively, it can be said that the capacity to perceive sound is called hearing. After striking the external ear, sound waves enter the external auditory meatus and strike the tympanic membrane and produce motion in the tympanic membrane. The tympanic membrane is coupled with the middle ear ossicles. The foot of stapes transmits vibration at the oval window to the perilymph in the scala vestibule. These pressure waves are transmitted to the perilymph in the scala tympani via the helicotrema. The secondary tympanic membrane at the round window bulges outward into the middle ear cavity. The pressure of sound waves causes the oval window to bulge inward. This pressure (after passing through incompressible fluids of the "scala vestibuli, scala media, and scala tympani) forces the round window to bulge outward. The foot of the

Table 8.1 Summary of the Mechanism of Hearing

- Sound waves strike the external ear
- Sound waves enter the external auditory meatus
- Sound waves strike the tympanic membrane
- Motion is produced in the tympanic membrane
- The foot of stapes transmits membrane vibrations to the oval window
- Pressure waves are transmitted to perilymph in the scala vestibuli and scala tympani
- Sound waves cause the oval window to bulge inward and the round window to bulge outward
- The foot of stapes moves inward, and the basilar membrane moves downward
- Stimulation/depolarization of hair cells occurs
- Sound is perceived by the auditory cortex area of the brain via the vestibulocochlear nerve

stapes moves inward", and the basilar membrane moves downward. This movement of the basilar membrane stimulates hair cells, which depolarize and generate action potentials in fibers of the cochlear nerve, which are transmitted to the auditory cortex in the brain [54, 56, 57] (Table 8.1).

8.5 The Auditory Pathway

The auditory neural pathway in the CNS carries sound signals from the inner ear to the brain. The pathway originates in the cochlear part of the vestibulocochlear nerve. Hair cells in the organ of Corti represent the auditory receptors situated in the internal ear. The first sensory neuron in the auditory pathway lies in the spiral ganglion of bipolar cells. The sensory neurons are situated in the canal around the modiolus at the base of the spiral lamina. Their peripheral processes are distributed to the organ of Corti and the central processes from the cochlear nerve, which ends in the dorsal and ventral cochlear nuclei. The second sensory neuron in the auditory pathway lies in the cochlear nuclei. Their axons form a trapezoid body in the lower pons, ascending in the brainstem as the lateral lemniscus on both sides. Some fibers of the lateral lemniscus terminate in the inferior colliculus of auditory reflex, and the rest of the lateral lemniscus relays in the medial geniculate body. The third and fourth neurons in the auditory pathway are the auditory radiation, a band of neurons that passes through the sub-lentiform part of the internal capsule and is finally relayed in the auditory cortex [54, 56, 57] (Table 8.2; Figure 8.4).

Table 8.2 Summary: Mechanism of Steps Involved Auditory Pathway

- Hair cells in the organ of Corti and spiral ganglion of the organ of Corti
- Vestibulocochlear nerve
- Dorsal and ventral cochlear nuclei in the upper medulla
- Trapezoid body
- Superior olivary nucleus
- Lateral lemniscus
- Fibers divide and go to the nucleus of lateral lemniscus higher center
- Inferior colliculi
- The medial geniculate nucleus in the thalamus
- Auditory cortex
- Hearing process complete

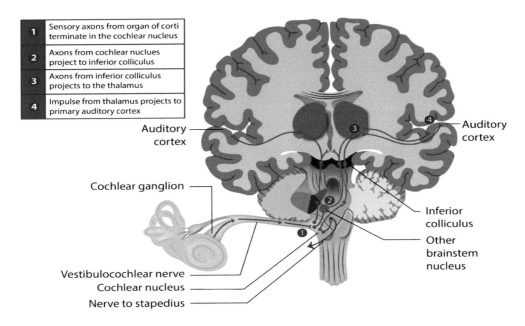

1	Sensory axons from organ of corti terminate in the cochlear nucleus
2	Axons from cochlear nuclues project to inferior colliculus
3	Axons from inferior colliculus projects to the thalamus
4	Impulse from thalamus projects to primary auditory cortex

Figure 8.4 The auditory pathway.

8.6 Global Prevalence and Economic Burden of Hearing Loss

In 1985, 42 million people worldwide (0.9% of the global population) had a hearing impairment [58]. In 1995, the approximate number of people with deteriorating hearing increased to more than 120 million (2.1% of the global population), with approximately 70 million adults and 8 million teenaged individuals from developing countries [58, 59]. In 2011, the number of cases of hearing impairment had increased to 360 million people, of which around 32 million were teenaged children, and 7.5 million were below five years of age [13]. The burden of hearing impairment among the elderly and children is highest in the Asian Pacific, Southern Asian, and sub-Saharan African regions [60].

The current WHO report [61] states that approximately 466 million people (6.1% of the global population) have hearing loss, of which 432 million are adults and 34 million are children. Furthermore, with the inclusion of mild and unilateral hearing loss, the estimated number of people worldwide with hearing loss rose from 1.2 billion (17.2% of the global population) in 2008 to 1.4 billion (18.7% of the worldwide population) in 2017 (Table 8.3). This rising prevalence has become a serious public health concern worldwide [62, 63]. Furthermore, an additional 1.1 billion people aged 12–35 years are at risk of hearing loss due to exposure to loud noise and environmental conditions. Hearing loss is estimated to have an annual global cost of approximately US$750 billion, according to a recent WHO report [61]. The major contributing factors to hearing loss are genetic, aging, birth complications, infectious diseases, acute or chronic ear infections, drugs, noise, air pollution, and electromagnetic field radiation.

Table 8.3 Global Burden of Disabling Hearing Loss

- Approximately 466 million people worldwide suffer from hearing loss
- 432 million adults and 34 million children (total 466 million)
- Approximately 1.1 billion people aged 12–35 years are at risk of hearing loss
- Globally, hearing loss results in an annual cost of approximately US$750 billion
- The prevalence of hearing loss predicted to reach over 900 million by 2050 [61]

8.7 Environmental Pollution and Hearing Impairment

Environmental pollution is a reported risk factor for various illnesses, including respiratory [7], cardiovascular [64], endocrine [65], and neurological disorders [66]. The association between environmental pollution and hearing disorders has been poorly studied. We, therefore, reviewed literature from the Institute of Science Information, Web of Sciences, National Library of Medicine, USA (PubMed), to evaluate the pathophysiological effects of environmental pollution, noise pollution, and RF-EMFR on hearing impairment.

8.7.1 Ambient Air Pollution and Hearing Impairment

Short- or long-term exposure to ambient air pollution results in adverse health consequences. Fine and ultrafine particulate matter with aerodynamic diameters of PM 2.5 μm and PM 0.1 μm, respectively, in addition to "carbon monoxide (CO), ozone (O_3), nitrogen dioxide (NO_2), and sulfur dioxide (SO_2)", are pollutants that are highly toxic to human health. These pollutants are predominantly emitted due to human activity and contribute to ambient air pollution at regional and international levels (Figure 8.5). In certain atmospheric conditions, air pollution can travel long distances across regional borders in 4–6 days, thereby affecting populations distant from their origin. Windblown dust from the desert regions of Africa, Central Asia, and China can carry large amounts of PM, fungal spores, and bacteria that impact health and air quality in distant regions [2, 3]. In August 2020, many US states were affected by PM from Californian wildfires.

Air pollution is a leading cause of hearing problems. The ears and hearing systems are particularly vulnerable to environmental air pollution, with many studies demonstrating the impact of environmental pollution on hearing.

Kuang-Hsi et al. [67] investigated the effect of weather and air pollution on sensorineural hearing loss (SHL). The authors recruited 75,767 subjects aged greater than 20 years without a history of SHL. Participants were consistently exposed to low, medium and high levels of "carbon monoxide (CO) and nitrogen dioxide (NO_2)" with exposure to medium and high levels of CO and NO_2 found to increase the incidence of SHL (Figure 8.6).

Similarly, Choi et al. [68] investigated the effects of climate change and air pollution on the onset of sudden sensorineural hearing loss (SSNHL). The authors selected 5,200 people,

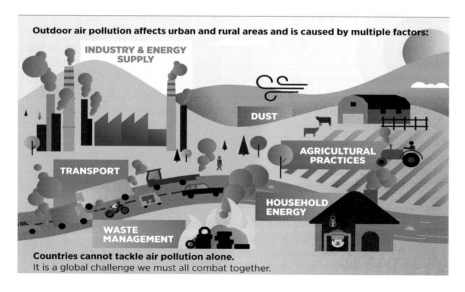

Figure 8.5 Sources of air pollution.

Source: Adapted with permission from WHO [2].

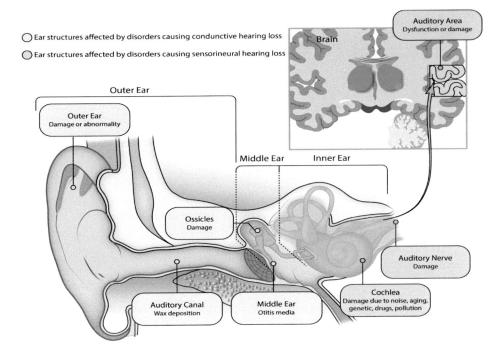

Figure 8.6 Conductive and sensorineural hearing impairment.

matched for age, gender, socioeconomic status, region, housing, and health conditions with 20,800 control subjects. Subjects with SSNHL, compared to the controls, demonstrated higher odds of NO_2 exposure. In another study, Lee et al. [69] assessed the effects of air pollution and particulate matter (PM) on the development of SSNHL, with a weak negative association observed between exposure to high PM2.5 and SSNHL. However, a substantial relationship was found between the daily number of patients admitted to hospitals with SSNHL and meteorological data, including PM levels (Tables 8.4 and 8.5; Figure 8.6).

Table 8.4 Possible Mechanisms Underlying the Association between Environmental Pollution and Hearing Impairment
Environmental pollution- via platelet dysfunction mechanism:
• Platelet dysfunction
• Thrombus formation
• Impair microcirculation
• Ischemia and hypoxia
• Inner ear dysfunction, cochlear damage
• Sensorineural hearing loss
• Hearing impairment [42–50]
Environmental pollution- via oxidative stress mechanism:
• Oxidative stress
• Increase superoxide radicals in the organ of Corti, spiral ganglion, & stria vascularis
• Inflammation
• Cochlear hair cells and neuronal damage
• Sensorineural hearing loss
• Hearing impairment [42–50]

Table 8.5 Summary: Causes of Conductive and Sensorineural Hearing Loss

Conductive Hearing Loss	Sensorineural Hearing Loss
External auditory canal obstruction	Cochlear pathology
Cerumen impaction	Presbycusis
External otitis	Ototoxicity
Congenital stenosis or atresia	Noise-induced hearing loss
Acquired stenosis	Meniere's disease
Neoplasm	Meningitis
Otitis media	Viral labyrinthitis
Tympanic perforation	Vestibulocochlear nerve damage
Otosclerosis	Acoustic neuroma and other
Ossicular trauma	Inner-ear or skull-base neoplasms
Environmental Pollution	Electromagnetic Field Radiation
Others	Environmental Pollution

Choi and Kim [70] examined 30,072 industrial workers and their exposure to work-related pollutants, heavy metals, organic solvents, and noise. They concluded that workers exposed to heavy metals or organic solvents from factories were more susceptible to hearing loss. In another study, Park et al. [71] examined the relationship between lead exposure and hearing impairment in 448 male subjects. The authors found that lead levels were associated with poorer hearing thresholds and hearing loss increased with higher lead exposure. Their study findings also suggested that prolonged low-level exposure to lead was a risk factor for age-related hearing loss.

Caciari et al. [72] conducted a study comprising 357 male outdoor workers exposed to urban noise with 357 control subjects and found differences in hearing thresholds in both ears between the groups. These findings may be due to the combined impact of exposure to both air pollutants and noise.

Park et al. [73] assessed the relationship between environmental pollution and the prevalence of otitis media (OM). The authors investigated five different air pollutants, including "particulate matter (PM 10 μm), nitrogen dioxide (NO_2), ozone (O_3), sulfur dioxide, and carbon monoxide". This analysis was based on 160,875 hospital visits of children aged less than 15 years for otitis media (OM). A positive correlation was between concentrations of the ambient air pollutants PM10, NO_2, CO, and O_3 and the incidence of OM.

Zemek et al. [74] studied the linkage between air pollution and OM based on 14,527 emergency department visits. The authors found an "association between the incidence of OM and air pollution due to carbon monoxide (CO), nitrogen dioxide (NO_2), ozone (O_3), sulfur dioxide (SO_2), and particulate matter of PM-10 and PM-2". Liu et al. [75] examined hearing ability, serum lead (Pb) and urinary cadmium (Cd) levels in 234 preschool children aged 3–7 years, with a higher prevalence of hearing loss observed in the exposed group compared to the reference group (28.8% vs. 13.6%) (Table 8.6).

Wilunda et al. [76] assessed the impact of maternal smoking during pregnancy on hearing impairment in children. They concluded that the risk of hearing impairment was higher in children whose mothers smoked during pregnancy, with a prevalence of hearing impairment in children at the age of 3 years of 4.6%. Abdel Rasoul et al. [77] found that exposure to lead can decrease cognitive function and hearing levels in school children. Exposure to lead damages the sensory, vascular, and neuronal components of the cochlea. Lead exposure can also cause oxidative stress in various tissues and organs by inducing lipid peroxidation and impairing antioxidant mechanisms [78, 79]. Recent studies have demonstrated that SHL and hearing impairment are multifactorial disorders with risk factors including environmental pollution, heavy metals, organic solvents, genetics, ageing, infections, smoking, RF-EMFR, and extensive exposure to noise pollution (Table 8.6).

Table 8.6 Effect of Environmental Pollution on Hearing

Author(s) and Year of Study	Sample Size	Age Years	Type of Study	Study Outcome
Kuang-Hsi et al. (2020) [67]	75,767	More than 20 years	Longitudinal Cohort	Air pollutants, CO, and NO_2 were associated with an increased risk of sensorineural hearing loss.
Choi et al. (2019) [68]	5,200 participants, 20,800 control	5–85 years	Longitudinal Cohort	Sensorineural hearing loss patients established high odds for NO_2 exposure than controls.
Lee et al. (2019) [69]	2.27 patients per day		Cross-Sectional	Establish a link between sensorineural hearing loss and exposure to particulate matter (PM) levels.
Choi and Kim (2014) [70]	30,072 workers from 1,935 industries	18–77 years	Longitudinal Cohort	Exposure to heavy metals or organic solvents: industrial workers were more vulnerable to hearing loss.
Park et al. (2010) [71]	2,264 participants	21–80 years	Multidisciplinary longitudinal study	Long-term exposure to low levels of Lead was linked with age-related hearing loss.
Caciari et al. (2013) [72]	357 workers, 357 unexposed		Cross-Sectional	**The hearing threshold results may be due to the combined impact of exposure to ototoxic air pollutants and noise.**
Park et al. (2018) [73]	160,875 hospital visits of children	1–90 years	National Cohort database	Relationship between ambient air pollutants and otitis media in children aged < 15 years.
Zemek et al. (2012) [74]	14,527 children	1–3 years	National Cohort database	Otitis media in children was linked to ambient air pollution.
Liu et al. (2018) [75]	234 preschool children	3–7 years	Cross-sectional	**Subjects exposed to Lead and cadmium, compared to controls, had a higher incidence of hearing loss: 28.8% vs. 13.6%**
Wilunda et al. (2018) [76]	50,734 children	About 3 years	Retrospective cohort	**Hearing impairment was 4.6% in children exposed to tobacco smoke, both prenatally and postnatally was linked to hearing impairment.**
Abdel Rasoul et al. (2012) [77]	190 primary school children	Primary school age	Cross-sectional	Exposure to Lead decreased intellectual function and hearing levels among children.

8.7.2 Noise Pollution and Hearing Impairment

Noise pollution confers deleterious physiological and psychological effects on human health. Environmental noise exposure-related hearing loss in the workplace represents a significant health issue [80]. Urbanization contributes to the destruction of natural habitats leading to increased air and noise pollution [81].

Noise is an unpleasant disturbance of sound or irregular fluctuation of sound waves and thus represents an increasingly problematic pollutant in modern societies. Noise pollution has adverse effects on human and animal health. Exposure to exceptionally loud noises from the environment, including industrial and occupational sectors, leads to noise-induced hearing loss [82]. Therefore, noise pollution is a common feature of all residential, industrial and urban localities, where it may pose both auditory and non-auditory hazards to human health [83, 84].

Hearing loss is ranked as the fourth most common disability globally, with noise exposure being its foremost risk factor [85]. The WHO has warned that approximately 1.1 billion people aged 12–35 years are at risk of developing hearing impairment due to hazardous and excessive use of personal audio systems and exposure to intense sound levels at entertainment venues [85]. Examples of environments with harmful sound levels include concerts, music venues, bars, nightclubs and festivals, centers for the practice and production of music, sports venues, and proximity to engine noise (mainly aircraft) [86, 87].

Armitage et al. [81] conducted a cross-sectional survey of 10,401 adults in the United Kingdom, demonstrating that recreational noise exposure is associated with hearing loss. Jamesdaniel et al. [88] demonstrated acquired hearing loss results from interactions between multiple, complex risk factors, including loud environments and environmental pollutants such as Lead, that are common in urban society. This study indicates that chronic exposure to Lead causes cochlear oxidative stress and leads to damage of parallel auditory pathways, thereby resulting in noise-induced hearing loss.

Similarly, Muhr et al. [89] evaluated the relationship between noise exposure and hearing impairment in 337 male Swedish military pilots with a mean age of 50 years. The authors assessed pure tone audiometry of pilots from the beginning until the end of their flight service, with a high prevalence of hearing thresholds greater than 20 dB observed in all age groups. Pilots with a greater annual amount of flying time or those flying fast jets were at increased risk of hearing disability. In another study, Lang et al. [90] demonstrated significantly higher hearing thresholds among pilots who had 20 or more service years. Contradictory to these findings, Kampel-Furman et al. [91] found no deterioration in hearing thresholds of Air Force pilots with over 20 years of military service compared to non-flying controls.

Śliwińska-Kowalska [80] reported low-quality evidence that long-duration exposure to loud music enhances the risk of hearing loss and worsening standard frequency audiometric thresholds. Similarly, Huang et al. [92] and Wouters et al. [93] reported that industrial, quarry, and brewery workers have early signs of noise-induced hearing loss.

While high-level noise exposure causes a permanent shift in the hearing threshold due to irreversible damage to the stereocilia of hair cells, moderate levels of noise exposure can cause a temporary shift in hearing thresholds due to reversible impairment of stereocilia. The potential for hearing damage depends on the type and intensity of noise in addition to the duration of exposure, as noise-induced hearing loss is typically related to the cumulative energy of the exposed sound [1].

Kim and Han [94] evaluated smartphones as potential risk factors for hearing impairment by measuring output levels at different volume levels. The volume level associated with risk among the six smartphones ranged from 84.1 to 92.4 dBA, with sound pressure found to increase with increasing volume. The loudest smartphone reached 113.1 dBA at the maximum volume level, with a 20–20,000 Hz frequency range. The sound pressure level peaked at 2,000 Hz for all smartphones. The authors concluded that sound pressure levels differed between smartphone models and genres of music. However, current smartphone models can produce volumes that can result in noise-induced hearing loss if individuals habitually listen to music at higher levels.

Hickox and Liberman [95] measured the responses in mice to "neuropathic' noise or a lower intensity 'non-neuropathic noise." The authors found that "neuropathic' noise exposure can result in rapid, permanent degeneration of the cochlear nerve despite a complete threshold recovery and a lack of hair cell damage". These results provide evidence that "acoustic injury is likely due to degeneration of cochlear nerve fibers rather than a direct impact of noise exposure on the central auditory system or damage to cochlear sensory cells" [96].

Noreña et al. [97] noted that sensory environments could influence the development of the nervous system. Continuous, long-term, and passive exposure to a spectrally enhanced acoustic environment (EAE) in juvenile cats caused a reorganization of the tonotopic map auditory cortex without hearing loss. The EAE induced a significant decrease in EAE frequencies a simultaneous over-representation of frequencies neighboring the EAE. The authors also observed lower appropriately tuned short-latency responses to EAE frequencies with more frequent long-latency responses tuned to EAE-neighboring frequencies.

Lin et al. [98] demonstrated that, despite the full recovery of cochlear thresholds and no loss of inner or outer hair cells, acoustic overexposure could result in a rapid and permanent loss of the peripheral "cochlear nerve terminals on inner hair cells (IHCs) and a slow degeneration of spiral ganglion cells." These results indicate that "irreversible neural degeneration is a common outcome following acoustic overstimulation in the mammalian ear even when threshold sensitivity returns to normal" [98].

The first of two possible mechanisms underlying hearing impairment due to noise pollution is the effect of noise on the cochlea. Scientific studies have indicated noise pollution may affect the cochlea in multiple ways, including loss of outer hair cells, diminished blood flow in the basal region, rupture of tight cell junctions, excitotoxicity of vestibulocochlear nerve fibers, cochlear synaptopathy, loss of IHCs, and damage to vestibulocochlear nerve fibers [99–101]. Exposure to loud noise can lead to hearing loss and hearing disorders such as recruitment, tinnitus, and hyperacusis [1].

The second pathophysiological mechanism is oxidative stress affecting the auditory system. Oxidative stress contributes to auditory impairment by activating cochlear cell death pathways in hearing loss associated with ageing. Oxidative stress may be induced by exposure to noise, organic solvents, heavy metals (such as cadmium), ototoxic drugs, and radiation [102–110]. Oxidative stress-induced damage in acquired hearing loss has been demonstrated in the sensory epithelium, modiolus, and lateral wall of the cochlea. These pathophysiological mechanisms underlie neural degeneration and hearing impairment following exposure to intense sound and noise [87].

8.7.3 Electromagnetic Pollution and Hearing Impairment

Modern technologies have become a major source of EMR, also known as electromagnetic pollution. Electromagnetic field pollution affects various aspects of the ecosystem and environment. Whether RF-EMFR pollution is hazardous to health hazards remains controversial as there is a lack of definitive evidence of its negative impact on human health. However, during the past decade, a significant body of scientific literature has been developed on the impact of EMR on the hearing system [111] (Table 8.7).

Chen and Pai [112] assessed the impact of numerous smartphone tasks, including calling, texting, music, and playing games, on inattentional blindness among 2,556 pedestrians. They found that smartphone overuse and unintentional deafness were common among people who listened to music on smartphones.

Velayutham et al. [113] assessed the effect of long-term usage of mobile phones on high-frequency hearing loss. The authors compared the results of audiometric measurements on the dominant ear and non-dominant ear. A total of 100 subjects were studied, including 53 males

Table 8.7 Effects of RF-EMFR on the Hearing System

Author(s) and Year of Study	Sample Size	Age Years	Type of Study	Study Outcome
Chen and Pai (2018) [112]	2,556 pedestrians	>18 years	Cross-sectional study	Inattentional deafness was common among music listeners.
Velayutham et al. (2014) [113]	100 subjects: M:53; F 47	Mean age 27 years	Prospective study	In mobile phone users, hearing loss was found in the dominant ear compared to the non-dominant ear.
Oktay et al. (2006) [114]	60 males		Cross-sectional study	Significant hearing loss was related to prolonged exposure to RF-EMFR generated from mobile phones.
Das et al. (2017) [115]	100 students	18–30 years	Cross-sectional	Air conduction and bone conduction hearing thresholds were increased in an exposed ear compared to a non-exposed ear.
Philip et al. (2017) [116]	150 health volunteers	21–45 years	Cross-sectional	Sensorineural hearing loss was observed in 10% of participants using mobile phones>1 h per day for a minimum period of 4 years. No sensorineural hearing loss was observed in control subjects.
Al-Dousary (2014) [118]	1 case	42 years	Case report	Sensorineural hearing loss was found due to the use of mobile phones
Panda et al. (2010) [119]	112 mobile phone users and 50 control	20–60 years	Cross-sectional	Audiologic abnormalities, high-frequency loss, and absent distortion of otoacoustic emissions were seen with excessive and long-term use of mobile phones.

Study	Sample	Age	Study type	Findings
Panda et al. (2011) [120]	125 mobile phone users 58 control	20–60 years	Cross-sectional	Excessive and prolonged use of mobile phones can damage the cochlea and auditory cortex.
Gupta et al. (2015) [121]	67 mobile phone users 33 control	18–30 years	Case-control Cross-sectional	There was no difference in latencies, interpeak latencies, and amplitudes of ABR waves between the groups.
Akdag et al. (2018) [122]	56 males	30–60 years	Case-control Cross-sectional	Mobile phone radiation caused DNA damage in follicles of hair cells in the ear canal, and damage was increased with the duration of exposure.
Godson et al. (2012) [123]	58 staff and 45 students	Students 23.8; staff 35.5 years	Cross-sectional	The audiometry findings showed that 22.2% of students and 28.0% of university staff had reported evidence of hearing deterioration.
Meo and Al-Dress (2005) [124]	873:M: 57.0% and F: 39.8%	18–46 years	Observational study	34.6% of subjects had complaints of impaired hearing, earache, and/or warmth on the ear.
Khan (2011) [125]	286	18–46 years	Observational study	Hearing problems were reported by 38.8% of mobile phone users; a feeling of warmth within the auricle and around the ear was found in 28.3% of mobile phone users.
Bhargav et al. (2017) [126]	61 healthy adults with a mean age of 17.4 years.		Cross-sectional	Energy levels decreased in multiple organs among subjects who used a mobile phone in the "on" mode compared to subjects used a mobile phone in the I" off" mode.

and 47 females with an average age of 27 years. The findings revealed that long-term mobile phone usage was associated with high-frequency hearing loss in the dominant ear but not the non-dominant ear.

Oktay et al. [114] determined the effect of radiation generated by mobile phones on hearing impairment. The study groups consisted of 20 males who used a cell phone for 2 hours per day for four years; 20 male subjects who used a cell phone for 10–20 minutes every day for four years; and 20 male subjects who never used a mobile phone (control group). "Brainstem evoked response audiometric and pure tone audiometric tests were conducted to measure the effects of exposure on the hearing function." The study found no damaging effects in moderate mobile phone users (10–20 minutes per day) or control subjects. However, subjects who used a mobile phone for 2 hours per day had higher thresholds than moderate users or control subjects. The study also found a high degree of hearing loss associated with long-term exposure to cell phones.

Das et al. [115] examined the effects of mobile phone usage on pure tone audiometry thresholds among 100 students with significant mobile phone usage. The findings demonstrated increased air and bone conduction hearing thresholds in exposed compared to non-exposed individuals.

Philip et al. [116] assessed the effect of mobile phone usage on auditory functions and patterns of hearing thresholds. The authors selected 150 healthy volunteers, with an age range of 21–45 years. Study participants were divided into three groups: 50 participants using their mobile phones for more than 1 hour per day (group A); 50 participants using their phones for less than 1 hour per day for a minimum period of 4 years (group B); and 50 subjects who were not using their mobile phones on a regular basis (less than 1 hour per week; group C). The incidence of SHL was 10% in group A and 2% in group B. However, no participants in group C had hearing loss. The authors concluded that prolonged and frequent use of mobile phones damages outer hair cells in the cochlea resulting in high-frequency hearing loss. Ragab et al. [117] demonstrated no increased risk between short-term exposure to mobile phone radiation and auditory system complaints.

Al-Dousary (2007) [118] reported the case of a 42-year-old male with SHL due to mobile phone usage. Panda et al. [119] conducted a case-control study that evaluated the effects of prolonged mobile phone usage on auditory function. The authors selected 112 subjects with heavy mobile phone use and 50 controls who never used a mobile phone. A trend of audio logic abnormalities, consisting of high-frequency loss and absent distortion product otoacoustic emissions, was observed with excessive and long-term use of mobile phones by individuals aged greater than 30 years. The authors concluded that intense and prolonged use of mobile phones might cause inner ear damage.

In another study, Panda et al. [120] evaluated hearing function at the level of the inner ear and central auditory pathway after chronic exposure to electromagnetic waves emitted from mobile phones. The authors selected 125 subjects who had used a mobile phone for more than one year and 58 controls who had never used a mobile phone. They found that mobile phone users, compared to controls, had a higher risk of having distortion product otoacoustic emissions. They also found that using a mobile phone with both ears for more than three years resulted in decreased earing in both ears. The authors concluded that long-term and excessive mobile phone usage could damage the cochlea and the auditory cortex.

Gupta et al. [121] performed a case-control study on the impact of prolonged mobile phone use on auditory brainstem evoked responses (ABRs). In this study, the authors selected 100 healthy subjects aged between 18 and 30 years, out of which 67 subjects were long-term mobile phone users and 33 were controls who had never used a mobile phone. No significant differences in latencies, inter-peak latencies, or ABR wave amplitudes were observed between groups.

Akdag et al. [122] investigated the impact of radiofrequency radiation (RFR) released from mobile phones on damage to DNA in the follicle cells of the ear in a study comprising 56 male subjects aged between 30 and 60 years. Mobile phone users were divided into three groups according to the duration of exposure: 0–30 minutes per day, 30–60 minutes per day, and more than 60 minutes per day. The study found RFR from mobile phones damaged DNA in follicle cells, and the amount of damage correlated with the daily duration of exposure.

Godson et al. [123] performed audiometry in 58 university staff members and 45 young adult students possessing mobile phones in Nigeria. The authors found that 28.0% of staff and 22.2% of students had evidence of hearing impairment. Meo and Al-Dress [124] investigated the association between mobile phone use and hearing problems, with 34.6% of participants reporting impaired hearing, earache, or warmth around the ear.

Similarly, Khan [125] conducted a study of 286 mobile phone users (73.8% male and 26.2% female) and found that 38.8% of mobile phone users had hearing problems, and 28.3% of participants had a sensation of warmth within the auricle and around the ear. Bhargav et al. [126] reported decreased metabolism, energy levels in multiple organs among subjects who used a mobile phone in the 'on' mode compared to control subjects who held a mobile phone in the 'off' mode. However, Sagiv et al. [127] did not find a relationship between the preferred ear for mobile phone usage and the onset of SSNHL.

Khullar et al. [128] determined the impact of mobile phones on auditory brainstem responses (ABRs). The authors recruited 60 subjects who were distributed into three groups of 20 subjects according to mobile phone use duration. Wave latency was significantly prolonged in subjects using mobile phones for 30 minutes per day for ten years compared to the control group. However, no differences in ABR parameters were observed between subjects using mobile phones for 30 minutes per day for five years compared to controls. The investigators concluded that prolonged exposure to mobile phones affects conduction in the peripheral auditory pathway.

The effects of RF-EMFR generated from mobile phones depend on the frequency, duration, and absorption of RF-EMFR. It has also been reported that over-ear use of mobile phones allows maximum levels of RF-EMFR to be absorbed by the brain's temporal lobe [129]. Previous studies have demonstrated same-side use of mobile phones is associated with the highest risk of malignant brain tumors and an increased risk of glioma and acoustic neuroma [130, 131].

Özgür et al. [132] performed a study on 12 adult Wistar albino rats. Rats were exposed to RF-EMFR over a period of 30 days, and ABR was recorded. The immunohistochemical and histopathologic studies demonstrated neuronal degeneration, augmented vacuolization in the cochlear nucleus, pyknotic cell appearances, and edema in the exposed group compared to the control group.

Similarly, Zuo et al. [133] investigated the sensitivity of spiral ganglion neurons of Sprague-Dawley rats aged between one and three days to RF-EMFR generated from the mobile phones at specific absorption rates of 2 and 4 watts per kilogram (W/kg). DNA damage and ultrastructure changes were detected after 24 hours of recurrent exposure at an absorption rate of 2 or 4 W/kg. The group exposed to 4 W/kg were found to have karyopyknosis, mitochondria vacuoles, and the presence of lysosomes and auto-phagosomes. Short-term 2 W/kg exposure to RF-EMFR did not result in DNA damage in spiral ganglion neurons; however, exposure to 4 W/kg RF-EMFR did result in cellular ultrastructure changes at the specific absorption rate. Studies regarding the effects of electromagnetic field radiation and RF-EMFR pollution on hearing provide essential knowledge and pertinent information regarding the hazardous effects of electromagnetic field radiation on SHL and hearing impairment. However, some studies did not find an association between RF-EMFR and hearing impairment.

8.8 Effect of RF-EMFR on Tumorigenesis and Hearing Impairment

The potential mechanism underlying the effect of RF-EMFR generated from mobile phone exposure on tumorigenesis remains controversial. There is ongoing debate regarding the impact of the RF-EMFR emitted from mobile phone use on the development of vestibular schwannomas, also known as acoustic neuromas. Some epidemiological studies have reported no significant increase in the risk of developing vestibular Schwannoma with mobile phone usage [134, 135]. These findings are supported by studies demonstrating that EMFs generated from mobile phones lack the energy required to distort chemical bonds or cause DNA damage [136, 137]. However, these studies did identify an increased risk of malignancy in the dominant or more frequently used ear while using mobile phones [138]. These findings are corroborated by studies reporting that RF-EMFR generated from mobile phones can penetrate the brain to a depth of approximately 4–5 cm, which can raise the temperature of the brain tissue by 0.1°C. This thermal effect may influence protein phosphorylation [139]. Other potential mechanisms are related to increased oxidative stress resulting from RF-EMFR, leading to increased carcinogenesis and apoptosis.

Lönn et al. [138] conducted a study in Sweden comprising 148 cases of acoustic neuroma and 604 controls aged 20–69 years. The authors found that the relative risk for acoustic neuroma increased with more than ten years of mobile phone use. However, the authors did not find an increased risk of acoustic neuroma in short-term mobile phone users. Similarly, Schoemaker et al. [139] performed a case-control study of 678 cases and 3,553 controls aged between 18 and 69 years. They found that the risk of ipsilateral tumor development was increased with ten or more years of mobile phone usage.

The Interphone study [140] reported that a cumulative ipsilateral mobile phone call time of greater than 1,640 hours was associated with the development of temporal lobe gliomas. Hardell et al. [141] conducted a case-control study on brain tumors comprising 233 patients aged 20–80 years. They found that ipsilateral use of a cell phone increased the risk of tumors in the temporal, temporoparietal, and occipital areas with the highest exposures to radiation generated from a mobile phone. Similarly, Corona et al. [142] identified radiation from cellular and cordless phones as risk factors for vestibulocochlear nerve schwannomas.

Hardell et al. [143] performed a "case-control study of acoustic neuromas among males and females aged 20–80 years. Using mobile phones on the same side of the head resulted in a higher risk of acoustic neuroma than contralateral use of mobile and cordless phones". The percentage of tumors increased per year of latency and 100 hours of cumulative phone use. Benson et al. [144] conducted a prospective study of mobile phone users and the prevalence of intracranial tumors in 791,710 middle-aged women in the United Kingdom. They reported an increased risk of acoustic neuroma in individuals who reported long-term use of mobile phones compared to people who never used mobile phones. This risk increased with the duration of exposure to mobile phones. However, mobile phone usage was not associated with an increased incidence of glioma, meningioma, or cancers of the nervous system.

Mild et al. [145] reported that long-term use of the mobile phone was associated with a rising risk of "acoustic neuromas and malignant brain tumors." The authors concluded that more than ten years of mobile phone use increased the risk of acoustic neuroma and glioma and that tumor occurrence was highest for ipsilateral exposure. They also reported that the use of analog cellular phones increased the risk of acoustic neuroma by approximately 5%. Moreover, this risk was increased per one-year use of analog phones by 10%, digital phones by 11%, and cordless phones by 8%. Similarly, Moon et al. [146] found Schwannoma tumors were more frequently seen in the ear used while talking on mobile phones and that tumor volume was associated with the duration of mobile phone use.

The International Agency for Research on Cancer of the WHO [147] has classified RF radiation in the frequency range 30 kHz to 300 GHz as a "possible human carcinogen." [147–150]. Epidemiologic, case-control, cohort, longitudinal, and case studies that have categorized RF-EMFR from mobile phones and other wireless devices as a possible human carcinogen (Group 2B) have been analyzed and summarized by Miller et al. [151]. However, several studies have reported contradictory findings with no association between RF-EMFR and tumorigenesis [152–154]. Previous case-control studies of the association between mobile phone use and cancer have been limited by systematic and randomization errors in selection bias [155].

Biological evidence of the hazardous effect of various environmental pollutants supports the hypothesis that environmental pollution and noise pollution can cause hearing impairment [73]. Choi et al. [156] reported the adverse effects of exposure to pollutants, including cadmium and Lead, on hearing outcomes both in subjects exposed to high and low noise levels. The basic mechanisms underlying the effects of environmental pollution and RF-EMFR on hearing impairment are oxidative stress, inflammation, and cochlear damage. Pollutants induce oxidative stress and systemic inflammation, which increase the risk of SHL [157]. Previous studies have also demonstrated that exposure to RF-EMFR is associated with increased ROS production, leading to DNA damage [158, 159]. Moreover, the cochlea has high aerobic metabolism, which can generate high levels of adenosine triphosphate (ATP) and ROS, with the cochlea thought to be particularly vulnerable to ROS [158, 159]. These potential mechanisms support the hypothesis that inner ear biology and hearing can be affected by increased ROS production due to environmental pollution, noise pollution, and REF-EMFR.

Conclusion

The advancement of technology, aimed at improving living standards, is undoubtedly a luxury but comes at a cost. Both the environment and human life pay this cost. Environmental pollution is increasingly becoming a major global health concern with its toxic effects affecting all life on the planet. Air pollution, noise pollution, and RF-EMFR pollution are all considered major components of environmental pollution. The auditory system, particularly the cochlea, hair cells, vestibular cochlear nerve, and vestibular system, is sensitive to various exogenous and endogenous agents. Human and animal model studies have demonstrated a relationship between environmental pollution, including noise and RF-EMFR pollution and SHL and hearing impairment. While some studies have shown a significant association between pollution and hearing loss, other studies have not found such an association. Therefore, to answer the question, "how dangerous is modern man-made environmental pollution?," members of the scientific community must collaborate to perform further data-generating studies and meta-analyses to further understand and reduce the harmful effects of environmental pollution on modern society.

References

[1] Rita Rosati, Samson Jamesdaniel. Environmental exposures and hearing loss. *Int. J. Environ. Res. Public. Health.* 2020; 17(13): 4879. doi.org/10.3390/ijerph17134879.

[2] World Health Organization (WHO). Air Pollution. Available at: www.who.int/westernpacific/health-topics/air-pollution. Cited date Nov 2, 2020.

[3] World Health Organization (WHO). Health effects of particulate matter. Policy implications for countries in eastern Europe, Caucasus and central Asia. 2013; ISBN 978 92 890 0001 7.

[4] Lim S.S., Vos T., Flaxman A.D., Danaei G., et al. A comparative risk assessment of burden of disease and injury attributable to 67 risk factors and risk factor clusters in 21 regions, 1990–2010: A systematic analysis for the Global burden of disease study 2010. *Lancet.* 2012; 380: 2224–60.

[5] Babatola S.S. Global burden of diseases attributable to air pollution. *J. Public. Health. Afr.* 2018; 9(3): 813. doi: 10.4081/jphia.2018.813.

[6] Rakhmanin Y.A., Mikhaylova R.I. Environment and health: Priorities for preventive medicine]. *Gig. Sanit.* 2014 Sep–Oct;(5): 5–10. Russian. PMID: 25831920.

[7] Meo S.A., Aldeghaither M., Alnaeem K.A., Alabdullatif F.S., Alzamil A.F., Alshunaifi A.I., Alfayez A.S., Almahmoud M., Meo A.S., El-Mubarak A.H. Effect of motor vehicle pollution on lung function, fractional exhaled nitric oxide and cognitive function among school adolescents. *Eur. Rev. Med. Pharmacol. Sci.* 2019; 23(19): 8678–86. doi: 10.26355/eurrev_201910_19185.

[8] Meo S.A., Almahmoud M., Alsultan Q., Alotaibi N., Alnajashi I., Hajjar W.M. Mobile phone base station tower settings adjacent to school buildings: Impact on students' cognitive health. *Am. J. Mens. Health.* 2019; 13(1): 1557988318816914. doi: 10.1177/1557988318816914.

[9] Tsay A.J., Giummarra M.J., Allen T.J., Proske U. The sensory origins of human position sense. *J. Physiol.* 2016; 594(4): 1037–49. doi: 10.1113/JP271498.

[10] Backman C., Crick M., Cho-Young D., Scharf M., Shea B. What is the impact of sensory practices on the quality of life of long-term care residents? A mixed-methods systematic review protocol. *Syst. Rev.* 2018; 7: 115. https://doi.org/10.1186/s13643-018-0783-9.

[11] Wallhagen M.I., Strawbridge W.J., Shema S.J., Kurata J., Kaplan G.A. Comparative impact of hearing and vision impairment on subsequent functioning. *J. Am. Geriatrics. Soc.* 2002; 49(8): 1086–92.

[12] Ohlenforst B., Zekveld A.A., Jansma E.P., Wang Y., Naylor G., Lorens A., Lunner T., Kramer S.E. Effects of hearing impairment and hearing aid amplification on listening effort: A systematic review. *Ear. Hear.* 2017; 38(3): 267–81.

[13] Le Prell C.G., Hammill T.L., Murphy W.J. Noise-induced hearing loss: Translating risk from animal models to real-world environments. *J. Acoust. Soc. Am.* 2019; 146: 3646.

[14] Hsu T.Y., Wu C.C., Chang J.G., Lee S.Y., Hsu C.J. Determinants of bilateral audiometric notches in noise-induced hearing loss. *Laryngoscope.* 2013; 123: 1005–10.

[15] Zhuang W., Wang C., Shi X., Qiu S., Zhang S., Xu B., Chen M., Jiang W., Dong H., Qiao Y. MCMV triggers ROS/NLRP3-associated inflammasome activation in the inner ear of mice and cultured spiral ganglion neurons, contributing to sensorineural hearing loss. *Int. J. Mol. Med.* 2018; 41: 3448–56.

[16] Chen H.C., Chung C.H., Wang C.H., Lin J.C., Chang W.K., Lin F.H., Tsao C.H., Wu Y.F., Chien W.C. Increased risk of sudden sensorineural hearing loss in patients with hepatitis virus infection. *PLoS. One.* 2017; 12: e0175266.

[17] Wu C.C., Hung C.C., Lin S.Y., Hsieh W.S. Tsao, P.N., Lee, C.N., Su, Y.N., Hsu, C.J. Newborn genetic screening for hearing impairment: A preliminary study at a tertiary center. *PLoS One.* 2011; 6: e22314.

[18] Lu J., Wang W., Liu H., Liu H., Wu H. Cisplatin induces calcium ion accumulation and hearing loss by causing functional alterations in calcium channels and exocytosis. *Am. J. Transl. Res.* 2019; 11: 6877–89.

[19] Jeong J., Lim H., Lee K., Hong C.E., Choi H.S. High risk of sudden sensorineural hearing loss in several autoimmune diseases according to a population-based national sample cohort study. *Audiol. Neurootol.* 2019; 24: 224–30.

[20] Reed N.S., Huddle M.G., Betz J., Power M.C., Pankow J.S., Gottesman R., Richey Sharrett A., Mosley T.H., Lin F.R., Deal J.A. Association of midlife hypertension with late-life hearing loss. *Otolaryngol. Head. Neck. Surg.* 2019; 161: 996–1003.

[21] Chen H.C., Chung C.H., Lu C.H., Chien W.C. Metformin decreases the risk of sudden sensorineural hearing loss in patients with diabetes mellitus: A 14-year follow-up study. *Diab. Vasc. Dis. Res.* 2019; 16: 324–7.

[22] Fang Q., Lai X., Yang L., Wang Z., Zhan Y., Zhou L., Xiao Y., Wang H., Li D., Zhang K., et al. Hearing loss is associated with increased stroke risk in the Dongfeng-Tongji Cohort. *Atherosclerosis.* 2019; 285: 10–6.

[23] Antonopoulos S., Balatsouras D.G., Kanakaki S., Dona A., Spiliopoulou C., Giannoulis G. Bilateral sudden sensorineural hearing loss caused by alcohol abuse and heroin sniffing. *Auris. Nasus. Larynx.* 2012; 39: 305–9.

[24] Kumar A., Gulati R., Singhal S., Hasan A., Khan A. The effect of smoking on the hearing status-a hospital-based study. *J. Clin. Diagn. Res.* 2013; 7(2): 210–4. doi: 10.7860/JCDR/2013/4968.2730.

[25] Kamenski G., Bendova J., Fink W., Sonnichsen A., Spiegel W., Zehetmayer S. Does COPD have a clinically relevant impact on hearing loss? A retrospective matched cohort study with a selection of patients diagnosed with COPD. *BMJ. Open.* 2015; 5: e008247.

[26] Chang K.H., Tsai S.C., Lee C.Y., et al. Increased risk of sensorineural hearing loss as a result of exposure to air pollution. *Int. J. Environ. Res. Public. Health.* 2020; 17(6): 1969. Published 2020 Mar 17. doi: 10.3390/ijerph17061969.

[27] Gagnaire F., Langlais C. Relative ototoxicity of 21 aromatic solvents. *Arch. Toxicol.* 2005; 79: 346–54.

[28] Fuente A., McPherson B. Central auditory damage induced by solvent exposure. *Int. J. Occup. Saf. Ergon.* 2007; 13: 391–7.

[29] Staudt A.M., Whitworth K.W., Chien L.C., Whitehead L.W., Gimeno Ruiz de Porras D. Association of organic solvents and occupational noise on hearing loss and tinnitus among adults in the U.S., 1999–2004. *Int. Arch. Occup. Environ. Health* 2019; 92: 403–13.

[30] Roggia S.M., de Franca A.G., Morata T.C., Krieg E., Earl B.R. Auditory system dysfunction in Brazilian gasoline station workers. *Int. J. Audiol.* 2019; 58: 484–96.

[31] Chen C.Y., Hung H.J., Chang K.H., Hsu C.Y., Muo C.H., Tsai C.H., Wu T.N. Long-term exposure to air pollution and the incidence of Parkinson's disease: A nested case-control study. *PLoS. One.* 2017; 12(8): e0182834.

[32] Peters R., Ee N., Peters J., Booth A., Mudway I., Anstey K.J. Air Pollution and dementia: A systematic review. *J. Alzheimers. Dis.* 2019; 70: S145–S163.

[33] Chang K.H., Chang M.Y., Muo C.H., Wu T.N., Hwang B.F., Chen C.Y., Lin T.H., Kao C.H. Exposure to air pollution increases the risk of osteoporosis: A nationwide longitudinal study. *Medicine (Baltimore).* 2015 May; 94(17): e733.

[34] Chang K.H., Hsu P.Y., Lin C.J., Lin C.L., Juo S.H., Liang C.L. Traffic-related air pollutants increase the risk for age-related macular degeneration. *J. Investig. Med.* 2019; 67(7): 1076–81.

[35] Nemmar A., Hoet P.H., Vanquickenborne B., Dinsdale D., Thomeer M., Hoylaerts M.F., Vanbilloen H., Mortelmans L., Nemery B. Passage of inhaled particles into the blood circulation in humans. *Circulation.* 2002 Jan 29; 105(4): 411–4.

[36] He X., Zhang H., Ma Y., Bai W., Zhang Z., Lu K., Ding Y., Zhao Y., Chai Z. Lung deposition and extrapulmonary translocation of nano-ceria after intratracheal instillation. *Nanotechnology.* 2010 Jul 16; 21(28): 285103.

[37] Oyewale M.M., Matlou I.M., Murembiwa S.M., Raymond P.H. Health outcomes of exposure to biological and chemical components of inhalable and respirable particulate matter. *Int. J. Environ. Res. Public. Health.* 2016; 13: 10.

[38] Liaoa B.Q., Liue C.B., Xiec B.Q., Liud Y., Denga Y.B., Hed S.W. Effects of fine particulate matter (PM2.5) on ovarian function and embryo quality in mice. *Environ. Int.* 2019; 135: 105338. doi: 10.1016/j.envint.2019.105338.

[39] Garcia G.J., Kimbell J.S. Deposition of inhaled nanoparticles in the rat nasal passages: Dose to the olfactory region. *Inhal. Toxicol.* 2009; 21(14): 1165–75.

[40] Maher B.A., Ahmed I.A., Karloukovski V., MacLaren D.A., Foulds P.G., Allsop D., Mann D.M., Torres-Jardón R., Calderon-Garciduenas L. Magnetite pollution nanoparticles in the human brain. *Proc. Natl. Acad. Sci. U S A.* 2016; 113(39): 10797–801.

[41] Calderón-Garcidueñas L., Reynoso-Robles R., Vargas-Martínez J., Gómez-Maqueo-Chew A., Pérez-Guillé B., Mukherjee P.S., Torres-Jardón R., Perry G., Gónzalez-Maciel A. Prefrontal white matter pathology in air pollution exposed Mexico City young urbanites and their potential impact on neurovascular unit dysfunction and the development of Alzheimer's disease. *Environ. Res.* 2016 Apr; 146: 404–17.

[42] Nemmar A., Hoet P.H., Dinsdale D., Vermylen J., Hoylaerts M.F., Nemery B. Diesel exhaust particles in lung acutely enhance experimental peripheral thrombosis. *Circulation.* 2003 Mar 4; 107(8): 1202–8.

[43] Mutlu G.M., Green D., Bellmeyer A., Baker C.M., Burgess Z., Rajamannan N., Christman J.W., Foiles N., Kamp D.W., Ghio A.J., Chandel N.S., Dean D.A., Sznajder J.I., Budinger G.R. Ambient particulate matter accelerates coagulation via an IL-6-dependent pathway. *J. Clin. Invest.* 2007 Oct; 117(10): 2952–61.

[44] Khandoga A., Stoeger T., Khandoga A.G., Bihari P., Karg E., Ettehadieh D., Lakatos S., Fent J., Schulz H., Krombach F. Platelet adhesion and fibrinogen deposition in murine microvessels upon inhalation of nanosized carbon particles. *J. Thromb. Haemost.* 2010 Jul; 8(7): 1632–40.

[45] Khandoga A., Stampfl A., Takenaka S., Schulz H., Radykewicz R., Kreyling W., Krombach F. Ultrafine particles exert prothrombotic but not inflammatory effects on the hepatic microcirculation in healthy mice in vivo. *Circulation.* 2004; 109(10): 1320–5.

[46] Tan H.H., Fiel M.I., Sun Q., Guo J., Gordon R.E., Chen L.C., Friedman S.L., Odin J.A., Allina J. Kupffer cell activation by ambient air particulate matter exposure may exacerbate the non-alcoholic fatty liver disease. *J. Immunotoxicol.* 2009; 6: 266–75.

[47] Sliwinska-Kowalska M., Zamyslowska-Szmytke E., Szymczak W., Kotylo P., Fiszer M., Wesolowski W., Pawlaczyk-Luszczynska M. Ototoxic effects of occupational exposure to styrene and co-exposure to styrene and noise. *J. Occup. Environ. Med.* 2003; 45: 15–24.

[48] Cappaert N.L., Klis S.F., Baretta A.B., Muijser H., Smoorenburg G.F. Ethyl benzene-induced oto-toxicity in rats: A dose-dependent mid-frequency hearing loss. *J. Assoc. Res. Otolaryngol.* 2000; 1: 292–9.

[49] Rolesi A.R., Paciello F., Eramo S.L.M., Grassi C., Troiani D., Paludetti G. Styrene enhances the noise-induced oxidative stress in the cochlea and affects differently mechanosensory and supporting cells. *Free Radic. Biol. Med.* 2016; 101: 211–25.

[50] Olivetto E., Simoni E., Guaran V., Astolfi L., Martini A. Sensorineural hearing loss and ischemic injury: Development of animal models to assess vascular and oxidative effects. *Hear. Res.* 2015 Sep; 327: 58–68.

[51] Shin S.A., Lyu A.R., Jeong S.H., Kim T.H., Park M.J., Park Y.H. Acoustic trauma modulates cochlear blood flow and vasoactive factors in a rodent model of noise-induced hearing loss. *Int. J. Mol. Sci.* 2019 Oct 25; 20(21).

[52] Ekdale E.G. Form and function of the mammalian inner ear. *J. Anat.* 2016; 228(2): 324–37. doi: 10.1111/joa.12308.

[53] Zhang Qing Z., DuanMao-li. Anatomy and physiology of peripheral auditory system and common causes of hearing loss. *J. Otol.* 2009; 4(1): 7–14.

[54] Timo Stöver, Marc Diensthuber. Molecular biology of hearing. *GMS Curr Top Otorhinolaryngol Head Neck Surg.* 2011; 10: Doc06. doi: 10.3205/cto000079.

[55] Nayagam B.A., Muniak M.A., Ryugo D.K. The spiral ganglion: Connecting the peripheral and central auditory systems. *Hear. Res.* 2011; 278(1–2): 2–20. doi: 10.1016/j.heares.2011.04.003.

[56] Boenninghaus Lenarz. Innenohr (Labyrinth) In: Boenninghaus Lenarz, editors. *HNO.* Heidelberg: Springer Medizin Verlag. 2007, pp. 15–20.

[57] Starlinger V., Masaki K., Heller S. Auditory physiology: Inner ear. In: Gulya A.J., Minor L.B., Poe D.S., editors. *Glasscock-Shambaugh's Surgery of the Ear*, 6th edition. Shelton, CT: People's Medical Publishing House. 2010, pp. 73–83.

[58] Smith A.W. Preventing deafness an achievable challenge. The WHO perspective. *Int. Con. Gr. Ser.* 2003; 1240: 183–91. http://dx.doi.org/10.1016/S0531-5131(03)00960-9.

[59] Smith A.W. WHO activities for prevention of deafness and hearing impairment in children. *Scand. Audiol. Suppl.* 2001; 53: 93–100.

[60] World Health Organization. *Global Estimates on the Prevalence of Hearing Loss. [Internet].* Geneva: World Health Organization. 2012. Available at: www.who.int/pbd/deafness/estimates. Cited date Jun 20, 2020.

[61] World Health Organization. Deafness and hearing loss. Available at: www.who.int/news-room/fact-sheets/detail/deafness-and-hearing-loss. Cited date Jun 20, 2020.

[62] GBD 2017 Disease and Injury Incidence and Prevalence Collaborators; GBD 2017 Disease and Injury Incidence and Prevalence Collaborators. Global, regional, and national incidence, prevalence, and years lived with disability for 354 diseases and injuries for 195 countries and territories, 1990–2017: A systematic analysis for the Global Burden of Disease study 2017. *Lancet.* 2018 11 10; 392(10159): 1789–858. http://dx.doi.org/10.1016/S0140-6736(18)32279-7 pmid: 30496104.

[63] Davis A.C., Hoffman H.J. Hearing loss: Rising prevalence and impact. *Bulletin. World. Health. Org.* 2019; 97: 646A. http://dx.doi.org/10.2471/BLT.19.224683.

[64] Meo S.A., Suraya F. Effect of environmental air pollution on cardiovascular diseases. *Eur. Rev. Med. Pharmacol Sci.* 2015 Dec; 19(24): 4890–7.

[65] Meo S.A., AlMutairi F.J., Alasbali M.M., Alqahtani T.B., AlMutairi S.S., Albuhayjan R.A., Al Rouq F., Ahmed N. Men's health in industries: Plastic plant pollution and prevalence of pre-diabetes and type 2 diabetes mellitus. *Am. J. Mens. Health.* 2018; 12(6): 2167–72. doi: 10.1177/1557988318800203.

[66] Meo S.A., Al-Khlaiwi T., Abukhalaf A.A., Alomar A.A., Alessa O.M., Almutairi F.J., Alasbali M.M. The nexus between workplace exposure for wood, welding, motor mechanic, and oil refinery workers and the prevalence of prediabetes and type 2 diabetes mellitus. *Int. J. Environ. Res. Public Health.* 2020; 17(11): 3992. doi: 10.3390/ijerph17113992.

[67] Kuang-Hsi Chang, Stella Chin-Shaw Tsai, Chang-Yin Lee, Ruey-Hwang Chou, Hueng-Chuen Fan, Frank Cheau-Feng Lin, Cheng-Li Lin, Yi-Chao Hsu. Increased risk of sensorineural hearing loss as a result of exposure to air pollution. *Int. J. Environ. Res. Public. Health.* 2020; 17(6): 1969. doi: 10.3390/ijerph17061969.

[68] Choi H.G., Min C., Kim S.Y. Air pollution increase the risk of SSNHL: A nested case-control study using meteorological data and national sample cohort data. *Sci. Rep.* 2019; 9: 8270. doi: 10.1038/s41598-019-44618-0.

[69] Lee H.M., Kim M.S., Kim D.J., Uhm T.W., Yi S.B., Han J.H., Lee I.W. Effects of meteorological factor and air pollution on sudden sensorineural hearing loss using the health claims data in Busan, republic of Korea. *Am. J. Otolaryngol.* 2019; 40(3): 393–9.

[70] Choi Y.H., Kim K.S. Noise-induced hearing loss in Korean workers: Co-exposure to organic solvents and heavy metals in nationwide industries. *PLoS. One.* 2014 May 28; 9(5): e97538. doi: 10.1371/journal.pone.0097538.

[71] Park S.K., Elmarsafawy S., Mukherjee B., Spiro A 3rd., Vokonas P.S., Nie H., Weisskopf M.G., Schwartz J., Hu H. Cumulative lead exposure and age-related hearing loss: The VA normative aging study. *Hear. Res.* 2010 Oct 1; 269(1–2): 48–55.

[72] Caciari T., Rosati M.V., Casale T., Loreti B., Sancini A., Riservato R., Nieto H.A., Frati P., Tomei F., Tomei G. Noise-induced hearing loss in workers exposed to urban stressors. *Sci. Total. Environ.* 2013 Oct 1: 302–8: 463–4. doi: 10.1016/j.scitotenv.2013.06.009.

[73] Park M., Han J., Jang M-J., Suh M-W., Lee J.H., Oh S.H. Air pollution influences the incidence of otitis media in children: A national population-based study. *PLoS. One.* 2018; 13(6): e0199296.

[74] Zemek R., Szyszkowicz M., Rowe B.H. Air pollution and emergency department visits for otitis media: A case-crossover study in Edmonton, Canada. *Environ. Health. Perspect.* 2010 Nov; 118(11): 1631–6. doi: 10.1289/ehp.0901675.

[75] Liu Y., Huo X., Xu L., et al. Hearing loss in children with e-waste lead and cadmium exposure. *Sci. Total. Environ.* 2018; 624: 621–7. doi: 10.1016/j.scitotenv.2017.12.091.

[76] Wilunda C., Yoshida S., Tanaka S., Kanazawa Y., Kimura T., Kawakami K. Exposure to tobacco smoke prenatally and during infancy and risk of hearing impairment among children in Japan: A retrospective cohort study. *Paediatr. Perinat. Epidemiol.* 2018; 32(5): 430–8.

[77] Abdel Rasoul G.M., Al-Batanony M.A., Mahrous O.A., Abo-Salem M.E., Gabr H.M. Environmental lead exposure among primary school children in Shebin El-Kom District, Menoufiya Governorate, Egypt. *Int. J. Occup. Environ. Med.* 2012; 3(4): 186–94.

[78] Ahamed M., Siddiqui M.K. Low-level lead exposure and oxidative stress: Current opinions. *Clin. Chim. Acta.* 2007 Aug; 383(1–2): 57–64.

[79] Roy A., Queirolo E., Peregalli F., Mañay N., Martínez G., Kordas K. Association of blood lead levels with urinary F_2-8α isoprostane and 8-hydroxy-2-deoxy-guanosine concentrations in first-grade Uruguayan children. *Environ. Res.* 2015 Jul; 140: 127–35.

[80] Śliwińska-Kowalska M., Zaborowski K. WHO environmental noise guidelines for the European region: A systematic review on environmental noise and permanent hearing loss and tinnitus. *Int. J. Environ. Res. Public. Health.* 2017; 14(10): 1139. doi: 10.3390/ijerph1410113.

[81] Armitage C.J., Loughran M.T., Munro K.J. Epidemiology of the extent of recreational noise exposure and hearing protection use: A cross-sectional survey in a nationally representative UK adult population sample. *BMC. Public. Health.* 2020 Oct 9; 20(1): 1529. doi: 10.1186/s12889-020-09602-8.

[82] Nelson D.I., Nelson R.Y., Concha-Barrientos M., Fingerhut M. The global burden of occupational noise-induced hearing loss. *Am. J. Ind. Med.* 2005; 48: 446–58.

[83] Allahverdy A., Jafari A.H. Non-auditory effect of noise pollution and its risk on human brain activity in different audio frequency using electroencephalogram complexity. *Iran. J. Public. Health.* 2016; 45(10): 1332–9.

[84] Basner M., Babisch W., Davis A., Brink M., Clark C., Janssen S., Stansfeld S. Auditory and non-auditory effects of noise on health. *Lancet.* 2014; 383: 1325–32.

[85] World Health Organization. Hearing loss due to recreational exposure to loud sounds: A review. 2015. Available at: https://apps.who.int/iris/handle/10665/154589. Cited Date Nov 3, 2020.

[86] Dehnert K., Raab U., Perez-Alvarez C., Steffens T., Bolte G., Fromme H., Twardella D. Total leisure noise exposure and its association with hearing loss among adolescents. *Int. J. Audiol.* 2015; 54: 665–73. doi: 10.3109/14992027.2015.1030510.

[87] Guest H., Dewey R.S., Plack C.J., Couth S., Prendergast G., Bakay W., Hall D.A. The noise exposure structured interview (NESI): An instrument for the comprehensive estimation of lifetime noise exposure. *Trends. Hear.* 2018; 22: Article no.: 2331216518803213. doi: 10.1177/2331216518803213.

[88] Jamesdaniel S., Rosati R., Westrick J., Ruden D.M. Chronic lead exposure induces cochlear oxidative stress and potentiates noise-induced hearing loss. *Toxicol. Lett.* 2018 Aug; 292: 175–80. doi: 10.1016/j.toxlet.2018.05.004.

[89] Muhr P., Johnson A.C., Selander J., Svensson E., Rosenhall U. Noise exposure and hearing impairment in air force pilots. *Aerosp. Med. Hum. Perform.* 2019 Sep 1; 90(9): 757–63. doi: 10.3357/AMHP.5353.2019. PMID: 31426890.

[90] Lang G.T., Harrigan M.J. Changes in hearing thresholds as measured by decibels of hearing loss in British army air corps lynx and Apache pilots. *Mil. Med.* 2012 Nov; 177(11): 1431–7. doi: 10.7205/milmed-d-11-00409. PMID: 23198527.

[91] Kampel-Furman L., Joachims Z., Bar-Cohen H., Grossman A., Frenkel-Nir Y., Shapira Y., Alon E., Carmon E., Gordon B. Hearing threshold shifts among military pilots of the Israeli air force. *J. R. Army. Med. Corps.* 2018 Feb; 164(1): 46–51. doi: 10.1136/jramc-2016-000758. Epub 2017 Sep 6. PMID: 28883024.

[92] Huang F.J., Hsieh C.J., Young C.H., Chung S.H., Tseng C.C., Yiin L.M. The assessment of exposure to occupational noise and hearing loss for stoneworkers in Taiwan. *Noise. Health.* 2018 Jul–Aug; 20(95): 146–51. doi: 10.4103/nah.NAH_45_17.

[93] Wouters N.L., Kaanen C.I., den Ouden P.J., Schilthuis H., Böhringer S., Sorgdrager B., Ajayi R., de Laat J.A.P.M. Noise exposure and hearing loss among brewery workers in Lagos, Nigeria. *Int. J. Environ. Res. Public. Health.* 2020 Apr 22; 17(8): 2880. doi: 10.3390/ijerph17082880.

[94] Kim G., Han W. Sound pressure levels generated at risk volume steps of portable listening devices: Types of smartphone and genres of music. *BMC. Public. Health.* 2018; 18: 481. https://doi.org/10.1186/s12889-018-5399-4.

[95] Hickox A.E., Liberman M.C. Is noise-induced cochlear neuropathy key to the generation of hyperacusis or tinnitus? *J. Neurophysiol.* 2014 Feb; 111(3): 552–64. doi: 10.1152/jn.00184.2013.

[96] Kujawa S.G., Liberman M.C. Adding insult to injury: Cochlear nerve degeneration after "temporary" noise-induced hearing loss. *J. Neurosci.* 2009; 29: 14077–85. doi.org/10.1523/JNEUROSCI.2845-09.2009.

[97] Noreña A., Gourévitch B., Aizawa N., et al. Spectrally enhanced acoustic environment disrupts frequency representation in cat auditory cortex. *Nat. Neurosci.* 2006; 9: 932–9. https://doi.org/10.1038/nn1720.

[98] Lin H.W., Furman A.C., Kujawa S.G., Liberman M.C. Primary neural degeneration in the Guinea pig cochlea after reversible noise-induced threshold shift. *J. Assoc. Res. Otolaryngol.* 2011; 12(5): 605–16. doi: 10.1007/s10162-011-0277-0.

[99] Wang Y., Hirose K., Liberman M.C. Dynamics of noise-induced cellular injury and repair in the mouse cochlea. *J. Assoc. Res. Otolaryngol.* 2002; 3: 248–68.

[100] Jamesdaniel S., Hu B., Kermany M.H., Jiang H., Ding D., Coling D., Salvi R. Noise-induced changes in the expression of p38/MAPK signaling proteins in the sensory epithelium of the inner ear. *J. Proteomics.* 2011; 75: 410–24.

[101] Kobel M., Le Prell C.G., Liu J., Hawks J.W., Bao J. Noise-induced cochlear synaptopathy: Past findings and future studies. *Hear. Res.* 2017; 349: 148–54.

[102] Böttger E.C., Schacht J. The mitochondrion: A perpetrator of acquired hearing loss. *Hear. Res.* 2013 Sep; 303: 12–9.

[103] Huth M.E., Ricci A.J., Cheng A.G. Mechanisms of aminoglycoside ototoxicity and targets of hair cell protection. *Int. J. Otolaryngol.* 2011; 2011: 937861.

[104] Jamesdaniel S., Rathinam R., Neumann W.L. Targeting nitrative stress for attenuating cisplatin-induced downregulation of cochlear LIM domain only 4 and ototoxicity. *Redox. Biol.* 2016 Dec; 10: 257–65.

[105] Kim S.J., Jeong H.J., Myung N.Y., Kim M.C., Lee J.H., So H.S., Park R.K., Kim H.M., Um J.Y., Hong S.H. The protective mechanism of antioxidants in cadmium-induced ototoxicity in vitro and in vivo. *Environ. Health. Perspect.* 2008 Jul; 116(7): 854–62.

[106] Poirrier A.L., Pincemail J., Van Den Ackerveken P., Lefebvre P.P., Malgrange B. Oxidative stress in the cochlea: An update. *Curr. Med. Chem.* 2010; 17(30): 3591–604.

[107] Samson J., Wiktorek-Smagur A., Politanski P., Rajkowska E., Pawlaczyk-Luszczynska M., Dudarewicz A., Sha S.H., Schacht J., Sliwinska-Kowalska M. Noise-induced time-dependent changes in oxidative stress in the mouse cochlea and attenuation by D-methionine. *Neuroscience.* 2008 Mar 3; 152(1): 146–50.

[108] Wong A.C., Ryan A.F. Mechanisms of sensorineural cell damage, death and survival in the cochlea. *Front. Aging. Neurosci.* 2015; 7: 58.

[109] Yang C.H., Schrepfer T., Schacht J. Age-related hearing impairment and the triad of acquired hearing loss. *Front. Cell. Neurosci.* 2015; 9: 276.

[110] Choi S.H., Choi C.H. Noise-induced neural degeneration and therapeutic effect of antioxidant drugs. *J. Audiol. Otol.* 2015 Dec; 19(3): 111–9.

[111] Redlarski G., Lewczuk B., Żak A., Koncicki A., Krawczuk M., Piechocki J., Jakubiuk K., Tojza P., Jaworski J., Ambroziak D., Skarbek Ł., Gradolewski D. The influence of electromagnetic pollution on living organisms: Historical trends and forecasting changes. *BioMed research international.* 2015: 234098. https://doi.org/10.1155/2015/234098.

[112] Chen P.L., Pai C.W. Pedestrian smartphone overuse, and inattentional blindness: An observational study in Taipei, Taiwan. *BMC. Public. Health.* 2018; 18(1): 1342. doi: 10.1186/s12889-018-6163-5.

[113] Velayutham P., Govindasamy G.K., Raman R., Prepageran N., Ng K.H. High-frequency hearing loss among mobile phone users. *Indian. J. Otolaryngol. Head. Neck Surg.* 2014 Jan; 66(Suppl 1): 169–72. doi: 10.1007/s12070-011-0406-4.

[114] Oktay M.F., Dasdag S. Effects of intensive and moderate cellular phone use on hearing function. *Electromagn. Biol. Med.* 2006; 25: 13–21. doi: 10.1080/15368370600572938.

[115] Das S., Chakraborty S., Mahanta B.A study on the effect of prolonged mobile phone use on pure tone audiometry thresholds of medical students of Sikkim. *J. Postgrad. Med.* 2017; 63(4): 221–5. doi: 10.4103/0022-3859.201419.

[116] Philip P., Bhandary S.K., Aroor R., Bhat V., Pratap D. The effect of mobile phone usage on hearing in an adult population. *Indian. J. Otol.* 2017; 23: 1–6.

[117] Ragab A., Salem M.E., Abd El Hady L.M. Mobile phone use and its risk on hearing. *Menoufia. Med. J.* 2014; 27: 432–9.

[118] Al-Dousary S.H. Mobile phone-induced sensorineural hearing loss Saudi. *Med J.* 2007; 28(8): 1283–9.

[119] Panda N.K., Jain R., Bakshi J., Munjal S. Audiologic disturbances in long-term mobile phone users. *J. Otolaryngol. Head. Neck. Surg.* 2010; 39(1): 5–11.

[120] Panda N.K., Modi R., Munjal S., Virk R.S. Auditory changes in mobile users: Is evidence forthcoming? *Otolaryngol. Head. Neck. Surg.* 2011; 144(4): 581–5. doi: 10.1177/0194599810394953.

[121] Gupta N., Goyal D., Sharma R., Arora K.S. Effect of prolonged use of mobile phone on brainstem auditory evoked potentials. *J. Clin. Diagn. Res.* 2015; 9(5): CC07–9. doi: 10.7860/JCDR/2015/13831.5976.

[122] Akdag M., Dasdag S., Canturk F., Akdag M.Z. Exposure to non-ionizing electromagnetic fields emitted from mobile phones induced DNA damage in human ear canal hair follicle cells. *Electromagn. Biol. Med.* 2018; 37(2): 66–75. doi: 10.1080/15368378.2018.1463246.

[123] Godson R.E.E., Ana A.E., Ukhun D.G., Shendell Patience A. Osisanya. Acute, repeated exposure to mobile phone noise and audiometric status of young adult users in a university community. *Int Sch Res Notices.* 2012; 2012: 241967. doi.org/10.5402/2012/241967.

[124] Meo S.A., Al-Dress A.M. Mobile phone related-hazards and subjective hearing and vision symptoms in the Saudi population. *Int. J. Occup. Med. Environ. Health.* 2005; 18(1): 53–60.

[125] Khan M.M. Adverse effects of excessive mobile phone use. *Int. J. Occup. Med. Environ. Health.* 2008; 21(4): 289–93. doi: 10.2478/v10001-008-0028-6.

[126] Bhargav H., Srinivasan T M., Bista S., Mooventhan A., Suresh S., Hankey A., Nagendra H.R. Acute effects of mobile phone radiations on subtle energy levels of teenagers using electro photonic imaging technique: A randomized controlled study. *Int. J. Yoga.* 2017; 10(1): 16–23. doi: 10.4103/0973-6131.186163.

[127] Sagiv D., Migirov L., Madgar O., Nakache G., Wolf M., Shapira Y. Mobile phone usage does not affect sudden sensorineural hearing loss. *J. Laryngol. Otol.* 2018; 132(1): 29–32. doi: 10.1017/S0022215117002365.

[128] Khullar S., Sood A., Sood S. Auditory brainstem responses and EMFs generated by mobile phones. *Indian. J. Otolaryngol. Head. Neck. Surg.* 2013; 65 (Suppl 3): 645–9.

[129] Cardis E., Deltour I., Mann S. Distribution of RF energy emitted by mobile phones in anatomical structures of the brain. *Phys. Med. Biol.* 2008; 53: 2771–83.

[130] Hardell L., Carlberg M., Hansson Mild K. Pooled analysis of two case-control studies on the use of cellular and cordless telephones and the risk for malignant brain tumours diagnosed in 1997–2003. *Int. Arch. Occup. Environ. Health.* 2006; 79: 630–9.

[131] Hardell L., Carlberg M., Hansson Mild K. Use of mobile phones and cordless phones is associated with increased risk for glioma and acoustic neuroma. *Pathophysiology.* 2013; 20: 85–110.

[132] Özgür A., Tümkaya L., Terzi S., Kalkan Y., Erdivanlı O.C., Dursun E. Effects of chronic exposure to electromagnetic waves on the auditory system. *Acta. Otolaryngol.* 2015; 135(8): 765–70. doi: 10.3109/00016489.2015.1032434.

[133] Zuo Yu-Juan H.U., Yang Yang, Xue-Yan Zhao, Yuan-Yuan Zhang, Wen Kong, Wei-Jia Kong. Sensitivity of spiral ganglion neurons to damage caused by mobile phone electromagnetic radiation will increase in lipopolysaccharide-induced inflammation in vitro model. *J. Neuroinflammation.* 2015; 12: 105. doi: 10.1186/s12974-015-0300-1.

[134] Christensen H.C., Schüz J., Kosteljanetz M., Poulsen H.S., Thomsen J., Johansen C. Cellular telephone use and risk of acoustic neuroma. *Am. J. Epidemiol.* 2004; 159(3): 277–83.

[135] Takebayashi T., Akiba S., Kikuchi Y., Taki M., Wake K., Watanabe S., Yamaguchi N. Mobile phone use and acoustic neuroma risk in Japan. *Occup. Environ. Med.* 2006 Dec; 63(12): 802–7. doi: 10.1136/oem.2006.028308.

[136] Repacholi M.H. Health risks from the use of mobile phones. *Toxicol. Lett.* 2001; 120(1–3): 323–31. doi: 10.1016/S0378-4274(01)00285-5.

[137] Moulder J.E., Foster K.R., Erdreich L.S., McNamee J.P. Mobile phones, mobile phone base stations, and cancer: A review. *Int. J. Radiat Biol.* 2005 Mar; 81(3): 189–203. doi: 10.1080/09553000500091097.

[138] Lönn S., Ahlbom A., Hall P., Feychting M. Mobile phone use and the risk of acoustic neuroma. *Epidemiology.* 2004; 15(6): 653–9. doi: 10.1097/01.ede.0000142519.00772.

[139] Schoemaker M.J., Swerdlow A.J., Ahlbom A., Auvinen A., Blaasaas K.G., Cardis Christensen H.C., Feychting M., Hepworth S.J., Johansen C., Klaeboe L., Lönn S., McKinney P.A., Muir K., Raitanen J., Salminen T., Thomsen J., Tynes T. Mobile phone use and risk of acoustic neuroma: Results of the Interphone case-control study in five North European countries. *Br. J. Cancer.* 2005 Oct 3; 93(7): 842–8. doi: 10.1038/sj.bjc.6602764.

[140] Interphone Study Group. Brain tumour risk in relation to mobile telephone use: Results of the INTERPHONE international case-control study. *Int. J. Epidemiol.* 2010; 39: 675–94.

[141] Hardell L., Mild K.H., Påhlson A., Hallquist A. Ionizing radiation, cellular telephones and the risk for brain tumours. *Eur. J. Cancer. Prev.* 2001; 10(6): 523–9. doi: 10.1097/00008469-200112000-00007.

[142] Corona A.P., Oliveira J.C., Souza F.P.A., Santana L.V., Rêgo M.A.V. Risk factors associated with vestibulocochlear nerve schwannoma: A systematic review. *Braz. J. Otorhinolaryngol.* 2009; 75(4): 593–615. doi: 10.1016/s1808-8694(15)30501-2.

[143] Hardell L., Carlberg M., Söderqvist F., Mild K.H. Pooled analysis of case-control studies on acoustic neuroma diagnosed 1997–2003 and 2007–2009 and use of mobile and cordless phones. *Int. J. Oncol.* 2013 Oct; 43(4): 1036–44. doi: 10.3892/ijo.2013.2025.

[144] Benson V.S., Pirie K., Schüz J., Reeves G.K., Beral V., Green J. Mobile phone use and risk of brain neoplasms and other cancers: A prospective study. *Int. J. Epidemiol.* 2013; 42(3): 792–802. doi: 10.1093/ije/dyt072.

[145] Mild K.H., Hardell L., Carlberg M. Pooled analysis of two Swedish case-control studies on the use of mobile and cordless telephones and the risk of brain tumours diagnosed during 1997–2003. *Int. J. Occup. Saf. Ergon.* 2007; 13(1): 63–71.

[146] Moon I.S., Kim B.G., Kim J., Lee J.D., Lee W. Association between vestibular schwannomas and mobile phone use. *Tumour. Biol.* 2014; 35(1): 581–7. doi: 10.1007/s13277-013-1081-8.

[147] IARC Working Group on the Evaluation of Carcinogenic Risks to Humans. Non-ionizing radiation, Part 2: Radiofrequency electromagnetic fields. *IARC Monogr Eval Carcinog Risks Hum.* 2013; 102(Pt 2): 1–460.

[148] Baan R., Grosse Y., Lauby-Secretan B., El Ghissassi F., Bouvard V., Benbrahim-Tallaa L., Guha N., Islami F., Galichet L., Straif K., WHO international agency for research on cancer monograph working group. Carcinogenicity of radiofrequency electromagnetic fields. *Lancet. Oncol.* 2011 Jul; 12(7): 624–6.

[149] Belpomme D., Hardell L., Belyaev I., Burgio E., Carpenter D.O. Thermal and non-thermal health effects of low-intensity non-ionizing radiation: An international perspective. *Environ. Pollut.* 2018 Nov; 242(Pt A): 643–58.

[150] Carlberg M., Hardell L. Evaluation of mobile phone and cordless phone use and glioma risk using the Bradford hill viewpoints from 1965 on association or causation. *Biomed. Res. Int.* 2017; 2017: 9218486.

[151] Miller A.B., Morgan L.L., Udasin I., Davis D.L. Cancer epidemiology update, following the 2011 IARC evaluation of radiofrequency electromagnetic fields (Monograph 102). *Environ. Res.* 2018 Nov; 167: 673–83. doi: 10.1016/j.envres.2018.06.043.

[152] Repacholi M.H., Lerchl A., Röösli M., Sienkiewicz Z., Auvinen A., Breckenkamp J., d'Inzeo G., Elliott P., Frei P., Heinrich S., Lagroye I., Lahkola A., McCormick D.L., Thomas S., Vecchia P. Systematic review of wireless phone use and brain cancer and other head tumors. *Bioelectromagnetics.* 2012 Apr; 33(3): 187–206. doi: 10.1002/bem.20716. Epub 2011 Oct 21. PMID: 22021071.

[153] Han Y.Y., Kano H., Davis D.L., Niranjan A., Lunsford L.D. Cell phone use and acoustic neuroma: The need for standardized questionnaires and access to industry data. *Surg. Neurol.* 2009; 72(3): 216–22; doi: 10.1016/j.surneu.2009.01.010.

[154] Balbani A.P., Montovani J.C. Mobile phones: Influence on auditory and vestibular systems. *Braz. J. Otorhinolaryngol.* 2008 Jan–Feb; 74(1): 125–31. doi: 10.1016/s1808-8694(15)30762-x. *Erratum in: Rev Bras Otorrinolaringol (Engl Ed).* 2008 Mar–Apr; 74(2): 319. PMID: 18392513.

[155] Vrijheid M., Deltour I., Krewski D., Sanchez M., Cardis E. The effects of recall errors and of selection bias in epidemiologic studies of mobile phone use and cancer risk. *J. Expo. Sci. Environ. Epidemiol.* 2006 Jul; 16(4): 371–84. doi: 10.1038/sj.jes.7500509. Epub 2006 Jun 14. PMID: 16773122.

[156] Choi Y.H., Hu H., Mukherjee B., Miller J., Park S.K. Environmental cadmium and lead exposures and hearing loss in U.S. Adults: The national health and nutrition examination survey, 1999 to 2004. *Environ. Health. Perspect.* 2012 Nov; 120(11): 1544–50.

[157] Masuda M., Kanzaki S., Minami S., Kikuchi J., Kanzaki J., Sato H., Ogawa K. Correlations of inflammatory biomarkers with the onset and prognosis of idiopathic sudden sensorineural hearing loss. *Otol. Neurotol.* 2012 Sep; 33(7): 1142–50.

[158] Yao K., Wu W., Yu Y., Zeng Q., He J., Lu D., Wang K. Effect of superposed electromagnetic noise on DNA damage of lenepithelial cells induced by microwave radiation. *Invest. Ophthalmol. Vis. Sci.* 2008; 49: 2009–2015.

[159] Avci B., Akar A., Bilgici B., Tunçel Ö.K. Oxidative stress induced by 1.8 GHz radiofrequency electromagnetic radiation and effects of garlic extract in rats. *Int. J. Radiat. Biol.* 2012; 88: 799–805.

Effects of Electromagnetic Field Radiation on the Eyes and Vision

9.1 Introduction

Advances in technology have led to a plethora of electronic devices, including computers, laptops, mobile phones, and devices using Wi-Fi, which generate RF-EMFR. This modern technology, especially the electromagnetic environment associated with it, has significantly altered human lives. The proliferation of electromagnetic fields (EMFs), also known as "electromagnetic smog," has become a major polluting element in the environment. There are without doubt benefits associated with the availability of these electronic devices. However, the unnecessary and excessive use of devices generating RF-EMFR is highly hazardous to the environment and human health.

The human sense organs are highly susceptible and provide awareness of light, sound, taste, touch and information about the internal and external environments. Humans also possess general senses, known as "somatosensation," which respond to stimuli like pain, pressure, temperature and vibration. Although each human organ is essential for a healthy life, vision is considered the most important sense during decision-making, planning interventions and all activities [1]. Given the importance of human senses, visual impairment has often been considered the highest form of disability. The major consequences of limited vision are an inability to function effectively and socialize, affecting different aspects of an individual's life [1, 2].

This chapter highlights the effects of RF-EMFR on the eyes and vision. Any impairment of the eyes and vision may profoundly reduce one's quality of life. Support from family and friends can facilitate positive adaptation to this, but it is still challenging to function and work at optimal capacity. The major etiopathological factors that impair the biology of the eyes and vision are associated with ageing, genetics, accidents, trauma, and environment-related factors. However, the role of mobile phone-generated RF-EMFR in the impairment of vision is not widely understood. Therefore, it is essential to highlight the scientific literature and obtained findings to improve our understanding of how excessive mobile phone usage impairs the biology of vision. Before these biological facts are discussed, this chapter first elaborates on the biological, anatomical, and physiological aspects of eyes and vision.

9.2 Biology of the Eyes and Vision

The eye is a paired, fluid-filled spherical and highly perceptive sensory organ. Among the sensory organs, vision is a dominant sense. The human eye is about 2.5 cm in diameter and is enclosed by three layers of tissue. The outer layer consists of a transparent, white fibrous tissue called the sclera. This dense outer layer is transformed into a highly specialized transparent tissue called the cornea at the front of the eye. It permits the entry of light into the eyes. The middle layer consists of three distinct continuous structures: iris, ciliary body and choroid. The iris is a flat, thin, ring-shaped membrane made of connective tissue. It lies behind the cornea with an adjustable circular opening in the center, the pupil. The iris gives the eye its color and contains two sets of muscles with opposing actions, which control the size of the pupil. The ciliary body is a ring-shaped structure that encircles and holds the lens in place to adjust the lens's refractive power. It is connected to the lens via a network of very delicate and small "ligaments called ciliary zonules or zonules of Zinn, which suspend the lens in place behind the pupil. The ciliary body also secretes the clear, colorless aqueous fluid that fills the eye's anterior segment. The choroid is composed of a rich capillary bed that serves as the primary source of blood supply for the photoreceptors of the retina. The retina contains neurons that are sensitive to light and are capable of transmitting visual signals to central targets" [3] (Table 9.1; Figure 9.1). The presence in the retina of a hundred million specialized photoreceptor cells known as rods and cones enables the conversion of light waves into electrochemical impulses, which are decoded by the brain. The higher centers in the brain interpret the image and respond accordingly. There are three types of circular layers of the eyeball.

The contents of the eyeball provide a refracting medium for the eye. These include aqueous humor, lens and vitreous body.

The sensation of vision is the result of complex processes in the eyes and brain. The light enters into the eye through the pupil and is projected onto the back of the inner part of the eye, known as the retina. The retina contains photoreceptor cells, known as cones and rods. The "photopigments in these two receptors absorb the light, resulting in a chemical-electrical signal that travels through a nerve into the visual cortex in the brain, where the visual sensation is invoked. The cone cells in the fovea have a one-to-one nerve connection to the brain, while the rod cells are located in the periphery of the retina" [3–5] (Table 9.2).

9.3 Global Burden of Vision Impairment and Blindness

As per a recent report from the WHO [6], about 1 billion people worldwide have vision impairment or blindness. These include people with moderate or severe distance vision

Table 9.1 Types of Layers of the Eyeball

1. **Fibrous/supporting layer**
   i. Sclera
   ii. Cornea
2. **Vascular/pigmented layer**
   i. Iris
   ii. Ciliary body
   iii. Choroid
3. **Nervous/retinal layer**
   i. Retina

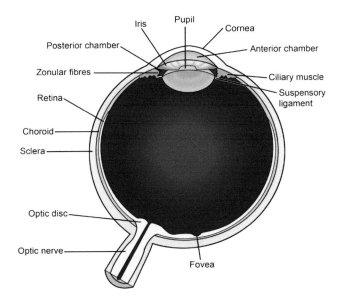

Figure 9.1 The eye.

Table 9.2 Pathway of Light Reflex

- Rods and cones
- Optic nerve
- Optic chiasma
- Optic tract
- The pretectal nucleus of the midbrain
- Edinger—Westphal nucleus
- Parasympathetic fibers in the oculomotor nerve
- Ciliary ganglion
- Ciliary nerve
- Pupil constricts

impairment or blindness due to unaddressed refractive error and near vision impairment caused by unaddressed presbyopia. Moreover, the leading causes of vision impairment all over the world are uncorrected refractive errors and cataracts. The majority of people with vision impairment are over the age of 50 years. [6]

It has also been reported that about 120 million are visually impaired because of uncorrected refractive errors, and 90% of visually impaired people reside in low- and middle-income countries. About 82% of people who are blind are over 50 years old, 28% of people with moderate and severe visual impairment are of working age, and 1.4 million children are blind [6].

The major causes of vision impairment and blindness among the above-mentioned 1 billion people include "unaddressed refractive error (123.7 million), cataract (65.2 million), glaucoma (6.9 million), corneal opacity (4.2 million), diabetic retinopathy (3 million), and trachoma (2 million), as well as near vision impairment caused by unaddressed presbyopia (826 million)" [6].

9.4 Electromagnetic Field Radiation (EMFR)

During the last three decades, the use of various electrical appliances has rapidly increased, especially with the use of wireless equipment and smartphones. Smartphone technology has left behind many other electronic devices that were essential and frequently used in the past, such as radio, telephone, telegraph, fax, and document scanners. With the rapid proliferation of wireless technologies, especially smartphones, and the installation of cell phone base-station towers in residential and densely populated areas, the public has become increasingly concerned about RF-EMFR generated from these novel devices. To better understand RF-EMFR and its impact on brain biology, it is vital to comprehend electromagnetic fields and their natural and human-made sources.

In the last two decades, the use of mobile phones has markedly increased among all age groups, in both urban and rural areas and developing and developed nations. Mobile phones are now a widely prevalent means of everyday communication [7]. Their role in daily life is not limited to calls but has spread to storing data, photography, gaming, listening to music, and watching videos. Moreover, modern cell phones provide internet access, online banking, transfers of documents, and geo-positioning system technology, allowing users to find locations and survey traffic conditions. In addition to the excessive usage of mobile phones, many people also use other information technology-related devices [7].

The scientific literature acknowledges numerous natural and synthetic sources of RF-EMFR. Mobile phones are widely used and are the most potent sources of human exposure to RF-EMFR [7]. The Global System for Mobile Communication (GSM) operated between 900 and 1,800 MHz. The third-generation (3G) or Universal Mobile Telecommunication System (UMTS) uses 1,900/2,100 MHz. The fourth-generation (4G) operates at 800/2,700 MHz. Furthermore, in 2019, the development of fifth-generation (5G) wireless technology was taking place, using frequencies of about 25–39 GHz. The literature also presents that the second-, third-, and fourth-generation mobile phones emit RF-EMFR in the range of about 800–2,700 MHz [8], while 5G high-frequency technology emits it at 24.25–52.6 GHz and 24–80 GHz [8].

Similar radiation is produced from other digital wireless systems such as data communication networks. The RF-EMF radiation causes various effects on human health [9] (Figure 9.2).

RF-EMFR affects human biology through thermal and non-thermal effects, which increase the temperature of tissues and organs, causing damage to cells. In 2011, the International Agency for Research on Cancer (IARC) classified the extremely low-frequency radiation, RF-EMFR, generated by electrical devices as possibly carcinogenic to humans [10, 11].

Figure 9.2 Exposure to RF-EMFR generated from mobile phones and mobile phone towers.

9.5 Mobile Phones and RF-EMFR

In the last decade, there has been an exponential evolution of mobile communication systems and associated devices, and as a result, the public is incidentally exposed to radiofrequency radiation. Along with ongoing public concerns regarding RF-EMFR, a large number of scientific studies have attempted to determine whether electromagnetic field radiation (EMFR) has any harmful effects on human health [12, 13]. All living organisms can absorb the RF-EMFR emitted by mobile phones and base-station towers. The absorption of RF-EMFR depends on the distance from the device, the frequency of the radiation, and the duration of exposure. The frequency is important since fifth-generation (5G) networks operate at higher frequencies [14, 15].

RF-EMFR has been one of the major impactors on health associated with multiple human-made sources such as mobile phones, mobile phone towers, tablets, laptops, computers, televisions, radars, microwave ovens, high power lines, transformers and various other wireless gadgets. The use of these devices that generate RF-EMF radiation has increased tremendously during the last three decades and exposure to them is widespread. Mobile phones generate various radio frequencies, including 900–1,800 and 2,600 MHz radiofrequency radiation. The RF-EMFR generated from mobile phones, Wi-Fi and electronic devices is a growing public concern regarding the hazardous effects of radiofrequency radiation on human health [16].

RF-EMFR has become a leading cause of persistent electromagnetic pollution that affects various environmental elements [17]. Mobile phones are the most common source of RF-EMF exposure to the brain. The number of mobile phones being used is more than the current world population (as many people have more than one mobile phone). Moreover, nowadays, children also frequently use mobile phones. Whether exposure to RF-EMFR can cause, cancer has been addressed in numerous case-control studies, with heterogeneous findings having been reported. However, recent case-control studies are reporting substantially increased risks [8].

9.6 Effect of RF-EMFR on Vision

Environmental stress, including the EMR generated from various electronic devices such as mobile phones, has a hazardous impact on the eyes and vision and is considered a risk factor for various eye problems. The eyes and vision are sensitive systems of the human body when determining safety standards. In the last three decades, the widespread use of mobile phones has raised various concerns about the possible hazardous health effects of the RF-EMFR emitted from mobile phones and their associated devices on eyes, vision, and the nervous system. When using cell phones, people tend to keep them close to the ear, head, eyes, and brain. RF-EMFR can affect vision and hearing by two main mechanisms. "These are thermal (heating) effects caused by holding mobile phones close to the body and possible non-thermal effects" [18]. In this chapter, while investigating the effect of RF-EMFR on vision, both human and animal model studies were reviewed to reach better conclusions (Tables 9.3 and 9.4; Figure 9.3).

Meo and Al-Dress (2005) [19] reported that mobile phone users have complaints of impaired hearing, earache, and warmth on the ear (34.59%), along with impaired vision or blurred vision (5.04%). In contrast [19], Davidson and Lutman (2007) [20] showed no hearing loss, tinnitus, or balance complaints among high or long-term users of mobile phones compared with low or short-term users. The most probable reason for this contradiction was the selection criteria. Harry and Mark (2007) [20] compared the findings between mobile users based on their usage. A better method is to compare the findings between mobile phone users and non-users. Considering that this study was conducted in 2007, the authors could easily have chosen individuals who were not avid mobile phone users.

Shokoohi-Rad et al. (2020) [21] explored the hazardous effects of RF-EMFR generated from cell phones on the intraocular pressure (IOP) in the eyes. The authors identified that the IOP in 42 glaucomatous eyes among people using a cell phone ipsilaterally before and after the intervention was 18.60 ± 6.70 and 23.50 ± 6.30, respectively. However, IOP in the control group without glaucoma (41 eyes) of people using a cell phone ipsilaterally, before and after the intervention, was 12.90 ± 3.50 and 13.40 ± 2.80, respectively. The authors found that acute effects of RF-EMFR emitted from mobile phones increased the IOP in glaucomatous eyes, while such changes were not observed in normal eyes. Moreover, Young-Hyun et al. (2017) [22] demonstrated visual fatigue caused by smartphone use that harmed balance function among 22 young, healthy adults.

Moon et al. (2016) [23] conducted a study on smartphone use in children. They found that smartphone use was linked to pediatric dry eye disease (DED). However, the outdoor activities of children were shown to play a role in protecting against pediatric DED. Older students who were residents of urban environments had a high risk of DED with a long duration of smartphone use and a short duration of outdoor activities.

Long et al. (2017) [24] investigated the eyestrain clinical characteristics in young adults reading from a smartphone for a period of 60 min. They found that eyestrain symptoms included tired eyes, uncomfortable eyes, and blurred vision, which increased significantly among the students who were watching and reading close to a smartphone for a period of 60 min. In addition, Golebiowski et al. (2020) [25] investigated the effect of 60 min of reading on a smartphone on "ocular symptoms, binocular vision, tear function, blinking and working distance." They found that eyestrain symptoms and ocular surface symptoms were increased after using a smartphone; these mainly involved discomfort, tiredness, and sleepiness, along with a decrease of binocular accommodative capability. The authors concluded that long-term use of smartphones has consequences for ocular health and binocular functions.

Küçer (2009) [26] conducted a study in Kocaeli, Turkey, focusing on clinical ocular symptoms. He found a significant increase in blurring of vision among those who had used a mobile phone for over two years compared with subjects who had used one for less than two years. Moreover, among the mobile phone users, more females complained of inflammation in the eyes than males.

Meo and Al-Dress (2005) [19] conducted a study on 873 (57.04% males and 39.86% females) subjects using mobile phones and found that 5.04% of users complained of decreased and blurred vision. There was also an association between the duration of calls and vision complaints among mobile phone users.

Table 9.3 RF-EMFR Generated from Mobile Phones and Its Effects on the Eyes and Vision (Findings Based on Human Model Studies).

Author(s) and Year of Study	Sample Size	Age (Years)	Type of Study	Outcome
Shokoohi-Rad et al. (2020) [21]	Glaucoma group 42, control group 41		Cross-sectional	Mobile phone radiation increased intraocular pressure in glaucomatous eyes, not normal eyes
Young-Hyun et al. (2017) [22]	22 adults (11 males and 11 females)	Young adults	Cross-sectional	Mobile phones caused visual fatigue and negatively impacted balance function
Moon et al. (2016) [23]	916 children		Case-control	Smartphone use was associated with pediatric dry eye disease (DED)
Long et al. (2017) [24]	18 adults with mean age 21.5 years	21.5 ± 3.3 years	Survey-based observational	Tired, uncomfortable eyes and blurred vision were increased after 60 min use of a smartphone
Golebiowski et al. (2020) [25]	12 young adults (18–23 years)		Survey-based observational	Binocular vision, tears, blinking, eyestrain symptoms were increased after 60 min use of a smartphone
Küçer (2009) [26]	229 (F:181, M:48)		Observational	Significant increase in blurring of vision among mobile phone users
Meo and Al-Dress (2005) [19]	873		Observational	Association between mobile phone use and vision complaints
Povolotskiy et al. (2019) [27]	2501 males and females	13–29 years	Retrospective cross-sectional	Injuries while using a phone affected head (33.1%); face, eyelid, eye area and nose (32.7%); and neck (12.5%)
Tawer et al. (2015) [28]	68 university students	18–25 years	Cross-sectional	90% showed complaints of defects in refraction for both eyes
Kan et al. (2016) [29]	Human model in vitro study	Human lens epithelial cells	Cross-sectional	Short- and long-term exposure to 50 Hz RF-EMFR at 0.4 mT could not induce DNA damage in human lens epithelial cells in vitro
Shuang et al. (2013) [30]	Human model in vitro study	In vitro study	Cross-sectional	Human lens epithelial cells exposed to 1.8 GHz showed increases in ROS and MDA levels in RF exposure group
Bhargav et al. (2017) [31]	61 healthy adults with a mean age of 17.40 years.	Adult age	Cross-sectional	Energy levels reduced in the left eye and left ear among subjects using mobile phone in ON mode after RF-EMF exposure compared with mobile phone in OFF mode group

Table 9.4 RF-EMFR Generated from Mobile Phones and Its Effects on the Eyes and Vision (Findings Based on Animal Model Studies).

Author(s) and Year of Study	Sample Size	Age (Years)	Type of Study	Outcome
Balci et al. (2009) [32]	40 albino Wistar rats		Animal model cross-sectional	Mobile phone radiation increased oxidative stress in cornea and lens tissues
Bormusov et al. (2008) [33]	58 bovine lenses of 1-year-old male calves	1-year-old male calves	Cross-sectional	RF-EMFR from mobile phones and other modern devices damaged eye tissue and lens optical quality
Eker et al. (2018) [34]	37 Wistar albino rats		Cross-sectional	EMF was a stress factor, activating caspase-3 and p38MAPK gene expression. RF-EMF could cause cellular damage in rat ocular cells
Talebnejad et al. (2017) [35]	40 adult white New Zealand rabbits		Cross-sectional	Histologically, cell phone radiation had no damaging effect on the retina but caused ciliary body congestion
Ahlers et al. (2014) [36]	Retinal cells		Cross-sectional	Exposure of RF-EMF at mobile phone frequencies (GSM-900, GSM-1800, MTS) and SARs up to 20 W/kg had no acute effects on retinal cells

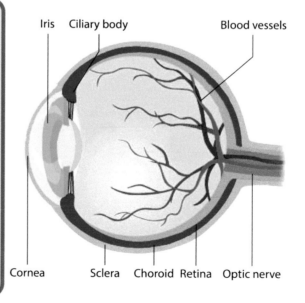

- Highly sensitive to heating effect
- Increase oxidative stress
- Inflammation
- Dry eye
- Increase intraocular pressure
- Change in vascular permeability
- Corneal lesion
- Impaired lens optical quality
- Neuronal injury
- Eye damage

Iris   Ciliary body   Blood vessels

Cornea   Sclera   Choroid Retina Optic nerve

Figure 9.3 Effect of RF-EMFR on eye and vision.

In another study, Povolotskiy et al. (2019) [27] assessed the incidence, types, and mechanisms of head/neck and eye injuries while using cell phones. In this study, the authors used the data from a regional database registry with head and neck injuries related to cell phone use. They identified a total of 2,501 patients aged 13–29 years who had presented with head and neck-related injuries during cell phone use. The most commonly reported injuries were in the head and neck region, including the head (33.1%), face including the eyelid, eye area, and nose (32.7%), and neck (12.5%). Laceration injuries constituted 26.3% of the total, contusions and abrasions were 24.5%, and internal organ injuries were 18.4%. These injuries occurred among people aged 13–29 years because of texting while walking.

Balci et al. (2009) [32] investigated the effects of RF-EMFR generated from mobile phones on oxidant/antioxidant balance in corneal and lens tissues. The findings confirmed that mobile phone radiation leads to oxidative stress in cornea and lens tissues, but antioxidants, such as vitamin C, avert these hazardous effects.

Siao et al. (2019) [37] examined the relationship between smartphone use and musculoskeletal/visual symptoms among 1,884 adolescents. Smartphone use was related to increased risk of discomfort in the neck, shoulders, upper back, arms, wrists, hands, and visual symptoms.

Bormusov et al. (2008) [33] conducted a study on the non-thermal effects of microwaves at the same frequency as used in cellular phones on the intact lens in organ culture. A dose similar to what the lens receives upon speaking on a cell phone was applied. The authors found that RF-EMFR damaged the lens' optical quality, and the effect was dose-dependent, while the damage to optical quality was reversed when the exposure was stopped. However, morphological changes to the epithelium were irreversible. The authors suggested that, while communicating on a cell phone, one should keep one's distance from it in order to minimise RF-EMFR exposure and thus reduce potentially harmful effects of the cell phone on the lens.

Tawer et al. (2015) [28] reported the effects of RF-EMFR emitted from cell phones on human eyes. They recruited 68 university students aged between 18 and 25 years old. They identified that about 90% of students had impaired refraction in both eyes, apart from the side of the head on which the cell phone was used. Similarly, Zhu et al. (2014) [38] reported the impact of low-frequency RF-EMFR on human fetal scleral fibroblasts. The growth rate of such fibroblasts decreased after exposure of only 24 hours to RF-EMFR. The study findings showed that RF-EMFR had biologically damaging effects on fetal scleral fibroblasts and could cause abnormalities in scleral collagen.

Radiofrequency (RF) energy has been reported to have various ocular effects on the lens, retina, cornea, and other ocular structures. It is reasonable to assume that temperatures equal to or greater than 41°C in or near the human eye lens can cause a cataract through thermal heating. It was also found that ocular effects, corneal lesions, retinal effects, and changes in vascular permeability developed after localized exposure of the eyes [39].

The available literature showed that exposure to RF-EMFR generated from a mobile phone can affect the eyes and vision through thermal and non-thermal mechanisms. RF-EMFR increases the intraocular temperature and damages visual tissues. The eye tissues cannot remove heat, and in the anterior part of the eye, active thermal transport does not occur. Moreover, blood flow through the retina serves to maintain a stable thermal environment for the photoreceptors. The literature also hypothesized that "poor heat dissipation capability from within the anterior segment of the eyes may lead to heat buildup and subsequent thermal damage" to ocular tissues. The eye is sensitive to heat within the anterior segment of the eye through the deep penetrating effect of RF-EMFR generated from various devices, including mobile phones [40]. Moreover, the human eye lens is highly vulnerable to RF-EMFR because of the decreased water content and the absence of vasculature. Hence, eye tissues such as the retina, iris vasculature, and corneal endothelium are sensitive to RF-EMFR [40].

Johansen et al. (2001) [41] performed a study on the trends in the prevalence of melanoma against the background of a rapidly increasing number of mobile phone users. The study findings did not provide support for a relationship between mobile phone use and ocular melanoma. Moreover, Eker et al. (2018) [34] "investigated the expression levels of heat shock protein 27 (Hsp27), p38 mitogen-activated protein kinase (p38MAPK), epidermal growth factor receptor, and caspase-3 genes in rat eye exposed to 1800 MHz RF-EMFR" [34]. After exposure of Wistar albino rats to 1,800 MHz RF-EMF for 2 hours per day for 8 weeks, the rats were sacrificed, the eyes were removed, and the four gene expression levels were investigated. Expression of the caspase-3 and p38MAPK genes was significantly upregulated in the ocular tissues following exposure to RF-EMF. Thus, it was found that RF-EMFR was a stress factor in response to which caspase-3 and p38MAPK gene expression were activated.

Moreover, the study findings also confirmed that RF-EMFR could cause cellular damage in ocular cells. In contrast, Kan et al. (2016) [29] investigated the effect of RF-EMFR on DNA damage and oxidative stress in human lens epithelial cells. The human lens cells were exposed or sham-exposed to 50-Hz ELF-MF at an intensity of 0.4 mT. The short- and long-term exposure to 50-Hz ELF-MF at 0.4 mT could not induce DNA damage in human lens epithelial cells in vitro.

Furthermore, Shuang et al. (2013) [30] explored the possible reasons for an increase in "reactive oxygen species (ROS) in human lens epithelial (HLE) cells exposed to 1.8 GHz radiofrequency fields (RF). The HLE cells were divided into RF exposure and RF sham-exposure groups. The ROS and MDA levels significantly increased in the RF exposure group, and cellular viability, mRNA expression of genes, and protein expression" markedly decreased compared with those in the sham-exposure group. It was hypothesized that "oxidative stress in HLE B3 cells exposed to 1.8 GHz and increased production of ROS might be due to the downregulation of genes encoding antioxidant enzymes induced by RF exposure" [30].

9.7 Transient Smartphone Blindness (TSB)

In smartphone users, transient smartphone blindness (TSB) has commonly emerged. "This refers to temporary monocular vision loss associated with smartphone usage while lying down in the dark. The symptoms reported are usually contralateral to the side on which they were lying. Blockage of light due to lying down caused one eye to undergo dark adaptation while the other (viewing) eye underwent light adaptation, causing temporary conflicting light adaptation between the two retinae [42]. When the blockage of light was removed, the light-adapted eye underwent monocular vision loss", lasting a couple of minutes in a patient; the reported time was 15 min [42]. This is a relatively unknown phenomenon with the potential for misdiagnosis as multiple sclerosis or ischemic vascular disease [43, 44]. Following the diagnosis, the patients started on therapy until further examinations ruled out the initial diagnosis. In this case, the patients used their smartphones while in bed [42–44].

The current risk of developing TSB is heightened by the ubiquitous use of smartphones, lack of awareness of the effects of smartphones among their users, and the intense blue light emitted by the latest smartphones. While TSB itself is considered relatively benign and a short-term pathology, the long-term implications arising from the disease have yet to be elucidated. Blue light emission from smartphones can contribute to various retinal pathologies, such as age-related macular degeneration [45]. The disruptions in the sleep cycle by blue light exposure may also be exacerbated by smartphone usage in bed and therefore promote TSB development. The danger may arise from the initiation of inappropriate treatments following misdiagnosis in patients with ambiguous test results [43]. Therefore, it is crucial for ophthalmologists and neurologists to include the history of smartphone usage of patients' daily practices before diagnosis.

To prevent TSB, people using smartphones should use them in well-lit rooms and in a position in which light does not enter only one eye. People should also try to avoid using their phones before going to bed in order to avoid disturbing their sleep cycles due to disruptive blue light emission. Mobile phone manufacturers can spread awareness through public service messages warning their clientele of possible side effects of using phones in bed and in poorly lit rooms. Ophthalmologists can also create awareness about TSB by counseling their patients, especially teens. Further interventions may be provided in the form of mobile applications that monitor blue light exposure and offer prompts for prolonged or intense exposure [46].

Geeraets et al. (1976) [47] described the mechanism by which RF-EMFR damages the retina. The most probable mechanisms involved are photochemical, thermal, and non-linear effects. The different wavelengths can be absorbed in the retinal layers. Thermal injury for minimal lesions produces penetrating central core damage surrounded by edema. It was identified that wavelengths have a role in the severity and origin of retinal damage upon long-term exposure.

Talebnejad et al. (2017) [35] evaluated the effects of RF-EMFR on the retina using 40 adult white rabbits. A cell phone simulator was used for irradiation. The rabbits were randomized into five groups of eight rabbits each. There was a trend toward greater changes in the irradiated eyes. However, histopathologically, cell phone-simulated irradiation had no significant detrimental effect on the retina. However, ciliary body congestion was observed in rabbits that received higher doses.

Suetov and Alekperov (2019) [48] investigated the clinical and morphological manifestations of ocular lesions resulting from acute exposure to microwave radiation. Twenty-four rabbits were included in the study, exposed for 15, 30, 45, or 60 s. The exposed (right) and paired (left) eyes were studied for clinical and morphological changes, as were the levels of pro-inflammatory cytokines IL-1β, IL-6, and TNF-α in the anterior chamber and the vitreous body after exposure to MR. Significant dose-dependent changes in the structure of the exposed eye were revealed. Partial or complete de-epithelialization, stromal edema, endothelial damage and inflammatory infiltration in the cornea, effusion of protein and cellular reaction in the aqueous humor were detected after MR exposure of 30, 45, and 60 s. In addition, the cellular reaction in the vitreous body was observed after exposure times of 45 and 60 s. Significantly higher levels of the pro-inflammatory cytokines IL-1β, IL-6, and TNF-α in the aqueous humor and vitreous body were revealed in animals exposed to MR for 45 and 60 s. Acute exposure of the organ of vision to electromagnetic microwave radiation can lead to adverse dose-dependent effects not only in the lens but also in other structures of the eye.

Bhargav et al. (2017) [31] investigated the effects of mobile phones on the subtle energy levels among 61 healthy adults with a mean age of 17.40 years. They found that subtle energy levels were significantly reduced in the pancreas, thyroid gland, cerebral cortex, cerebral vessels, left ear and left eye, liver, right kidney, spleen, and immune system among the mobile phone in ON mode [MPON] group compared with the levels in the mobile phone in OFF mode group.

A recent study reported by McCully et al. (2020) [49] demonstrated that environmental pollution, including RF-EMFR, is a major cause of vision impairment and blindness. They showed that environmental pollutants, including dust, fumes, chemicals, herbicides, pesticides, additives to food and water, and electromagnetic fields (EMFs), promote numerous chronic diseases. These environmental pollutants cause adverse health outcomes by impairing mitochondrial function to produce oxidative stress and loss of the active site complex for oxidative phosphorylation, thereby preventing thioretinaco ozonide oxygen nicotinamide adenine dinucleotide phosphate from opening the mitochondrial permeability transition pore. EMFs are a pollutant promoting loss and decomposition of the active site for oxidative phosphorylation, producing mitochondrial dysfunction and oxidative stress. Adverse health effects of EMFs include hearing loss, cataracts, blindness and retinal hemorrhage.

The thermal effect of RFEMF could be responsible for the increased IOP changes in the eyes. Temperature changes may regulate the secretion, exertion, and flow dynamics of aqueous humor and thus affecting its balance. As the temperature rises in the anterior segment, vasodilation and increased metabolic processes occur in the ciliary body, caused by increased blood flow, resulting in higher secretion. However, the opposite mechanism is observed after temperature decreases due to activation of the sympathetic system and vasoconstriction. The metabolic process is enhanced by increased temperature, which activates the inflammatory pathways and oxidative stress. The pathophysiological responses increase the cellularity of the anterior segment structures and disturb the trabecular meshwork drainage while also impairing vision [50, 51].

Conclusions

Current evidence-based on the available literature, including that on animal and human model studies and epidemiological literature, indicates the hazardous effects of RF-EMFR generated from mobile phones on the various structures and functions of the eyes, including the lens, cornea, sclera, retina, and nerves, leading to visual impairment. Similar to biological stress responses at the body organ level, cellular stress responses at the cellular levels confer protection to the cell against external and internal stressors. RF-EMFR has been demonstrated to stimulate stress responses at the cellular level in various structures of the eyes. RF-EMFR generated from mobile phones can produce various ocular effects, including cataract, corneal edema, lacrimation of eyes, dry eye disease, and vision impairment. Health officials must implement directives to counsel individuals, especially the young, on avoiding the excess usage of phones, especially before sleeping hours, and adopt other measures to minimize public exposure to RF-EMFR by avoiding MPBSTs installation in or near densely populated residential areas.

References

[1] Wallhagen M.I., Strawbridge W.J., Shema S.J., Kurata J., Kaplan G.A. Comparative impact of hearing and vision impairment on subsequent functioning [journal]. *J Am Geriatr Soc*. 2001; 49(8): 1086–92. doi: 10.1046/j.1532-5415.2001.49213.x, PMID 11555071.

[2] Wahl H., Tesch-Romer C. Aging, sensory loss, and social functioning. In: Charness N., Parks D.C., Sabel B.A., editors. *Communication, Technology, and Aging*. New York: Springer. 2001, pp. 108–26.

[3] Purves D., Augustine G.J., Fitzpatrick D., editors. Sunderland, MA: Neuroscience. 2nd edition. Available at: www.ncbi.nlm.nih.gov/books/NBK11120.

[4] Mauser M.W. Exploring the anatomy of your own eye. *Am Biol Teach*. 2011; 73(1): 28–33. doi: 10.1525/abt.2011.73.1.6.

[5] Sudha A. Anatomy, physiology, histology and normal cytology of eye. *J Cytol*. 2007; 24(1): 16–9. doi: 10.4103/0970-9371.42084.

[6] World Health Organization. Available at: www.who.int/news-room/fact-sheets/detail/blindness-and-visual-impairment. Cited date July 12, 2020.

[7] Al-Khlaiwi T., Meo S.A. Association of mobile phone radiation with fatigue, headache, dizziness, tension, and sleep disturbance in Saudi population. *Saudi Med J*. 2004; 25(6): 732–6. PMID 15195201.

[8] Röösli M., Lagorio S., Schoemaker M.J., Schüz J., Feychting M. Brain, salivary gland tumors, and mobile phone use evaluate evidence from various epidemiological study designs. *Annu Rev Public Health*. 2019; 40: 221–38. doi: 10.1146/annurev-publhealth-040218-044037, PMID 30633716.

[9] Bhatt C.R.R., Thielens A., Billah B., Redmayne M., Abramson M.J., Sim M.R., Vermeulen R., Martens L., Joseph W., Benke G. Assessment of personal exposure radiofrequency-electromagnetic fields in Australia and Belgium using on-body calibrated expositors. *Environ Res*. 2016; 151: 547–63. doi: 10.1016/j.envres.2016.08.022, PMID 27588949.

[10] World Health Organization (WHO), International Agency for Research on cancer (IARC). Classifies radiofrequency electromagnetic fields as possibly carcinogenic to humans. 2018. Available at: www.iarc.fr/wp-content/uploads/2018/07/pr208_E.pdf. Cited date Jan 2.

[11] WHO. Clarification of mooted relationship between mobile telephone base stations and cancer. 2018. Available at: www.who.int/mediacentre/news/statements/statementemf/en/. Cited date 16 Nov.

[12] Feychting M., Ahlbom A., Kheifets L. EMF and health. EMF and health. *Annu Rev Public Health*. 2005; 26(1): 165–89. doi: 10.1146/annurev.publhealth.26.021304.144445, PMID 15760285.

[13] Röösli M. Radio frequency electromagnetic field exposureand non-specific symptoms of ill health: A systematic review. *Environ Res*. 2008; 107(2): 277–87. doi: 10.1016/j.envres.2008.02.003, PMID 18359015.

[14] Colombi D., Thors B., Tornevik C. Implications of emf exposure limits on output power levels for 5-g devices above 6 ghz. *IEEE Antennas Wirel Propag Lett*. 2015; 14: 1247–9. doi: 10.1109/LAWP.2015.2400331.

[15] Pi Z., Khan F. An introduction to millimeter-wave mobile broadband systems. *IEEE Commun Mag*. 2011; 49(6): 101–7. doi: 10.1109/MCOM.2011.5783993.

[16] Alkis M.E., Bilgin H.M., Akpolat V., Dasdag S., Yegin K., Yavas M.C., Akdag M.Z. Effect of 900–1800, and 2100-MHz radio-frequency radiation on DNA and oxidative stress in brain. *Electromagn Biol Med*. 2019; 38(1): 32–47. doi: 10.1080/15368378.2019.1567526, PMID 30669883.

[17] Redlarski G., Lewczuk B., Żak A., Koncicki A., Krawczuk M., Piechocki J., Jakubiuk K., Tojza P., Jaworski J., Ambroziak D., Skarbek Ł., Gradolewski D. The influence of electromagnetic pollution on living organisms: Historical trends and forecasting changes. *BioMed Res Int*. 2015; 2015: 234098. doi: 10.1155/2015/234098.

[18] Westerman R., Hocking B. Diseases of modern living: Neurological changes associated with mobile phones and radiofrequency radiation in humans. *Neurosci Lett*. 2004; 361(1–3): 13–6. doi: 10.1016/j.neulet.2003.12.028, PMID 15135881.

[19] Meo S.A., Al-Dress A.M. Mobile phone related-hazards and subjective hearing and vision symptoms in the Saudi population. *Int J Occup Med Environ Health*. 2005; 18(1): 53–7. PMID 16052891.

[20] Davidson H.C., Lutman M.E. Survey of mobile phone use and their chronic effects on the hearing of a student population. *Int J Audiol*. 2007; 46(3): 113–8. doi: 10.1080/14992020600690472, PMID 17365064.

[21] Shokoohi-Rad S., Ansari M.R., Sabzi F., Saffari R., Rajaei P., Karimi F. Comparison of intraocular pressure changes due to exposure to mobile phone electromagnetics radiations in normal and glaucoma eye. *Middle East Afr J Ophthalmol*. 2020; 27(1): 10–3. doi: 10.4103/meajo.MEAJO_20_19, PMID 32549718.

[22] Park Y-H., An C.M., Moon S.J 1. Effects of visual fatigue caused by smartphones on balance function in healthy adults. *J Phys Ther Sci*. 2017 Feb; 29(2): 1. doi: 10.1589/jpts.29.221, PMID 28265143.

[23] Moon J.H., Kim K.W., Moon N.J. Smartphone use is a risk factor for pediatric dry eye disease according to region and age: A case control study. *BMC Ophthalmol*. 2016 Oct 28; 16(1): 188. doi: 10.1186/s12886-016-0364-4, PMID 27788672.

[24] Jennifer Long J., Cheung R., Simon Duong. *Clin Exp Optom*. 2017 Mar; 1., Lisa Asper. Viewing distance and eyestrain symptoms with prolonged viewing of smartphones. *Rosemary Paynter*; 100(2): 1, 133–7. doi: 10.1111/cxo.12453.

[25] Golebiowski B., Long J., Harrison K., Lee A., Chidi-Egboka Ngozi, Asper L. Smartphone use and effects on tear film, blinking and binocular vision. *Curr Eye Res*. 2020 Apr; 45(4): 428–34. doi: 10.1080/02713683.2019.1663542, PMID 31573824.

[26] Küçer N. Some ocular symptoms experienced by users of mobile phones. *Electromagn Biol Med*. 2008; 27(2): 205–9. doi: 10.1080/15368370802072174, PMID 18568938.

[27] Povolotskiy R., Gupta N., Leverant A.B., Kandinov A., Paskhover B. Head and neck injuries associated with cell phone use. *JAMA Otolaryngol Head Neck Surg*. 2020; 146(2): 122–7. doi: 10.1001/jamaoto.2019.3678, PMID 31804678.

[28] Tawer Kafi S.T., Ahmed A.M., Ismail B.A., Nayel E.A., Awad A.R., Alhassan E.A. Effects of RF radiation emitted from cellphones on human eye function (vision acuity/refraction) [journal]. of. *Electr Eng*; 3(3): 128–33. doi: 10.17265/2328-2223/2015.03.003.

[29] Zhu K., Lv Y., Cheng Q., Hua J., Zeng Q. Extremely low frequency magnetic fields do not induce DNA damage in human lens epithelial cells in vitro. *Anat Rec (Hoboken)*. 2016 May; 299(5): 688–97. doi: 10.1002/ar.23312, PMID 27079842.

[30] Ni S., Yu Y., Zhang Y., Wu W., Lai K., Yao K. Study of oxidative stress in human lens epithelial cells exposed to 1.8 GHz radiofrequency fields. *PLoS One*. 2013; 8(8): e72370. doi: 10.1371/journal. pone.0072370, PMID 23991100.

[31] Bhargav H., Srinivasan T.M., Bista S., Mooventhan A., Suresh V., Hankey A., Nagendra H.R. Acute effects of mobile phone radiations on subtle energy levels of teenagers using electrophotonic imaging technique: A randomized controlled study. *Int J Yoga*. 2017; 10(1): 16–23. doi: 10.4103/0973-6131.186163, PMID 28149063.

[32] Balci M., Devrim E., Durak I. Effects of mobile phones on oxidant/antioxidant balance in cornea and lens of rats. *Curr Eye Res*. 2007; 32(1): 21–5. doi: 10.1080/02713680601114948, PMID 17364731.

[33] Bormusov E., Andley U.P., Sharon N., Schächter L., Lahav A., Dovrat A. Non-thermal electromagnetic radiation damage to lens epithelium. *Open Ophthalmol J*. 2008; 2: 102–6. doi: 10.2174/1 874364100802010102, PMID 19517034.

[34] Eker E.D., Arslan B., Yildirim M., Akar A., Aras N. The effect of exposure to 1800 MHz radiofrequency radiation on epidermal growth factor, caspase-3, Hsp27 and p38MAPK gene expressions in the rat eye. *Bratisl Lek Listy*. 2018; 119(9): 588–92. doi: 10.4149/BLL_2018_106, PMID 30226071.

[35] Talebnejad M.R., Sadeghi-Sarvestani A., Nowroozzadeh M.H., Mortazavi S.M.J., Abbas Alighanbari A., Khalili M.R. The effects of microwave radiation on rabbit's retina. *J Curr Ophthalmol*. 2017; 30(1): 74–9. doi: 10.1016/j.joco.2017.08.010.

[36] Ahlers M.T, Ammermüller J. No influence of acute RF exposure (GSM-900, GSM-1800, and UMTS) on mouse retinal ganglion cell responses under constant temperature conditions. *Bioelectromagnetics*. 2014 Jan; 35(1): 16–29. doi: 10.1002/bem.21811. PMID 24115076.

[37] Toh S.H., Coenen P., Howie E.K., Mukherjee S., Mackey D.A., Straker L.M. Mobile touch screen device use and associations with musculoskeletal symptoms and visual health in a nationally representative sample of Singaporean adolescents. *Ergonomics*. 2019 Jun; 62(6): 778–93. doi: 10.1080/00140139.2018.1562107, PMID 30575442.

[38] Zhu Jie H., Wang J.C., Fan Xianqun. Effects of extremely low frequency electromagnetic fields on human fetal scleral fibroblasts. *Toxicol Ind Health*. 2014; 32: 1042–51.

[39] Elder J.A. Ocular effects of radiofrequency energy. *Bioelectromagnetics*. 2003 (Suppl 6): S148–S61. doi: 10.1002/bem.10117, PMID 14628311.

[40] D'Andrea J.A., Chalfin S. Effects of microwave and millimeter wave radiation on the eye. In: Klauenberg B.J., Miklavčič D., editors. *Radio Frequency Radiation Dosimetry and its Relationship to the Biological Effects of Electromagnetic Fields. NATO Science Series (Series 3: High Technology)*. Vol. 82. Dordrecht: Springer. 2000. doi: 10.1007/978-94-011-4191-8_43.

[41] Johansen C., Boice J.D., McLaughlin J.K., Christensen H.C., Olsen J.H. Mobile phones and malignant melanoma of the eye. *Br J Cancer*. 2002 Feb 1; 86(3): 348–9. doi: 10.1038/sj.bjc.6600068, PMID 11875697.

[42] Alim-Marvasti A., Bi W., Mahroo O.A., Barbur J.L., Plant G.T. Transient smartphone blindness. *N Engl J Med*. 2016; 374(25): 2502–4. doi: 10.1056/NEJMc1514294.

[43] Sathiamoorthi S., Wingerchuk D.M. Transient smartphone blindness: Relevance to misdiagnosis in neurologic practice. *Neurology*. 2017; 88(8): 809–10. doi: 10.1212/WNL.0000000000003639.

[44] Irshad F., Adhiyaman V. Transient smartphone blindness. *Can J Ophthalmol*. 2017; 52(3): e107–8. doi: 10.1016/j.jcjo.2016.10.017. PMID 28576218.

[45] Tosini G., Ferguson I., Tsubota K. Effects of blue light on the circadian system and eye physiology. *Mol Vis*. 2016; 22: 61–72. PMID 26900325.

[46] Hasan C.A., Hasan F., Mahmood Shah S.M. Transient smartphone blindness: Precaution needed. *Cureus*. Oct 24 2017; 9(10): e1796. doi: 10.7759/cureus.1796. PMID 29282440.

[47] Geeraets W.J., Geeraets R., Goldman A.I. Electromagnetic radiation damage to the retina (author's transl) author's transl. *Albrecht. Von. Graefes. Arch. Klin. Exp. Ophthalmol*. 1976; 200(3). Albrecht Von Graefes Arch Klin Exp Ophthalmol. 1976; 200(3): 263–78. doi: 10.1007/BF01028543, PMID 1086617.

[48] Suetov A.A., Alekperov S.I. Acute ocular lesions after exposure to electromagnetic radiation of ultrahigh frequency (an experimental study) (an experimental study. *Vestn. Oftalmol*. 2019; 135(4): 41–9. doi: 10.17116/oftalma201913504141. PMID 31573556.

[49] McCully K.S. Environmental pollution, oxidative stress and Thioretinaco ozonide: Effects of glyphosate, fluoride and electromagnetic fields on mitochondrial dysfunction in carcinogenesis, atherogenesis and aging. *Ann Clin Lab Sci*. 2020 May; 50(3): 408–11. PMID 32581036.

[50] Saccà S.C., Centofanti M., Izzotti A. New proteins as vascular biomarkers in primary open angle glaucomatous aqueous humor. *Invest Ophthalmol Vis Sci.* 2012; 53(7): 4242–53. doi: 10.1167/iovs.11-8902, PMID 22618596.

[51] Saccà S.C., Pulliero A., Izzotti A. The dysfunction of the trabecular meshwork during glaucoma course. *J Cell Physiol.* 2015; 230(3): 510–25. doi: 10.1002/jcp.24826. PMID 25216121.

Environmental Pollution and Brain Damage

10.1 Introduction

Environmental pollution is a leading public health concern that threatens global health-care systems and economies. Environmental pollution generated from different natural, human-made and industrial sources causes various health hazards to physiological systems, including respiratory [1], coronary [2], endocrine [3, 4], and nervous systems [1]. Environmental pollution consists of a complex mixture of coarse, fine, and ultrafine particles, particulate matter (PM) of various sizes, along with noxious substances including dust, fumes, and gases. It also includes VOCs, including benzene, toluene, and xylene and metals such as lead, manganese, vanadium, and iron, along with biological materials [5]. These pollutants possess different biological, physical, and chemical characteristics and have numerous pathophysiological aspects upon penetrating the lungs, blood circulation, and brain.

The major sources of environmental pollution include industry, power stations, oil refineries, plastic refineries and the chemical, metallurgical and fertilizer industries. Moreover, planes, ships, cars, buses, trucks, trains, combustion of wood, wildfires, and volcanoes are also sources of environmental pollution [6]. Urbanization, human activities and the industrial revolution resulted in the production of a vast amount of pollutants with adverse effects on the environment by polluting the air, water, and soil. Anthropogenic air pollution accounts for approximately nine million deaths per annum [7].

A majority of developing nations are trying to achieve rapid economic growth through industrialization. To achieve this, many countries have undergone poorly planned industrialization and urbanization. This rapid change has had a significant environmental impact and

DOI: 10.1201/9781003212461-10   201

Figure 10.1 Wildfire: a source of fine and ultrafine particulate matter.

caused air and marine pollution. Moreover, motor vehicles and electromagnetic field radiation are becoming sources of environmental pollution of increasing concern globally. All of these factors are evolving major threats to the atmosphere and human health [8]. Pollutants can be defined as natural or anthropogenic, biodegradable, or non-biodegradable [9, 10], and harm the environment upon exceeding normal limits. These pollutants pose health hazards through impairing various physiological functions and systems, including respiratory [1], coronary [2], endocrine [3, 4], and nervous systems [1]. Air pollution impairs not only human health but also poses economic challenges. The weather conditions and climate change are markedly affecting both urban and rural areas. Wildfires are also a major source of environmental pollution (Figure 10.1).

10.2 Environmental Pollution: Particulate Matter (PM)

Environmental pollution is not restricted by international borders. Once produced, PM components remain in the environment for an extended period and travel a long distance, sometimes from one country to another and even from one continent to another [11]. PM can also form a haze, an atmospheric phenomenon whereby dust, smoke, and PM combine. Pollutants are sometimes transported with the wind, and chemical reactions occur in the atmosphere and clouds, creating a haze. These change the biological and chemical nature of the pollutants before they are deposited elsewhere [11, 12]. The effects of contaminants on human physiological systems depend on the type and biological, chemical and physical nature of the pollutants and the duration of exposure [13].

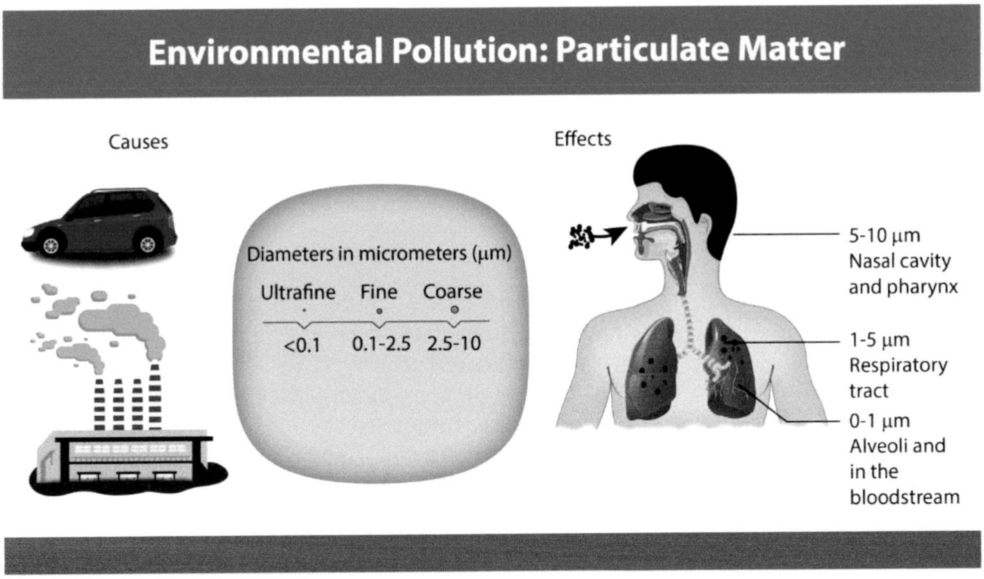

Figure 10.2 Particulate matter: coarse, fine, and ultrafine pollutants.

The PM component of air pollution is defined and classified by its aerodynamic diameter, which ranges across three different categories: (i) coarse particulate matter (PM10; <10 μm); (ii) fine particulate matter (PM2.5; <2.5 μm); and (iii) ultrafine PM <100 nm; PM <0.1 μm) [14–16]. These types of PM enter the human body and exert effects dependent on their aerodynamic diameter (Figure 10.2) [17].

The human respiratory system is particularly vulnerable to air pollution. The body reflex mechanisms such as coughing and sneezing and the respiratory tract mucociliary system play significant roles in removing pollutants and protecting the body. However, PM's entry, deposition, and toxicity depend on the biological, physical, and chemical nature and size of the pollutant and the duration of exposure [18].

As per a WHO report, about 91% of the global population and 97% of the inhabitants of low- and middle-income countries reside in places where air pollution exceeds the recommended air quality standards [7]. Environmental pollution affects human health through multiple pathways [19, 20]. The respiratory system is the system that is most vulnerable to air pollutants. Pollutants enter the human body through the respiratory, gastrointestinal and olfactory systems. Pollutants with an aerodynamic diameter of 10 μm are deposited in the upper respiratory tract, mainly the nose and pharynx. The particles between 2.5 and 10 μm in aerodynamic diameter can be distributed throughout the tracheobronchial tree, while particles between 0.5 and 2.5 μm are frequently deposited in the respiratory zone, mostly the alveoli. Particles smaller than 0.1 μm remain in the air stream and are exhaled [21] (Figure 10.2).

PM, mainly in the categories of PM2.5 and ultrafine PM, are highly toxic because of their [22, 23] biological, physical and chemical characteristics. It has also been reported that particles equal to or smaller than 0.10 μm enter the general circulation and can be deposited in various tissues and organs, resulting in a range of health problems [18].

Such health problems can arise as a result of the physiological failure of one or more protective mechanisms. If dusty air is continuously breathed in, mixing between the inspired air and the dead-space air will cause the dust to reach the terminal airways. The dust particles accumulate in the alveoli, subsequently entering the circulatory system and eventually the brain.

10.3 Pollution and the Brain

Air pollutants of small size nanoparticles are widespread in the environment. These nanoparticles enter the general circulation through the respiratory and gastrointestinal systems, being readily deposited in extrapulmonary tissues [24, 25] (Figure 10.3). Ultrafine PM consists of nanoparticles, which are highly toxic air pollution components. They penetrate deep into the lung alveoli to reach the lung capillaries and circulating cells.

The ultrafine particles swiftly translocate from the lungs into the cells, tissues, and blood and quickly enter the systemic circulation [26]. About 50% of inhaled ultrafine PM are deposited in pulmonary alveolar regions of the lungs, passing through the alveolar-capillary barrier and accessing the pulmonary interstitium. The particles can cross the endothelial cells, move into the blood circulation, and are then deposited in various

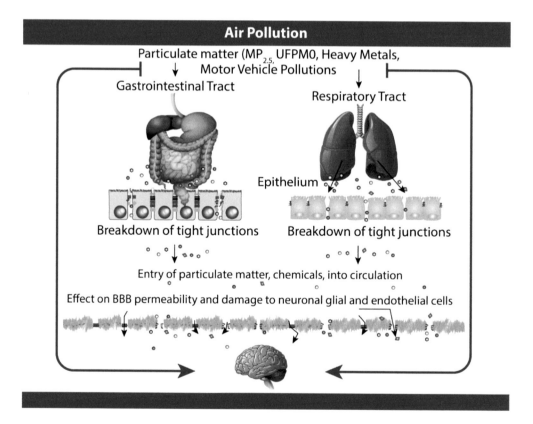

Figure 10.3 Entry of pollutants into the nervous system.

organs, including the liver, heart, and nervous system, where they can have serious consequences for health [27, 28].

The entry of pollutants into the brain is directly through the nasal olfactory pathway or indirectly through the lungs and circulatory system. Pollutants can cause inflammation in the respiratory system and lead to the release of pro-inflammatory cytokines into the circulation, causing systemic inflammation. Moreover, these pollutants activate the hypothalamus–pituitary–adrenal (HPA) axis and enhance the secretion of stress hormones, including cortisol in the blood. The contaminants, pro-inflammatory cytokines and cortisol can reach the brain and cause neuroinflammation and impair various neuronal functions [29–32]. These pollutants can, in turn, cause stroke, Alzheimer's disease, Parkinson's disease, and neurodevelopmental disorders [32].

Ultrafine PM is considered a highly reactive toxic element of air pollution that can penetrate into the brain directly through the olfactory nerve transport mechanism [22] and indirectly through the respiratory and gastrointestinal systems and blood circulation. The upper respiratory tract nasal cavity provides another transport pathway for the direct entry of some inhaled PM into the brain through the olfactory neurons and olfactory bulbs. The literature reveals that PM and ultrafine particles move through the nasal cavity and enter the brain across the olfactory epithelium [27, 28].

PM can also cross the synapses within the olfactory pathway and travel via the secondary and tertiary neurons to various areas in the brain [33]. It has also been reported that metal compounds such as "manganese, iron, cadmium, thallium, mercury, cobalt and zinc, and carbon particles move into the brain following inhalation or intranasal tracheal" instillation [34, 35]. The suggested mechanism through which transport of these particles occurs is through the olfactory nerve transport system.

10.4 Pollution and Brain Disorders

Fine and ultrafine PM enters the nervous system and brain through the respiratory system and general circulation. Another pathway is directly through the olfactory pathway into the brain [29]. Either way, once the fine and ultrafine PM has entered general circulation and the brain, these pollutants alter the BBB function, injure endothelial cells and damage the BBB and neurons in the brain [36]. "Long-term exposure to air pollution causes damage to the BBB, endothelial cell damage in the cerebral vasculature, and entry of pollutants into the brain," especially in the higher brain centers [37]. Upon entering the brain, environmental pollutants can cause various nervous system diseases (Table 10.1: Figure 10.4).

Table 10.1 Environmental Pollution Exposure and Brain Damage	
Environmental Pollution Exposure	Effect
Acute exposure	Acute effect Cerebro vascular accident by precipitating platelets activation, blood coagulation and thrombus formation
Chronic exposure	Chronic effect Development of atherosclerosis

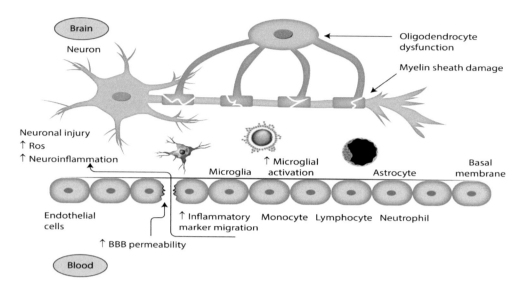

Figure 10.4 Environmental pollution and injury to the blood–brain barrier and neurons.

Environmental pollution has a substantial effect on various parts of the CNS. The growing evidence on this topic has revealed links between air pollution and brain diseases, behavioral impairments, neuroinflammation and neurodegeneration [33]. Recent literature in the field of epidemiology has raised concerns about the health hazards of air pollution on brain biology, the associated brain inflammation and white matter abnormalities, conferring risks of "autism spectrum disorders, lower IQ, neurodegenerative diseases, Parkinson's disease, Alzheimer's disease" and stroke [38]. The literature also asserted that people living or working in urban areas with high air pollution levels are more likely to have impaired cognitive function and brain pathology [39, 40] (Figure 10.4).

Air pollution affects the brain biology by producing pro-inflammatory mediators [41]. The biochemical arrangement of pollutants demonstrates both spatial and temporal variations that reflect local sources, including motor vehicle pollutants, industry and natural biological processes. Outdoor and indoor pollutants can penetrate the body and reach the brain, resulting in adverse impacts on the nervous system and neurodegenerative disorders [23, 26].

Exposure to indoor and outdoor environmental contamination is a major health issue worldwide. Environmental pollution contains noxious compounds, PM of various physical, biological and chemical characteristics. Inhaled, fine, and ultrafine PM can cross into the blood circulation; these PMs are taken by cells and cause oxidative stress and mitochondrial damage [42]. The ultrafine PMs are the most toxic PM form and may penetrate into the higher centers, including the brain, directly through the olfactory nerve [29]. These environmental pollutants are mainly responsible for neurodegenerative disorders with long-lasting complications.

10.5 Pollution: Dementia/Alzheimer's Disease

Neurodegenerative disorders are multifactorial complex conditions connected with genetic, lifestyle-related, and environmental factors (Figure 10.5). The literature reported that exposure to air pollution enhances the risk of dementia and Alzheimer's disease. Alzheimer's disease is a

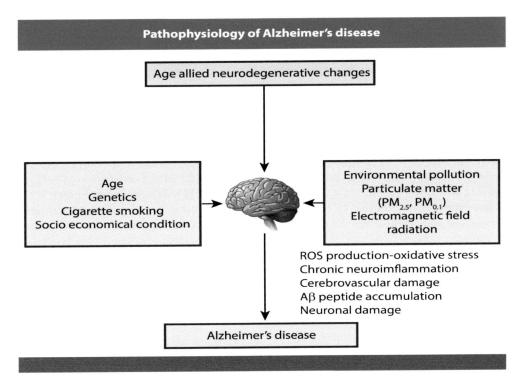

Pathophysiology of Alzheimer's disease

Age allied neurodegenerative changes

Age
Genetics
Cigarette smoking
Socio economical condition

Environmental pollution
Particulate matter
($PM_{2.5}$, $PM_{0.1}$)
Electromagnetic field
radiation

ROS production-oxidative stress
Chronic neuroimflammation
Cerebrovascular damage
$A\beta$ peptide accumulation
Neuronal damage

Alzheimer's disease

Figure 10.5 Pathophysiology of Alzheimer's disease.

progressive and irreversible neurodegenerative disorder of older adults. It affects a large number of people worldwide, mainly in urban and polluted areas. In dementia, there are impairments of memory, cognitive behavior, and the ability to perform everyday activities [43, 44].

Alzheimer's disease is a complex, progressive and irreversible neurodegenerative disease characterized by gradual complaints of progressive dementia beginning with memory loss and decline in cognitive function. The WHO demonstrated that approximately 50 million people have dementia worldwide, and about 10 million new cases of dementia develop every year. Alzheimer's disease is the most common form of dementia, contributing to 60–70% of cases. Alzheimer's disease and dementia cause disability and dependency, with long-lasting physical, social, psychological, and economic impacts on individuals, families, and society as a whole [43].

Prolonged exposure to environmental pollution, including air pollutants, PM, nanoparticles, metals, chemicals, pesticides, and gases, can accelerate the development of AD. The most probable mechanisms are an increase in amyloid-β ($A\beta$) peptide, along with tau phosphorylation and the development of senile plaques and neurofibrillary tangles. These are resulting in neuronal damage [45] and ultimately neurodegenerative diseases, including Alzheimer's disease (Figure 10.5).

Dementia is a disabling, degenerative disease of the brain and is becoming a global issue of growing importance [46, 47]. There is plausible evidence for a relationship between environmental pollution and an increased risk of dementia [29]. The literature demonstrates that neurodegenerative disorders, including AD, are moderately mediated by oxidative stress. Oxidative stress results in redox imbalance when the level of ROS surpasses the capacity of antioxidant defense mechanisms to deal with them [48].

Air pollution can enhance ROS generation, thus posing a risk for AD by increasing oxidative stress. These biological and biochemical alterations in the CNS can cause neurodegenerative disorders, including Alzheimer's disease (Figure 10.5).

In the scientific literature, it is demonstrated that people with long-term exposure to "high levels of air pollution show damage in the olfactory mucosa, olfactory bulb and frontal cortex, similar to that observed in AD" brains [29, 49, 50].

People exposed to urban air pollution exhibit the accumulation of "amyloid-beta (Aβ) peptide fibrils known as amyloid plaques" and neurofibrillary tangles (NFTs) in the frontal cortex. This demonstrates a link between chronic exposure to high concentrations of air pollution and oxidative stress, neuroinflammation and neurodegeneration [29, 50]. In another study, it was also reported that air pollution increased the accumulation of amyloid-beta-42 (Aβ-42), which is a known cause of neuronal dysfunction [51, 52]. Moreover, human studies have shown that exposure to air pollution impairs cognitive function and induces cerebrovascular damage [53, 54]. All of these findings support a plausible association between air pollution and AD [55–60]. Although AD is a multifactorial disorder, there is a piece of sufficient evidence proving that air pollution is a risk factor for neurodegenerative diseases. The relationship between ageing, genetic predisposition, and air pollution can cause AD (Figure 10.6).

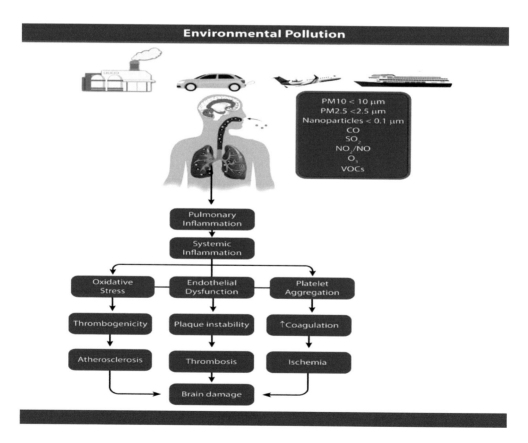

Figure 10.6 Environmental pollution: blood coagulation and brain damage.

10.6 Pollution and Degenerative Changes

The brain is a well-protected, highly organized, complex system. However, it is vulnerable to ultrafine ambient PM. The nanoparticles can cross the BBB, enter the brain and damage the neurobiological and physiological organization [61]. PM is associated with autism, mental retardation, and neurodegenerative disorders [62]. Exposure to pollutants such as lead and manganese is also implicated in neuronal damage and brain disease. The neurotoxicity of the lead includes impaired intellectual function and attention, encephalopathy, and convulsions [63]. In the literature, it is asserted that exposure to air pollutants can cause damage to the sensory neurons in the olfactory epithelium [64]. Exposure to mixtures of air pollution such as ozone has been reported to cause cerebral edema [65], neurodegeneration in the hippocampus, striatum, and substantia nigra [66], and altered human behavior and mood.

The most probable mechanism mediating air pollution-induced adverse effects on the human brain is oxidative stress. The particles escape through the respiratory tract and enter the circulatory system, ultimately reaching the brain. Moreover, ultrafine PM can enter the brain through the olfactory system [27, 28]. The literature reveals that the brain tissue of the individuals residing in highly polluted areas shows an increase in CD68-, CD163- and HLA-DR-positive cells, indicating infiltrating monocytes, elevated pro-inflammatory markers, interleukin-1β (IL-1β) and cyclooxygenase 2 (COX2). There is also an increase in Aβ42 deposition, a hallmark of Alzheimer's disease, BBB damage [67], upregulation of pro-inflammatory markers, and brain lesions in the prefrontal lobe [68–72]. These studies clearly indicate that air pollution has notable adverse impacts on the nervous system, with numerous degenerative changes and brain damage (Figures 10.6 and 10.7).

10.7 Pollution and Atherosclerosis

In recent years, air pollution has attracted attention from the scientific community and broader society concerning its impact on human health in terms of vasculature and atherosclerosis [73]. PM is heterogeneous material suspended in the air of various "sizes, surface areas, concentrations and chemical compositions [74]. PM2.5 and ultrafine particles penetrate deep into the lungs, reach the smallest airways and alveoli, and spread into the blood circulation through the alveolar/capillary" membrane [75].

The literature reveals epidemiological evidence of a relationship between atherosclerosis and ambient air pollutant PM2.5 in human subjects has been presented by Kunzli et al. [76]. They reported a 5.9% increase in carotid intima-medial thickness for every 10 µg/m^3 rise in the PM2.5 level. Similarly, PM2.5 was also found to accelerate the progression of coronary atherosclerosis, as assessed by coronary artery calcification scores [77].

Araujo and Nel [78] and Yao et al. [79] conducted an animal model study in which they used an intravenous "injection of PM2.5 suspension to study the effect of PM2.5 on the progression of atherosclerosis. They claimed that, after only 24 h, PM2.5 injection changed the plasma lipid profile, with reductions in triglyceride and HDL cholesterol and increased total cholesterol and LDL cholesterol. These changes were associated with a reduction of superoxide dismutase (SOD) and increases of malondialdehyde (MDA), TNF-α" and high-sensitivity CRP. The most probable mechanism by which PM2.5 exposure affects atherogenesis is the generation of oxidative reaction products by the reaction of pollutants with lipids or cellular membranes in the airways and lung alveoli. Then, circulating products trigger atherogenic mechanisms such as lipid peroxidation and high-density-lipoprotein dysfunction. [75]. The continuous and insistent activation of this pathway is related to the development of atherosclerosis [80–81].

10.8 Pollution and Cerebrovascular Accident (Stroke)

The term cerebrovascular accident (stroke) is described as an episode of acute neurological dysfunction caused by ischemia or hemorrhage. Stroke is a leading cause of mortality and morbidity worldwide, accounting for about 9% of all deaths globally and 10–12% of deaths in high-income countries. The WHO reported that, due to secondary effects of ambient air pollution on the lungs and brain, air pollution may cause about 4.2 million deaths globally [82].

Hahad et al. (2020) [83] reported that exposure to ambient air pollution is a determinant of human health and a cause of various debilitating diseases. The Lancet Commission on pollution [46] and health concluded that air pollution is the leading environmental cause of global illness and premature death. There is also growing evidence linking air pollution with cardiorespiratory, cerebrovascular, and neuropsychiatric disorders. Moreover, accumulating evidence shows that exposure to air pollutants, particularly fine particles smaller than PM2.5, affects the "brain health, contributing to increased risks of stroke, dementia, Parkinson's disease, cognitive dysfunction, neurodevelopmental disorders, depression and other conditions. The underlying molecular mechanisms involved are inflammation and oxidative stress, which are crucial factors in the pathogenesis of air pollution-induced brain disorders, driven by the enhanced production of pro-inflammatory mediators and ROS in response to exposure to various air pollutants" [46] (Figure 10.7).

Li et al. (2020) [84] conducted a time-series multicity study in developing states to differentiate the acute and chronic adverse effects of ambient nitrogen dioxide (NO_2) on the years of life lost (YLL) from different subtypes of stroke. It was found that a 10 µg/m³ increase in ambient NO_2 concentration was liked to lead to increments in percentage change of the YLL from ischemic stroke (0.82%) than hemorrhagic stroke (0.46%). The ischemic stroke relationship was significantly stronger in the low-education population than in the high-education population. Moreover, for both subtypes of stroke, a substantially stronger linkage was identified in south China than in north China. These facts confirm the relation between air pollution, ambient NO_2 concentration, and ischemic and hemorrhagic stroke [84].

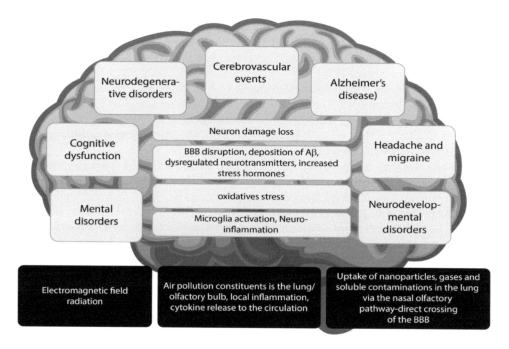

Figure 10.7 Air pollution contributes to neurological disorders [83].

Similarly, Rocha et al. (2020) [85] demonstrated that short-term exposure to air pollution was associated with Alzheimer's disease, epilepsy, ischemic stroke, and migraine. However, long-term exposure to air pollution was associated with Alzheimer's disease, amyotrophic lateral sclerosis, dementia, and ischemic stroke. Moreover, Zhang et al. (2019) [86] and Ljungman et al. (2019) [87] also reported that long-term exposure to air pollutants and PM is associated with stroke incidence. These studies clearly showed that short- or long-term exposure to air pollution, mainly PM2.5, can cause a stroke.

Ischemic stroke may be caused by a thrombus or an embolus, resulting in reduced blood supply to part of the brain, damaging the brain tissue. A hemorrhagic stroke may be due to a primary intracerebral hemorrhage or a subarachnoid hemorrhage. Exposure to air pollution PM2.5 has been associated with an increased risk of cerebrovascular accident (stroke). The literature also identified that short-term exposure to ambient PM2.5 could cause an increased risk of hospital admission for stroke. Each $10\,\mu g/m^3$ increase in long-term exposure to PM2.5 was associated with a 13% increase in the incidence of stroke. In addition, the incidence of both ischemic and hemorrhagic stroke was found to be related to long-term exposure to PM2.5 [47, 88–92].

It was hypothesized that exposure to PM2.5 might result in platelet activation, promoting blood coagulation and the formation of thrombi. In addition, long-term exposure to PM2.5 may induce systemic inflammation through the increased release of plasma cytokines, leading to an increased risk of atherosclerosis [92]. It has also been reported that PM2.5 could increase the risk of ischemic or hemorrhagic stroke through pollution-associated pathophysiological changes by arterial vasoconstriction and an increase in blood pressure [75]. The endothelial dysfunction caused by air pollution might increase the vulnerability of brain vessels to rupture and cause a cerebrovascular accident [85].

Another hypothesis that links this association is that the "brain is highly vulnerable to oxidative stress injury because of its high metabolic activity, low activity of antioxidant enzymes and low content of endogenous radical scavengers such as vitamin C." Moreover, the high cellular content of lipids and proteins and high levels of redox metals can act as a potent catalyst for ROS production [88, 89, 93]. Environmental pollutants have been consistently linked to oxidative stress, which has been associated with neurodegeneration insult. All of these findings revealed a link between air pollution-associated oxidative stress and brain damage (Figure 10.8).

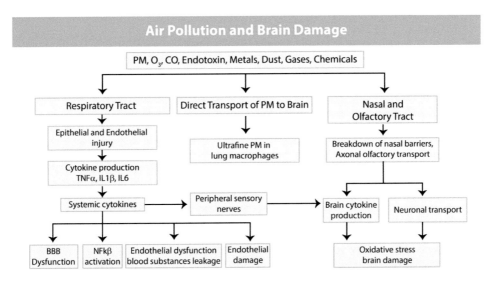

Figure 10.8 Pathophysiology of air pollution and brain damage.

Conclusions

Environmental pollution is a mixture of numerous coarse, fine and ultrafine particles of dust, gases and chemicals with various biological, physical and chemical characteristics. PM moves from the atmosphere into the brain through the respiratory, olfactory, gastrointestinal, and blood circulatory systems and can be deposited in various parts of the brain. These pollutants cause brain damage through a complex interaction of cellular and molecular mechanisms, resulting in neuronal injury. Air pollution can cause pulmonary and systemic inflammation, activate platelet aggregation, thrombogenicity, atherosclerosis, ischemia, and brain damage. Other mechanisms involved are oxidative stress, neuroinflammation, neurodegeneration, cerebral vascular damage, white matter abnormalities, and cerebrovascular accident (stroke). These findings presented in this paper strongly support the adverse impact of pollution on human health. To protect humans from such effects, the scientific community and broader society must come together and encourage efforts to protect the environment, preserve mother nature, and promote a healthier world.

References

[1] Meo S.A., Aldeghaither M., Alnaeem K.A., Alabdullatif F.S., Alzamil A.F., Alshunaifi A.I., et al. Effect of motor vehicle pollution on lung function, fractional exhaled nitric oxide and cognitive function among school adolescents. *Eur Rev Med Pharmacol Sci*. 2019; 23(19): 8678–86.

[2] Meo S.A., Suraya F. Effect of environmental air pollution on cardiovascular diseases. *Eur Rev Med Pharmacol Sci*. 2015 Dec; 19(24): 4890–7.

[3] Meo S.A., AlMutairi F.J., Alasbali M.M., Alqahtani T.B., AlMutairi S.S., Albuhayjan R.A., et al. Men's health in industries: Plastic plant pollution and prevalence of pre-diabetes and type 2 diabetes mellitus. *Am J Mens Health*. 2018; 12(6): 2167–72.

[4] Meo S.A., Memon A.N., Sheikh S.A., Rouq F.A., Usmani A.M., Hassan A., et al. Effect of environmental air pollution on type 2 diabetes mellitus. *Eur Rev Med Pharmacol Sci*. 2015 Jan; 19(1): 123–8.

[5] Phalen R.F., Mendez L.B., Oldham M.J. New developments in aerosol dosimetry. *Inhal Toxicol*. 2010; 22(Suppl 2): 6–14.

[6] Manisalidis I., Stavropoulou E., Stavropoulos A., Bezirtzoglou E. Environmental and health impacts of air pollution: A review. *Front Public Health*. 2020; 8: 14.

[7] World Health Organization (WHO). Air pollution. 2018. Available at: www.who.int/newsroom/air-pollution. Cited date Dec.

[8] Otsuka K. Shift in China's commitment to regional environmental governance in NorthEast Asia? [journal]. *Asian Stud*. 2018; 7(1): 16–34. doi: 10.1080/24761028.2018.1504643.

[9] Colbeck I., Lazaridis M. Aerosols and environmental pollution. *Sci Nat*. 2009; 97: 117–31. doi: 10.1007/s00114-009-0594-x.

[10] Incecik S., Gertler A., Kassomenos P. Aerosols and air quality. *Sci Total Environ*. 2014; 488–9: 355. doi: 10.1016/j.scitotenv.2014.04.012, PMID 24834897.

[11] Xu F., Xiang N., Higano Y. How to reach haze control targets by air pollutants emission reduction in the Beijing-Tianjin-Hebei region of China? *PLoS One*. 2017; 12(3): e0173612. doi: 10.1371/journal.pone.0173612, PMID 28282464.

[12] Bottenheim J.W., Dastoor A., Sun-Ling G., Kaz Higuchi. YI-Fanli. Long Range Transp Air Pollut Arct. 2004: 1–15. doi: 10.1007/b94522.

[13] Meo S.A., Al-Drees A.M., Al Masri A.A., Al Rouq F., Azeem M.A. Effect of duration of exposure to cement dust on respiratory function of non-smoking cement mill workers. *Int J Environ Res Public Health*. 2013 Jan 16; 10(1): 390–8.

[14] Oberdörster G., Ferin J., Lehnert B.E. Correlation between particle size, in vivo particle persistence, and lung injury. *Environ Health Perspect*. 1994; 102(Suppl 5): 173–9.

[15] Li N., Sioutas C., Cho A., Schmitz D., Misra C., Sempf J., Wang M., et al. Ultrafine particulate pollutants induce oxidative stress and mitochondrial damage. *Environ Health Perspect*. 2003; 111(4): 455–60.

[16] Kumar P., Morawska L., Birmili W., Paasonen P., Hu M., Kulmala M., et al. Ultrafine particles in cities. *Environ Int.* 2014; 66: 1–10.

[17] Deng Q., Deng L., Miao Y., Guo X., Li Y. Particle deposition in the human lung: Health implications of particulate matter from different sources. *Environ Res.* 2019; 169: 237–45.

[18] Meo S.A., Azeem M.A., Ghori M.G., Subhan M.M. Lung function and surface electromyography of intercostal muscles in cement mill workers. *Int J Occup Med Environ Health.* 2002; 15(3): 279–87. PMID 12462455.

[19] Oberdörster G., Utell M.J. Ultrafine particles in the urban air: To the respiratory tract-and beyond? *Environ Health Perspect.* 2002; 110(8): A440–1.

[20] Oberdörster G., Sharp Z., Atudorei V., Elder A., Gelein R., Kreyling W., et al. Translocation of inhaled ultrafine particles to the brain. *Inhal Toxicol.* 2004; 16(6–7): 437–45.

[21] Sheppard D., Hughson W.G, Shellito J. Occupational lung diseases. In: La Dou J., editor. *Occupational Medicine.* Norwalk, CT: Appleton & Lange. 1990, pp. 221–36.

[22] Walter J.B., Israel M.S. The body's defense against infection. In: *General Pathology.* 6th edition. Edinburgh: Churchill Livingstone. 1987, pp. 102–3.

[23] Li Q., Song C., Mao H. Particulate matter and public health. *Encyclopedia. of. Environ Health.* 2019; 31–35. doi: 10.1016/B978-0-12-409548-9.10988-1.

[24] Kreyling W.G., Semmler-Behnke M., Möller W. Ultrafine particle-lung interactions: Does size matter? *J Aerosol Med.* 2006; 19(1): 74–83.

[25] Kreyling W.G., Semmler M., Erbe F., Mayer P., Takenaka S., Schulz H., et al. Translocation of ultrafine insoluble iridium particles from lung epithelium to extrapulmonary organs is size dependent but very low. *J Toxicol Environ Health A.* 2002; 65(20): 1513–30.

[26] Gurgueira S.A., Lawrence J., Coull B., Murthy G.G., González-Flecha B. Rapid increases in the steady-state concentration of reactive oxygen species in the lungs and heart after particulate air pollution inhalation. *Environ Health Perspect.* 2002; 110(8): 749–55.

[27] Lewis J., Bench G., Myers O., Tinner B., Staines W., Barr E., et al. Trigeminal uptake and clearance of inhaled manganese chloride in rats and mice. *Neurotoxicology.* 2005; 26(1): 113–23.

[28] Elder A., Gelein R., Silva V., Feikert T., Opanashuk L., Carter J., et al. Translocation of inhaled ultrafine manganese oxide particles to the central nervous system. *Environ Health Perspect.* 2006; 114(8): 1172–8.

[29] Block M.L., Calderón-Garcidueñas L. Air pollution: Mechanisms of neuroinflammation and CNS disease. *Trends Neurosci.* 2009; 32(9): 506–16.

[30] Genc S., Zadeoglulari Z., Fuss S.H., Genc K. The adverse effects of air pollution on the nervous system. *J Toxicol.* 2012; 2012: 782462.

[31] Maher B.A., Ahmed I.A., Karloukovski V., MacLaren D.A., Foulds P.G., Allsop D., et al. Magnetite pollution nanoparticles in the human brain. *Proc Natl Acad Sci U S A.* 2016; 113(39): 10797–801.

[32] Chen C., Nakagawa S. Planetary health and the future of human capacity: The increasing impact of planetary distress on the human brain. *Challenges.* 2018; 9(2): 41. doi: 10.3390/challe9020041.

[33] Leavens T.L., Rao D., Andersen M.E., Dorman D.C. Evaluating transport of manganese from olfactory mucosa to striatum by pharmacokinetic modeling. *Toxicol Sci.* 2007; 97(2): 265–78.

[34] Lucchini R.G., Dorman D.C., Elder A., Veronesi B. Neurological impacts from inhalation of pollutants and the nose-brain connection. *Neurotoxicology.* 2012; 33(4): 838–41. doi: 10.1016/j.neuro.2011.12.001, PMID 22178536.

[35] Sunderman F.W. Nasal toxicity, carcinogenicity, and olfactory uptake of metals. *Ann Clin Lab Sci.* 2001; 31(1): 3–24.

[36] Chen L., Yokel R.A., Hennig B., Toborek M. Manufactured aluminum oxide nanoparticles decrease expression of tight junction proteins in brain vasculature. *J Neuroimmune Pharmacol.* 2008; 3(4): 286–95.

[37] Calderón-Garcidueñas L., Solt A.C., Henríquez-Roldán C., Torres-Jardón R., Nuse B., Herritt L., et al. Long-term air pollution exposure is associated with neuroinflammation, an altered innate immune response, disruption of the blood–brain barrier, ultrafine particulate deposition, and accumulation of amyloid beta-42 and alpha-synuclein in children and young adults. *Toxicol Pathol.* 2008b; 36(2): 289–310.

[38] Allen J.L., Klocke C., Morris-Schaffer K., Conrad K., Sobolewski M., Cory-Slechta D.A. Cognitive effects of air pollution exposures and potential mechanistic underpinnings. *Curr Environ Health Rep.* 2017; 4(2): 180–91.

[39] Weuve J., Puett R.C., Schwartz J., Yanosky J.D., Laden F., Grodstein F. Exposure to particulate air pollution and cognitive decline in older women. *Arch Intern Med.* 2012; 172(3): 219–27.

[40] Finkelstein M.M., Jerrett M. A study of the relationships between Parkinson's disease and markers of traffic-derived and environmental manganese air pollution in two Canadian cities. *Environ Res.* 2007; 104(3): 420–32.

[41] Block M.L., Elder A., Auten R.L., Bilbo S.D., Chen H., Chen J.C., et al. The outdoor air pollution and brain health workshop. *Neurotoxicology.* 2012; 33(5): 972–84.

[42] Li N., Sioutas C., Cho A., Schmitz D., Misra C., Sempf J., Wang M., et al. Ultrafine particulate pollutants induce oxidative stress and mitochondrial damage. *Environ Health Perspect.* 2003; 111(4): 455–60.

[43] World Health Organization. Dementia. 2020. Available at: www.who.int/news-room/fact-sheets/detail/dementia. Cited date Sept 22.

[44] Moulton P.V., Yang W. Air pollution, oxidative stress, and Alzheimer's disease [journal]. *J Environ Public Health.* 2012; 2012: 472751. doi: 10.1155/2012/472751.

[45] Mir R.H., Sawhney G., Pottoo F.H., Mohi-Ud-Din R., Madishetti S., Jachak S.M., et al. Role of environmental pollutants in Alzheimer's disease: A review. *Environ Sci Pollut Res Int.* 2020; 27(36): 44724–42. doi: 10.1007/s11356-020-09964-x.

[46] Landrigan P.J., Fuller R., Acosta N.J.R., Adeyi O., Arnold R., Basu N.N., et al. The Lancet commission on pollution and health. *Lancet.* 2018; 391(10119): 462–512.

[47] Shah A.S., Lee K.K., McAllister D.A., Hunter A., Nair H., Whiteley W., et al. Short term exposure to air pollution and stroke: Systematic review and meta-analysis. *BMJ.* 2015; 350: h1295.

[48] Poon H.F., Calabrese V., Scapagnini G., Butterfield D.A. Free radicals and brain aging. *Clin Geriatr Med.* 2004; 20(2): 329–59.

[49] Doty R.L. The olfactory vector hypothesis of neurodegenerative disease: Is it viable? *Ann Neurol.* 2008; 63(1): 7–15.

[50] Calderón-Garcidueñas L., Kavanaugh M., Block M. Neuroinflammation, Alzheimer's disease-associated pathology, and down-regulation of the prion-related protein in air pollution exposed children and young adults[journal]. Of. *J Alzheimers Dis.* 2011; 28(1): 93–107.

[51] Campbell A. Inflammation, neurodegenerative diseases, and environmental exposures. *Ann N Y Acad Sci.* 2004; 1035(1): 117–32.

[52] Peters A., Veronesi B., Calderón-Garcidueñas L., Gehr P., Chen L.C., Geiser M., et al. Translocation and potential neurological effects of fine and ultrafine particles a critical update. *Part Fibre Toxicol.* 2006; 3: 13.

[53] Ranft U., Schikowski T., Sugiri D., Krutmann J., Krämer U. Long-term exposure to traffic-related particulate matter impairs cognitive function in the elderly. *Environ Res.* 2009; 109(8): 1004–11.

[54] Craig L., Brook J.R., Chiotti Q., Croes B., Gower S., Hedley A., et al. Air pollution and public health: A guidance document for risk managers [journal]. *J Toxicol Environ Health A.* 2008; 71(9–10): 588–698.

[55] Calderón-Garcidueñas L., Reed W., Maronpot R.R., Henríquez-Roldán C., Delgado-Chavez R., Calderón-Garcidueñas A., et al. Brain inflammation and Alzheimer's-like pathology in individuals exposed to severe air pollution. *Toxicol Pathol.* 2004; 32(6): 650–8.

[56] Calderón-Garcidueñas L., Solt A.C., Henríquez-Roldán C., Torres-Jardón R., Nuse B., Herritt L., et al. Long-term air pollution exposure is associated with neuroinflammation, an altered innate immune response, disruption of the blood–brain barrier, ultrafine particulate deposition, and accumulation of amyloid beta-42 and alpha-synuclein in children and young adults. *Toxicol Pathol.* 2008; 36(2): 289–310.

[57] Sutherland G.T., Siebert G.A., Kril J.J., Mellick G.D. Knowing me, knowing you: Can knowledge of risk factors for Alzheimer's disease prove useful in understanding the pathogenesis of Parkinson's disease? [journal]. *J Alzheimers Dis.* 2011; 25(3): 395–415.

[58] Markesbery W.R. Oxidative stress hypothesis in Alzheimer's disease. *Free Radic Biol Med.* 1997; 23(1): 134–47.

[59] Markesbery W.R., Carney J.M. Oxidative alterations in Alzheimer's disease. *Brain Pathol.* 1999; 9(1): 133–46.

[60] Tuppo E.E., Forman L.J., Spur B.W., Chan-Ting R.E., Chopra A., Cavalieri T.A. Sign of lipid peroxidation as measured in the urine of patients with probable Alzheimer's disease. *Brain Res Bull.* 2001; 54(5): 565–8.

[61] Lockman P.R., Koziara J.M., Mumper R.J., Allen D.D. Nanoparticle surface charges alter blood–brain barrier integrity and permeability. *J Drug Target*. 2004; 12(9–10): 635–41.

[62] Meyer U., Feldon J., Fatemi S.H. In-vivo rodent models for the experimental investigation of prenatal immune activation effects in neurodevelopmental brain disorders. *Neurosci Biobehav Rev*. 2009; 33(7): 1061–79.

[63] Mendola P., Selevan S.G., Gutter S., Rice D. Environmental factors associated with a spectrum of neurodevelopmental deficits. *Ment Retard Dev Disabil Res Rev*. 2002; 8(3): 188–97.

[64] Tonelli L.H., Postolache T.T. Airborne inflammatory factors: From the nose to the brain. *Front Biosci (Schol Ed)*. 2010; 2: 135–52.

[65] Crețu D.I, Sovrea A., Ignat R.M., Filip A., Bidian C., Crețu A. Morpho-pathological and physiological changes of the brain and liver after ozone exposure. *Rom J Morphol Embryol*. 2010; 51(4): 701–6.

[66] Pereyra-Muñoz N., Rugerio-Vargas C., Angoa-Pérez M., Borgonio-Pérez G., Rivas-Arancibia S. Oxidative damage in substantia nigra and striatum of rats chronically exposed to ozone. *J Chem Neuroanat*. 2006; 31(2): 114–23.

[67] Calderón-Garcidueñas L., Azzarelli B., Acuna H., Garcia R., Gambling T.M., Osnaya N., et al. Air pollution and brain damage. *Toxicol Pathol*. 2002; 30(3): 373–89.

[68] Calderón-Garcidueñas L., Solt A.C., Henríquez-Roldán C., Torres-Jardón R., Nuse B., Herritt L., et al. Long-term air pollution exposure is associated with neuroinflammation, an altered innate immune response, disruption of the blood–brain barrier, ultrafine particulate deposition, and accumulation of amyloid beta-42 and alpha-synuclein in children and young adults. *Toxicol Pathol*. 2008; 36(2): 289–310.

[69] Calderón-Garcidueñas L., Mora-Tiscareño A., Ontiveros E., Gómez-Garza G., Barragán-Mejía G., Broadway J., et al. Air pollution, cognitive deficits and brain abnormalities: A pilot study with children and dogs. *Brain Cogn*. 2008; 68(2): 117–27.

[70] Campbell A., Oldham M., Becaria A., Bondy S.C., Meacher D., Sioutas C., et al. particulate matter in polluted air may increase biomarkers of inflammation in mouse brain. *Neurotoxicology*. 2005; 26(1): 133–40.

[71] Kleinman M.T., Araujo J.A., Nel A., Sioutas C., Campbell A., Cong P.Q., et al. Inhaled ultrafine particulate matter affects CNS inflammatory processes and may act via MAP kinase signaling pathways. *Toxicol Lett*. 2008; 178(2): 127–30.

[72] Calderón-Garcidueñas L., Gónzalez-Maciel A., Vojdani A., Franco-Lira M., Reynoso-Robles R. The intestinal barrier in air pollution-associated neural involvement in Mexico City residents: Mind the gut, the evolution of a changing paradigm relevant to Parkinson disease risk. *J Alzheimers Dis Parkinsonism*. 2015; 5: 179.

[73] Lee H.C., Lin T.H. Air pollution, particular matter, and atherosclerosis. *Acta Cardiol Sin*. 2017 Nov; 33(6): 646–7.

[74] Brook R.D., Franklin B., Cascio W., Hong Y., Howard G., Lipsett M., et al. Air pollution and cardiovascular disease: A statement for healthcare professionals from the expert panel on population and prevention science of the American heart association. *Circulation*. 2004; 109(21): 2655–71.

[75] Brook R.D., Rajagopalan S., Pope C.A., Brook J.R., Bhatnagar A., Diez-Roux A.V., et al. Particulate matter air pollution and cardiovascular disease: An update to the scientific statement from the American heart association. *Circulation*. 2010; 121(21): 2331–78.

[76] Künzli N., Jerrett M., Mack W.J., Beckerman B., LaBree L., Gilliland F., et al. Ambient air pollution and atherosclerosis in Los Angeles. *Environ Health Perspect*. 2005; 113(2): 201–6.

[77] Hoffmann B., Moebus S., Möhlenkamp S., Stang A., Lehmann N., Dragano N., et al. Residential exposure to traffic is associated with coronary atherosclerosis. *Circulation*. 2007; 116(5): 489–96.

[78] Araujo J.A., Nel A.E. Particulate matter and atherosclerosis: Role of particle size, composition, and oxidative stress. *Part Fibre Toxicol*. 2009; 6: 24.

[79] Yao H.M., Lv J.Y. Statin attenuated myocardial inflammation induced by PM2.5 in rats. *Acta Cardiol Sin*. 2017; 33(6): 637–45. doi: 10.6515/ACS20170518A, PMID 29167617.

[80] Cosselman K.E., Navas-Acien A., Kaufman J.D. Environmental factors in cardiovascular disease. *Nat Rev Cardiol*. 2015; 12(11): 627–42.

[81] World Health Organization. Ambient air pollution: Health Impacts. Available at: impacts.tps. Vol. 12, 2002. Available at: www.who.int/airpollution/ambient/health-impacts/en/. Cited date Sept 12, 2020.

[82] Lemprière S. Air pollution linked to multiple sclerosis and stroke. *Nat Rev Neurol.* 2020; 16(3): 127. doi: 10.1038/s41582-020-0322-x.

[83] Hahad O., Lelieveld J., Birklein F., Lieb K., Daiber A., Münzel T. Ambient air pollution increases the risk of cerebrovascular and neuropsychiatric disorders through induction of inflammation and oxidative stress. *Int J Mol Sci.* 2020 Jun 17; 21(12): 4306. doi: 10.3390/ijms21124306.

[84] Li J., Huang J., Wang Y., Yin P., Wang L., Liu Y., et al. Years of life lost from ischaemic and haemor-rhagic stroke related to ambient nitrogen dioxide exposure: A multicity study in China. *Ecotoxicol Environ Saf.* 2020 Oct 15; 203: 111018. doi: 10.1016/j.ecoenv.2020.111018, PMID 32888591.

[85] Rocha I.I., Narasimhalu K., De Silva D.A. Impact of air pollution and seasonal haze on neurolog-ical conditions. *Ann Acad Med Singap.* 2020; 49(1): 26–36.

[86] Zhang H.W., Kok V.C., Chuang S.C., Tseng C.H., Lin C.T., Li T.C., et al. Long-term ambient hydrocarbons exposure and incidence of ischemic stroke. *PLoS One.* 2019 Dec 4; 14(12): e0225363. doi: 10.1371/journal.pone.0225363.

[87] Ljungman P.L.S., Andersson N., Stockfelt L., Andersson E.M., Nilsson Sommar J., Eneroth K., et al. Long-term exposure to particulate air pollution, black carbon, and their source components in relation to ischemic heart disease and stroke. *Environ Health Perspect.* 2019; 127(10): 107012.

[88] Lin H., Tao J., Du Y., Liu T., Qian Z., Tian L., et al. Differentiating the effects of characteristics of PM pollution on mortality from ischemic and hemorrhagic strokes. *Int J Hyg Environ Health.* 2016; 219(2): 204–11.

[89] Huang K., Liang F., Yang X., Liu F., Li J., Xiao Q., et al. Long term exposure to ambient fine par-ticulate matter and incidence of stroke: A prospective cohort study from the China-PAR project. *BMJ.* 2019; 367(367): 16720.

[90] Yorifuji T., Kawachi I., Sakamoto T., Doi H. Associations of outdoor air pollution with hemor-rhagic stroke mortality. *J Occup Environ Med.* 2011; 53(2): 124–6.

[91] Liu H., Tian Y., Xu Y., Huang Z., Huang C., Hu Y., et al. Association between ambient air pollution and hospitalization for ischemic and hemorrhagic stroke in China: A multicity case-crossover study. *Environ Pollut.* 2017; 230: 234–41.

[92] Migliore L., Coppedè F. Environmental-induced oxidative stress in neurodegenerative disorders and aging. *Mutat Res.* 2009; 674(1–2): 73–84.

[93] Jomova K., Vondrakova D., Lawson M., Valko M. Metals, oxidative stress, and neurodegenerative disorders. *Mol Cell Biochem.* 2010; 345(1–2): 91–104.

Glossary

Acoustic encoding: The act of encoding sounds, words in the memory system.

Action potential: Also known as "spike potential, neuron firing," occurs when a sufficient stimulus produces electrical activity in the nerve or muscle cells. The action potential is a swift, momentary, and proliferating change in the resting membrane potential [1].

Air pollution: Air contains harmful biological, physical, and chemical particles emitted into the environment [2–4].

Air pollution episode: A specific period in which air pollutants concentration increases [2–4].

Alzheimer's disease: This is an irreversible, progressive neurodegenerative disease or brain disorder that gradually impairs memory and thinking, and ultimately, the subject cannot perform tasks related to memory.

Amygdala: An essential component of the brain's limbic system, which plays a role in emotional responses, such as happiness or anxiety.

Amyloid plaque: The extracellular deposition of amyloid-beta proteins is mainly in the gray matter around the neurons and synapses. These proteins are found in people with old age and Alzheimer's disease.

Amyloid-beta (Aβ) protein: Abnormal clumps of protein form the amyloid plaques, a biomarker of Alzheimer's disease. These proteins are toxic to brain cells and have a causal link to Alzheimer's disease [5].

Android: Android is a Linux-based platform for smartphones.

Animal model study: A study findings based on the laboratory animals.

Anxiety: A condition in which the subject feels persistent worry and stress about everyday situations.

Apoptosis: A programmed cell death occurs as part of average growth and development.

Artificial intelligence: Computer-based technology performs tasks that usually require human intelligence, such as learning, decision-making, and problem-solving behaviors.

Astrocytes: Are star-shaped glial cells in the central nervous system. They have many processes which envelop synapses. A single astrocyte can interact with about 2 million synapses. These cells also modulate neuronal activities.

Ataxia: A degenerative disease of the nervous system. The main clinical feature includes slurred speech, stumbling, reeling wide-based gait, and incoordination. These symptoms are caused by damage to the cerebellum.

Autism spectrum disorder (ASD): A complex neurodevelopmental disorder characterized by "deficits in socialization, communication, and repetitive or unusual behavior."

Autonomic nervous system: An essential part of the central nervous system that controls internal organ functions, for example, heart rate, respiratory rate, blood pressure, intestinal function; its functions are not under the control of a subject (involuntary).

Axons: A component of a neuron that conducts impulses away from the neuron's cell body.

Axon terminals: Are club- or button-shaped, small swellings found at the terminal ends of axons. The electrochemical signals transmit through the synapse to neighboring cells under neurotransmitters control mechanisms.

Biodiversity: Biological diversity, variety of life on the planet, animals, and microorganisms.

Biomass: Fuel source firmed from the living and plant materials such as wood and leaves.

Biosphere: Part of the earth and its atmosphere that can support life.

Blood–brain barrier (BBB): This is a highly selective semipermeable endothelial cells layer. It is a protective barrier that prevents the brain from the entry of toxic substances. It promotes neuronal functions.

Brain tumor: A mass or growth of abnormal cells found in the brain.

Brain waves: Are also called neural oscillations. These waves are the periodic patterns of neural activity in the central nervous system.

Brain-derived neurotrophic factor (BDNF): This is a protein of the neurotrophins family. It promotes the growth, maintenance, survival, and differentiation of neuronal populations during neuron development.

Broca's area: A minor key component area in the left frontal lobe. It plays an essential role in complex speech production networks.

Carbon dioxide (CO2): A colorless gas composed of one carbon and two oxygen atoms; present on the earth are in a low concentration and acts as a greenhouse gas.

Carbon monoxide (CO): A colorless noxious gas formed by incomplete fossil fuel combustion. The most significant sources of CO are vehicles or machinery that burn fossil fuels.

Central nervous system: A controlling and coordination system of the body consists of two parts, the brain, and spinal cord.

Cerebellum: Sensorimotor part of the hindbrain with biochemical and neuronal networks links the brainstem and spinal cord [6].

Cerebrospinal fluid (CSF): The clear, colorless liquid found in the brain and spinal cord. It provides a cushion-like function to the brain and spinal cord and serves as a means of nutrient transport and waste elimination system [6].

Cerebrum: The most significant part of the brain, has left and right cerebral hemispheres, and four lobes.

Climate: The pattern of weather conditions in a particular region over a while.

Climate change: A natural or human-made change at a regional environment and weather conditions levels.

Coarse particles: Particles with diameters between 2.5 and 10 µm are called coarse particles [2–4].

Cochlea: This is a hollow, spiral-shaped part of the inner ear that plays a significant role in hearing mechanisms.

Cognition: The cognition or cognitive process includes thinking, observing, distinguishing, considering, judging, reasoning, and conception.

Cognitive neuroscience: The study of neuroscience deals with biological processes, including attention, memory, and other facets of cognition.

Concentration: The process of paying focused, full attention to things.

Contamination: The presence of potentially damaging substances, including biological, physical, chemical toxic substance which contaminates and adversely affect the environment.

Deforestation: The permanent removal of the trees from the forest.

Dementia: A term used to describe various clinical features of cognitive impairment such as impairment in memory, communication, thinking, forgetfulness.

Dendrites: The neuronal projections which receive information from other neurons.

Deoxyribonucleic acid (DNA): The two strands molecule that carries genetic instructions in all living things.

Depression: A attitude or emotional disorder characterized by lack of motivation and sadness. The condition is mainly associated with disruptions in the brain's neurotransmitter systems, including dopamine and serotonin.

Desert: A unfertile landscape with less humidity and living conditions is hostile to plant and animal life.

Domestic waste: The daily basis home waste.

Dopamine: A neurotransmitter involves in pleasure, motivation, learning, and also other brain functions.

Dose–response relationship: The measurable association between different doses, time, or biological or ecological response.

Down syndrome: A genetically abnormal condition in which a person has an extra chromosome. The syndrome is characterized by cognitive function impairment and physical abnormalities.

Dump: A place or area used to dispose of solid wastes without environmental controls.

Dysarthria: Unable to express the words accurately, and slurred and inappropriate phrasing.

Dysdiadochokinesia: Unable to perform rapid alternating movements.

Ecology: The relationships between organisms and environment.

Ecosystem: Biological environment where animals, plants, and other organisms live.

Electric and magnetic fields (EMFs): Invisible energy, also referred to as radiation, is linked to electrical power, mobile phones, and various forms of natural and human-made electrical devices [7].

Electric fields: The electric field created by differences in voltage; increases the voltage stronger will be the resultant field. It exists even when no current flowing [7].

Electroencephalography (EEG): An investigation that measures electrical activity in the brain.

Electromagnetic energy: Energy allied to photons containing an electromagnetic spectrum such as ultraviolet and infrared energy [7].

Electromagnetic spectrum: The wide range of frequencies of electromagnetic fields. It has an "extremely low frequency (ELF), very low frequency (VLF), radio frequency (RF), microwave, visible light, and ionizing radiation." [7].

Emission: The discharge of pollutants, dust, gases, chemicals into the environment. It contributes to poor air quality and global warming.

Encoding: The process of receiving and transference of information, items, things into memory. This is an input of data into the memory.

Environmental risk: A hazard or possibility of disease damage due to exposure to some environmental factors.

Environmental toxicology: The study of ecological factors influencing dust, fumes, chemicals, or hazardous substances that are potentially toxic [2–4].

Epigenetics: The study about environmental changes which affect the means of gene's work. The epigenetic modifications are reversible and do not change the DNA sequence.

Epilepsy: A neurological condition that is characterized by abnormal brain electrical activity [8].

Familiar path: The act in which locations along a path used as points while developing linkage.

Fine particles: Particles generally with 2.5 μm in diameter or smaller. These particles are inhalable coarse particles [2–4].

Fossil fuels: The formation of fuels, coal, gases in the soil from dead plants and animals over a long time.

Fourth generation: Fourth generation of digital mobile handsets, the speed at 100 Mbit/s for high-mobility communication.

Gamma-aminobutyric acid (GABA): A neurotransmitter plays a role in brain biology.

Glioma: A tumor that arises from the brain's glial tissue.

Global Standard for Mobile Communications (GSM): The standard protocols define the digital cellular networks used by mobile devices.

Global warming: The steady increase in the temperature caused by human activities generates high carbon dioxide levels and other gases.

Gamma-aminobutyric acid (GABA): Gamma amino acid neurotransmitters inhibit the electrochemical activity.

Golgi apparatus: Folded structure, located near to nucleus, with clusters of flattened cisternae.

Greenhouse effect: The rapid rising of temperatures, weather conditions of the earth's atmosphere caused by increasing levels of gases such as "carbon dioxide, methane, and nitrous oxide (NO2)."

Habitat: The place or environment where plants grow and animals live in specific environmental conditions for their survival.

Hazardous waste: The waste that can pose a risk to human health or the environment.

Hippocampus: A primitive brain structure critical for memory and learning.

Humidity: This is the level of water molecules in the atmosphere [8–9].

Hygroscopicity: The quality of absorption of moisture from the air.

Hypothalamus: The small area in the brain located at the base of the brain, near the pituitary gland.

Interneurons: Interneurons, also known as "relay neurons," connect and carry impulses between sensory or motor neurons.

Ionizing radiation: High-level radiation has possible effects on cellular and DNA damage.

Liquid waste: The urban areas' waste includes sewage, industrial, and household fluids.

Long-term memory: Permanent storage of memories with unlimited capacity.

Machine learning: The artificial intelligence algorithm that can learn the processor identifying the diagnostic criteria of brain imaging or genetic information.

Magnetic field: A vector field that defines the magnetic impact on moving electric charges, currents, and magnetic materials.

Memory: The process of retaining items, ideas, things learned or experienced [10].

Memory retrieval: This is a process of remembering information stored in long-term memory.

Memory storage: Act of getting information into memory system and retention of the data.

Memory trigger: A stimuli that facilitate the recall of memory, such as sensory or contextual [10–11].

Meninges: The three delicate layers surround and protect the brain and spinal cord. These layers are the dura mater, arachnoid mater, and pia mater.

Mitochondria: Powerhouse of the cell, spherical or rod-shaped in appearance.

Motor neurons: Motor neurons integrate neural inputs into an output signal. Transmits nerve impulses from the nervous system, spinal cord to effector structures such as skeletal muscles [10–11].

Nerve impulse: The process of "nerve signal" in which a neuron communicates with other cells through the electrochemical signals.

Nervous system: The chief controlling and coordinating system regulates all body functions.

Neurodevelopmental disorder: The conditions that impair, delay the development and maturation of the brain.

Neuroglia: Supports neurons and other structures.

Neurons: The basic structural and functional unit of the nervous system.

Neurotransmitters: Chemical substances released at synapses regulate the various function.

Noise pollution: A loud noise that disturbs the people and environment.

Non-ionizing: A low-level radiation, which is generally perceived as harmless to humans. It is divided into "extremely low-frequency fields (ELF fields), intermediate frequency fields (IF fields), and radiofrequency fields (RF fields)."

Olfactory: The biology of sense of smell. The term can be used for the olfactory nerve, olfactory mechanism, or olfactory pathway.

Optic nerve: The second cranial nerve transmits information from the retina to the brain.

Oxidizing smog: Air pollution containing ozone, peroxyacetyl nitrate, and other oxidant gases [10–11].

Ozone (O3): The gas consists of "three atoms of oxygen, formed by chemical reactions between oxides of nitrogen (NOx) and volatile organic compounds (VOCs). This occurs once pollutants emitted by motor vehicles and other sources chemically react in the presence of sunlight."

Parkinson's disease: A progressive neurodegenerative disorder mainly affects dopamine-producing ("dopaminergic") neurons in a substantia nigra. The main characteristics are muscle stiffness, tremors, slowed movement, impaired posture, balance, and speech changes.

Particulate matter: The fine particles, including dust, smoke, gases, pollen, and soil particles suspended in the air and polluted the environment [2–4].

Particulate matter (PM10): These are inhalable coarse particles with a diameter of 10 μm [2–4].

Particulate matter (PM2.5): Fine particles with an aerodynamic diameter of 2.5 μm [2–4].

Particulate matter pollution: Pollution due to particulate matter solids, liquid droplets are suspended in the air and contaminate the environment [2–4].

Phonetic: The sounds associated with speech.

Plasma cell membrane: Covers of cell organelles and made up of lipids bilayer and proteins.

Plasticity: The brain's ability to change and adapt to developmental forces, learning processes, or aging.

Pollution: The exposure of dust, fumes, gases, chemicals, organisms, or radiation energy in quantities that exceed their required limit causing toxicity or other ecological damages.

Radiation: A form of energy transmitted as rays, waves, or streams of energetic particles. The term is frequently used concerning the emission of radiation from rays, mobile phones, X-rays, etc.

Radioactive: Material that can generate radiation.

Renewable energy: Energy collected from renewable resources such as wind power, solar energy, or biomass.

Residue: Persistent concentrations of pesticides or other chemicals in the environment.

Secondary pollutants: Pollutants that are not directly produced or released in the environment by chemical reactions but made due to involving emitted substances or chemicals.

Sensory memory: The memory process for swiftly forming insights.

Sensory neurons: Transmits impulses from sensory receptors to the spinal cord and brain.

Sewage: The liquid wastes may be a mixture of domestic or industrial waste.

Short-term memory: Memory storage for a short period of 20–30 seconds.

Sludge: A semi-solid or solid or precipitate that settles during the polluted water treatment. It is produced mainly during sewage and industrial facilities.

Smog: Air pollution consists of smoke, and fog occurs mainly in urban and industrial areas.

Synapses: Communicating site or junction between the neurons or their extensions.

Synaptic cleft: The space gap that can establish a junction between the neurons. It facilitates to transmission of the nerve impulse from one neuron to the other.

Synaptic transmission: The process of transmission of nerve-to-nerve impulse in the central nervous system.

Thermal pollution: Release of heated water from industrial processes affect the life processes.

Ultrafine particles: The ultrafine particles are nanoparticles with diameters less than 0.1 μm.

Urbanization: The development of large metropolitan cities and towns on natural lands.

Visual encoding: This is the input or receiving of images into the memory system.

Volatile: Substance that evaporates swiftly.

Wernicke's area: Brain area in the left temporal lobe responsible for speech comprehension.

Wi-Fi: "Wireless Fidelity": Wireless networking allows computers, phones, laptops, and other devices to communicate over a wireless signal.

Zero emissions: The engine does not produce any harmful gases into the environment.

References

[1] Bean, B. The action potential in mammalian central neurons. *Nat Rev Neurosci.* 2007; 8: 451–65. https://doi.org/10.1038/nrn2148.

[2] Lijima S. Electron microscopy of small particles. *J. Electron Microsc.* 1985; 34(4): 249.

[3] Environmental protection agency: Available at: www.epa.ie/footer/a-zglossaryofenvironmentalte rms/. Cited date Apr 2, 2021.

[4] The United States Environmental Protection Agency. Available at: www.epa.gov/pmcourse/ glossary-air-pollution-terms-particle-pollution-and-your-patients-health. Cited date Apr 2, 2021.

[5] Weller J., Budson A. Current understanding of Alzheimer's disease diagnosis and treatment. *F1000Res.* 2018; 7: F1000 Faculty Rev-1161. Published 2018 July 31. doi: 10.12688/f1000research.14506.1.

[6] Telano L.N., Baker S. Physiology, cerebral spinal fluid. In: *StatPearls [Internet].* Treasure Island, FL: StatPearls Publishing. 2021 Jan. Available at: www.ncbi.nlm.nih.gov/books/NBK519007/. Cited date July 14, 2020.

[7] World Health Organization. Radiation: Electromagnetic fields. Available at: www.who.int/news-room/q-a-detail/radiation-electromagnetic-fields. Cited date Apr 2, 2021.

[8] Bogdan Lewczuk, Grzegorz Redlarski, Arkadiusz Żak, Natalia Ziółkowska, Barbara Przybylska-Gornowicz, Marek Krawczuk. Influence of electric, magnetic, and electromagnetic fields on the circadian system: Current stage of knowledge. *BioMed Res. Int.* 2014: 169459. doi. org/10.1155/2014/169459.

[9] Raymond Richard Neutra. Glossary of terms used when discussing exposure to electric and magnetic fields. *J Epidemiol Community Health.* 2005; 59: 546–50. doi: 10.1136/jech.2003.019075.

[10] Ker terms glossary: Dana Foundation. Available at: https://dana.org/explore-neuroscience/brain-basics/key-brain-terms-glossary/. Cited date Apr 2, 2021.

[11] Glossary: Available at: www.brainfacts.org/Glossary. Cited date Apr 2, 2021.

Index